Lewis Mumford AND *Patrick Geddes*

Lewis Mumford AND Patrick Geddes

THE CORRESPONDENCE

Edited and introduced by

Frank G. Novak, Jr.

LONDON AND NEW YORK

First published 1995
by Routledge
11 New Fetter Lane, London EC4P 4EE

Simultaneously published in the USA and
Canada by Routledge
29 West 35th Street, New York, NY 10001

© 1995 Frank G. Novak, Jr.

Typeset in Sabon by Solidus (Bristol) Limited
Printed and bound in Great Britain by
T.J. Press (Padstow) Ltd, Padstow, Cornwall

British Library Cataloguing in Publication Data
A catalogue record for this book is available
from the British Library

*Library of Congress Cataloguing in Publication
Data*
Mumford, Lewis
Lewis Mumford and Patrick Geddes:
the correspondence/
[edited by] Frank G. Novak, Jr.
p. cm.
Includes bibliographical references (p.) and
index.
1. Mumford, Lewis, 1895–1990
Correspondence. 2. Geddes, Patrick, Sir,
1854–1932 – Correspondence. 3. City
planners – United States – Correspondence.
4. City planners – Great Britain –
Correspondence. 5. Architects – United States
– Correspondence. 6. Sociologists – Great
Britain – Correspondence. 7. Social reformers
– United States – Correspondence. I. Geddes,
Patrick, Sir, 1854–1932. II. Novak, Frank
G. III. Title.
CT275.M734A4 1995
711.4'092'2–dc20   94-41551

ISBN 0-415-11906-5

CONTENTS

ACKNOWLEDGEMENTS

Work on this project was supported by an Arnold L. and Lois S. Graves Award in the Humanities, grants from the Seaver College Reassigned Time Committee, and grants from the Pepperdine University Research Council. I am indebted to the following for making the manuscripts of the letters and related materials available to me: Nancy Shawcross and the staff of the University of Pennsylvania Van Pelt Library Special Collections Department; Elspeth Yeo, Olive Geddes, and the staff of the National Library of Scotland Manuscripts Division; Tor Blekastad of the University of Oslo Library; James McGrath of the Strathclyde University Archives. It was a privilege to work with Tristan Palmer and Caroline Cautley, kind and efficient editors at Routledge. I am grateful to John Banks for his work on the manuscript. I owe special thanks to Sophia (Mrs. Lewis) Mumford for her interest and assistance. Affectionate gratitude goes to my family – Cynthia, Laura, and Alison – who provided good-humored patience and encouragement throughout my work on the project.

A NOTE ON THE TEXT

The Lewis Mumford Collection at the University of Pennsylvania Van Pelt Library contains most of Geddes' letters to Mumford (folders 5827–5835). In the late 1960s, Mumford sent a number of Geddes' letters to his friend Phillip Boardman, in Norway, as Boardman was researching the revised edition of his Geddes biography. A few of these letters were apparently never returned to Mumford, probably having been lost or destroyed, but copies remain in the collection of Boardman papers at the University of Oslo. Nearly all of Geddes' letters are handwritten, and a number of words and passages are very difficult to decipher. Geddes often used abbreviations and an idiosyncratic shorthand in his writing. The following transcriptions spell out these abbreviations and translate the shorthand. Geddes was inconsistent and sometimes careless in his spelling, punctuation, and use of accents on foreign words. The transcribed letters correct the obvious errors, and an effort has been made to be consistent in regularizing mechanical elements without altering meaning or compromising Geddes' distinctive style. Where the standard form is not obvious or appropriate, the letter has been transcribed verbatim; in other words, one should assume that any unusual or nonstandard forms contained in the transcriptions printed here also exist in the originals. Geddes' letters contain British spellings, which Geddes used in almost all instances. In order to retain something of the flavor of Geddes' swift, often impulsive writing, his use of the ampersand (&) is retained throughout.

The Patrick Geddes Collection at the National Library of Scotland in Edinburgh contains most of Mumford's letters to Geddes (manuscript 10575). One Mumford letter is in the collection of Geddes papers at the University of Strathclyde Archives, Glasgow. Nearly all of Mumford's letters are typed. Geddes occasionally refers to letters from Mumford that have apparently been lost; but, in view of Geddes' peripatetic career and disorderly habits, it is amazing that so many have survived. Mumford kept copies of several of his earliest letters to Geddes, whose originals have been lost, and those copies (from the University of Pennsylvania Lewis Mumford Collection) are reprinted here. The relatively few obvious errors in Mumford's letters have been corrected. Following the original manuscripts, Mumford's letters are transcribed using American spellings.

This volume contains transcriptions of all the Geddes–Mumford letters at the National Library of Scotland, the University of Pennsylvania, the University of Oslo, and the University of Strathclyde. Mumford included excerpts from a few of his letters to Geddes in *Findings and Keepings: Analects for an Autobiography* (New York: Harcourt Brace Jovanovich, 1975) and in *My Works and Days: A Personal Chronicle* (New York: Harcourt Brace Jovanovich, 1979).

The letters are generally arranged in chronological order. When logically appropriate, the order varies to take into account the sequence in which each letter was received and responded to – although one or both correspondents may have written subsequent letters while previous ones were still in transit. For example, Mumford's letter of 5 December 1920 is followed by Geddes' of 10 January 1921 (as a direct and detailed response to the 5 December letter); Mumford's of 3 January 1921 follows Geddes' of 10 January.

Certain parts of the letters are not transcribed because they are missing or illegible. These omissions are indicated in the text by the bracketed ellipsis: […].

The following abbreviations are used in the identification which appears in brackets beneath the date of each letter:

Ms: Handwritten
Ts: Typescript
NLS: National Library of Scotland, Edinburgh
Oslo: Philip Boardman Papers, University of Oslo Library, Norway
UP: Lewis Mumford Collection, University of Pennsylvania Van Pelt Library, Philadelphia

PATRICK GEDDES:
A CHRONOLOGY

1854   Born at Ballater, West Aberdeenshire, Scotland, on 2 October.

1857   Family moves to Perth, Scotland.

1874–1878   Studies biology under Thomas Huxley at the Royal School of Mines in London.

1879   During a scientific expedition in Mexico suffers temporary blindness and develops the graphic method using folded squares of paper.

1880–1888   Biology demonstrator and zoology lecturer at the University of Edinburgh.

1886   Marries Anna Morton; resides at James Court, in old Edinburgh, where he instigates urban improvement projects and establishes University Hall as a student residence.

1887–1899   Directs the Edinburgh Summer Meetings, a summer school program visited by many distinguished intellectuals.

1888   Unsuccessful applicant for the Regius Chair of Botany at the University of Edinburgh.

1888–1919   Holds chair of botany at University College in Dundee, summer terms only, endowed for Geddes by J. Martin White.

1890   Acquires Outlook Tower in Edinburgh to develop as regional museum and "sociological laboratory."

1897	Directs agricultural development project in Cyprus.

1899–1900   Visits America, December–March; meets intellectual luminaries and education leaders in Philadelphia, Chicago, and New York.

1900   Conducts a study program at the Exposition Universelle held in Paris.

1903   Helps to found the Sociological Society of London.

1904   Publishes *City Development: A Study of Parks, Gardens, and Culture-Institutes: A Report to the Carnegie Dunfermline Trust.*

1910   Cities and Town Planning Exhibition first displayed.

1912   Declines offer of knighthood.

1914   Travels to India at the invitation of Lord Pentland, Governor of Madras, to show the Cities Exhibition and to assist municipalities as a town-planning consultant. Continues to work in India until 1924, returning to Dundee for the summer term (through 1919). In October 1914, the Cities Exhibition, en route to India, is lost when the freighter carrying it is sunk by a German raider.

1915   Publishes *Cities in Evolution: An Introduction to the Town Planning Movement and to the Study of Civics.*

1917   In April, his son Alasdair is killed in action on the Western Front. In May, his wife Anna dies in Calcutta. Receives the first letter from Lewis Mumford, probably in August.

1919   Commissioned by the Zionist Federation to plan the new Hebrew University in Jerusalem; appointed to the Chair of Sociology at the University of Bombay, which he holds through 1924, and founds the Department of Civics and Sociology at Bombay.

1923   On 7 May, he arrives in New York and meets Lewis Mumford. He gives a course of lectures at the New School for Social Research, 25 June–4 August.

1924   Seriously ill, in April he travels to Montpellier, France, to recuperate. In September, he begins work on the Collège des Ecossais at Montpellier.

1925   In April, he travels to Jerusalem for the inauguration ceremony of the Hebrew University. In September, Lewis Mumford visits Geddes at Edinburgh.

1928   Marries Lillian Brown in January.

1932   Awarded knighthood in January.

1932   Dies at Montpellier on 17 April.

LEWIS MUMFORD:
A CHRONOLOGY

1895   Born in Flushing, New York, on 19 October.

1912   Enrolls in the City College of New York.

1914   First reads works by Geddes in the fall.

1915   Writes to the Outlook Tower. Reads Geddes' *City Development* and *Cities in Evolution*.

1916   Inspector for the Joint Board of the Dress and Waist Industry in New York.

1917   Works in Pittsburgh as a cement tester for the Bureau of Standards, May–July. Receives the first letter from Geddes, dated 8 August.

1918   Joins the Navy; is stationed first at Newport, Rhode Island, then assigned to the Radio Training School at Cambridge, Massachusetts.

1919   Discharged from the Navy in February. Serves as an associate editor for *The Dial*, March–November.

1920   Lives in London where he serves as editor of the *Sociological Review*, April–October.

1921   Marries Sophia Wittenberg and lives in Greenwich Village.

1922   Travels in Europe with Sophia, July–November. *The Story of Utopias*. Moves to Brooklyn Heights.

1923   Helps to found the Regional Planning Association of America and serves as the organization's secretary. Meets Geddes for the first time when he visits New York to lecture at the New School for Social Research.

1924   *Sticks and Stones.*

1925   Son Geddes born on 5 July. Lectures on American literature at the International Summer School in Geneva. Visits Geddes in Edinburgh in September. Moves to Sunnyside Gardens, Queens.

1926   Spends the summer at Troutbeck, Joel Spingarn's estate near Amenia, New York. *The Golden Day.*

1929   Visiting professor at Dartmouth College. *Herman Melville.*

1931   Begins writing "The Sky Line" column for *The New Yorker,* which he continued to write through 1963. *The Brown Decades.*

1932   Study trip to Europe.

1934   *Technics and Civilization,* first volume in the Renewal of Life series.

1935   Daughter Alison born on 28 April.

1936   Settles permanently in the Leedsville community, near Amenia, New York.

1938   *The Culture of Cities.*

1942   Serves as head of the School of Humanities at Stanford University (resigned in 1944).

1944   Son Geddes killed in combat in Italy. *The Condition of Man.*

1951   Visiting professor at the University of Pennsylvania, an appointment that he held for ten years. *The Conduct of Life.*

1956   *The Transformations of Man.*

1961   *The City in History* wins the National Book Award.

1967   *The Myth of the Machine: I. Technics and Human Development.*

1970   *The Myth of the Machine: II. The Pentagon of Power.*

1982   *Sketches from Life*, his autobiography.

1990   Dies at his Amenia home on 26 January.

INTRODUCTION:
MASTER AND DISCIPLE

═══════════

THE CORRESPONDENTS

Patrick Geddes exerted a profound and decisive influence on Lewis Mumford's thought and writing. Beginning in 1915, Geddes' writings on the city, his vitalistic philosophy, and his role as a "professor of things in general" were crucial factors in shaping Mumford's interests and purposes at the outset of his literary career. Although he was separated from the peripatetic Scotsman by oceans and continents and met him on only two occasions, Mumford repeatedly acknowledged Geddes as the mentor of his youth and as his most important teacher. In 1930, the cumulative effect of Geddes' teachings and example ignited in Mumford what he termed an "explosion of energy" that propelled his signal achievement in American letters, and Geddes continued to be a vital influence throughout his long and voluminous career. The spirit of the man Mumford addressed as "master" permeates his work – from his first book, *The Story of Utopias* (1922), to his last, *Sketches from Life* (1982).

Lewis Mumford wrote his first letter to Patrick Geddes in 1917, and they corresponded regularly until the latter's death in 1932. Given Geddes' remarkable impact on Mumford, their correspondence is particularly important for understanding the development of this prolific and versatile American critic of culture. Mumford's letters to Geddes tell the story of his intellectual and literary coming-of-age, and they document the evolution of his early career. The letters also tell the story of a personal relationship that had a sad, if not tragic, conclusion. Mumford's youthful dream of studying and working with his intellectual hero never came to pass; nor was Geddes' hope that Mumford become his surrogate son, amanuensis, biographer, and popularizer ever realized. The letters chart the progressive changes in Mumford's relationship with his master, from initial adulation, through early disillusionment and rebellion, to reconciliation and continued, though tempered, admiration.

Their relationship was in certain important respects that of father and son: in its early stages, Mumford looked to Geddes as his mentor and role model; yet within a few years Mumford found himself growing increasingly restive under the insistent demands and dogmatism of the man whom he had acknowledged his "intellectual parent" (LM 17 November 1922). After 1925, the aging Geddes' pleas that Mumford join him for collaboration at Montpellier, France, became progressively more urgent and unrealistic. Concurrently, Mumford was rapidly gaining literary prestige and was enjoying an expanding demand for his articles and books. In the final phase of their relationship, the weak and ill Geddes desperately, even pathetically, begged Mumford to edit his chaotic "middens" of papers and to continue his work. However, by 1930 it was too late, for the "disciple" had become a confident and wide-ranging intellectual as well as a prominent and successful writer, pursuing a career quite independent of his master; Mumford had reached the threshold of the major phase of his mature career. At that point, Geddes reluctantly began to recognize that he was at the end of his career and that his grandiose ambitions would probably never be realized. Nevertheless, less than a year before his death, Geddes still nurtured the dream that Mumford would join him for the long-anticipated, rousing "Pallas owl-flight" together and would assume the role of intellectual heir and "literary legatee" (PG 17 May 1931).

Although they shared many of the same interests and values, the Geddes–Mumford relationship linked two contrasting personalities who possessed very different temperaments and talents. Geddes was a man of action who was incessantly forming new contacts, travelling to new places, and promoting new schemes; he planned and renovated cities, founded societies and colleges, organized conferences and educational programs. He was notoriously impatient with those who balked at his frenetic pace and precipitate impulsiveness. Continually in financial difficulty, Geddes typically turned to new projects before previous ones had been completed. He was a man of the Victorian Age who had studied under Thomas Huxley, had met Charles Darwin, and had encountered his hero Thomas Carlyle walking in Chelsea. Mumford, on the other hand, was disciplined, methodical, and consistently productive. He aspired above all else to be a great writer and did his main work in the library and at his writing desk. He was single-mindedly, sometimes even ruthlessly, committed to his writing and the success of his literary career. Growing up in turn-of-the-century New York, he served in the U.S. navy during World War I and lost his son in World War II; he witnessed the "roaring twenties," the Great Depression, the atomic bomb, the Vietnam War, and the election of Ronald Reagan. Geddes was imperious, impatient, and impulsive while Mumford was contemplative and cautious; Geddes was more the extrovert, Mumford the introvert. Geddes liked to express his ideas in an idiosyncratic, "intellectual shorthand," while Mumford preferred carefully crafted, eloquently written prose. Despite the deep affection and respect each held for the other, these fundamental differences in age, experience, temperament, and work habits undermined their hopes to establish a collaborative partnership.

Geddes and Mumford were prolific correspondents, who both typically wrote several letters each day. The approximately 160 letters they exchanged represent only a small fraction of the thousands of letters they wrote to and received from others. Mumford conducted extensive correspondences with numerous people, including Frank Lloyd Wright, Henry A. Murray, Van Wyck Brooks, and Frederic J. Osborn.

Geddes' many correspondents included Jagadis Bose, Henri Bergson, James Mavor, and Mahatma Gandhi.[1]

While he is renowned for his writings on architecture and the city, Mumford considered his letters to be an important aspect of his literary achievement and to contain some of his best writing. His two autobiographical "miscellanies," *Findings and Keepings* and *My Works and Days*, include extensive selections from his correspondence. He claimed that his letters "tell as much about my literary as about my intellectual and personal development."[2] And he believed that they possess merits that distinguish them from the writing he produced for publication: "I have always said that my letters were better than any formal work I had written: my correspondents drew out something that an anonymous audience never received."[3] Mumford's letters to Geddes do, in fact, portray a persona different from that projected in his more formal published writings. The letters are more discursive and self-revealing, often more fluid and graceful than the massive volumes of cultural history and criticism on which his reputation rests. The letters more resemble in style and tone his autobiographical works, *Green Memories*, a memoir of his son Geddes, and *Sketches from Life*, his autobiography.[4]

Although he wrote many books and articles, Geddes never completed the "opus syntheticum," the comprehensive statement of his philosophy he hoped to write with Mumford's assistance.[5] As for what Geddes did publish, Mumford warned that readers are likely to be "put off by his crabbed, somewhat Carlylean style, by his incomplete thoughts, by his impatient shortcuts and his willful exaggerations." Mumford believed that Geddes' published writings inadequately reflect the depth and richness of the man's mind and personality; he stated that they do not fully convey "his stern common sense, his massive practical grasp, his astonishing breadth of scholarship, his relentless confrontation of reality."[6] Mumford insisted that to appreciate the true genius of Geddes one had to encounter the man in person, but he also noted that "the essential Geddes from time to time would spring up in his letters."[7] Accordingly, while their correspondence is more important for what it reveals about Mumford's intellectual development and early career, Geddes' letters provide vivid glimpses of a mind and personality that his published writings rarely offer. The letters reveal "the essential Geddes" in flashes of memorable eloquence, original insight, and ruthlessly honest personal confession; they reflect his impressive learning and zest for life; they show his generosity, passion, and humor. Moreover, Mumford found in Geddes' letters an intimacy, particularly in his sensitive responses to Mumford's requests for guidance, that their personal encounters lacked: "Paradoxically, I was closest to Geddes when we were spatially far apart and had to communicate by letter" (WD 103).

Geddes' enduring renown must be attributed, in large part, to his relationship with Mumford. Throughout his major works and in a number of essays and autobiographical writings, Mumford frequently and generously acknowledges Geddes' influence. Discussions of Mumford almost invariably mention the dramatic and profound impact Geddes had on his young disciple, and several scholars have given particular attention to this influence.[8] When he first read Geddes' writings on the city in 1915, Mumford experienced what he described as a sudden and dramatic intellectual awakening. Because of Geddes' influence, Mumford eventually expanded his research and writing beyond the streets and architecture of his native New York, beyond

American art and literature, to a broader field that encompassed the history of the city and technological development in Western civilization and, ultimately, the evolution of the human mind and cultural expression from pre-history to the present. Just as important as the Geddesian ideas and outlook that the young Mumford found so stimulating was the master's role as a model of the intellectual vocation. This example inspired Mumford to develop a wide-ranging approach, a style of "audacious insurgency," and a literary career that were uniquely his own. Although the disciple rebelled and eventually supplanted his master, much of Mumford's work can be read as a response to and extension of Geddes' thought. Moreover, Mumford has paid homage to his master by commemorating him as an ideal, symbolic personality. In fact, given the master's decisive and continuing influence on his productive and distinguished disciple, Geddes might well have claimed Lewis Mumford as his greatest achievement.

Geddes' published writing only partially accounts for the enormous impact he had on Mumford, and their two personal encounters were brief, disappointing, and unproductive. The correspondence was their consistent and most important means of contact and dialogue: Geddes and Mumford established, sustained, and defined their relationship by means of the letters they wrote to one another. Moreover, Mumford's dated annotations on Geddes' letters indicate that he read and re-read them throughout the half-century after his master's death. The Geddes–Mumford letters are notable not only because they document an important Anglo-American intellectual friendship, but also because they describe a fascinating psychological relationship that linked two very different, very powerful personalities; theirs was a complex, multi-faceted relationship both intellectually and personally. In order to appreciate the significance of the correspondence, it is important to understand something of Geddes' continuing, long-term influence on Mumford as well as how Mumford commemorated the thought and personality of his master. For the letters are interesting because of what they conceal as well as what they disclose.

DISCOVERY

Mumford discovered the work of Patrick Geddes in the fall of 1914, at the age of eighteen, when he was a student in the Evening Session of the City College of New York. As part of his studies in a biology course, he read *The Evolution of Sex* (1899) and *Evolution* (1911), works co-authored by Geddes and his long-time associate J. Arthur Thomson. Mumford was arrested by those sections of the books which bore the distinctive Geddesian imprint, "his Meredithian style and his vigorous and entirely refreshing point of view" (SL 144).[9] Within a year he had also read Geddes' *City Development* (1903), and in early 1916 he imported *Cities in Evolution* (1915) from the Outlook Tower in Edinburgh, Geddes' quondam research headquarters where he had established an "index museum" and "sociological laboratory."[10] Mumford recalls that he experienced a feeling of "elation equal only to that upon the first publication of my own first book" when he received the package containing *Cities in Evolution* and other writings by Geddes (SL 152).

Occurring at a critical juncture in his development, Mumford's discovery of Geddes' writings on the city was a dramatic and decisive experience for a young man who aspired to be a great writer. At the time Mumford was grappling with the decision as to whether he should leave the stimulating Evening Session of City College – which enrolled primarily mature, working adults – and enroll in the regular Day Session to complete his undergraduate degree. He had vague intentions of taking a Ph.D. in philosophy in preparation for an academic career. However, as a regular student, Mumford quickly grew impatient with what he considered the arbitrary and demeaning requirements of the undergraduate curriculum. The courses he was required to take seemed irrelevant to his literary ambitions, and he found the day students less serious and mature than those he had known in the Evening Session. At this decisive point, Geddes' writings prompted Mumford to leave college and to embark on the independent program of self-education which laid the foundation for his lifelong intellectual interests and literary career. As a result of his initial contact with Geddes, Mumford writes, "I reconstructed my personal activities as to get for myself the essentials of a genuine education."[11] This "genuine education" was not to be had at City College or any conventional college or university: Mumford recalls that Geddes' "thought and example turned me away from a routine academic education, and caused me to use the city itself, with its museums and its art galleries, with its busy economic and political life, as a field of study."[12] At this fateful moment, Mumford embraced Geddes as his intellectual master; and the disciple's commitment was enthusiastic and unqualified. By 1917 Mumford had abandoned plans to complete his undergraduate degree and to pursue an academic career. This decision was inspired by the independent and iconoclastic spirit of his intellectual hero, Patrick Geddes, who had "refused on principle to take examinations or stand for a degree ... to be entangled in [the] formalities, legalisms, stale traditions, and tepid conventions" of academic life.[13]

In an unpublished note written in January 1916, Mumford describes the dramatic impact of discovering Geddes' *City Development* in the City College library the previous fall:

> The one book which I had always eagerly half-anticipated momentously entered my life.... Geddes' *City Development* capped my educational climax. I felt, and still feel, and hope I shall continue to feel that in the planning, or at least in the enrichment of city life my serious work in future lay. Here I saw a means of utilizing my aptitudes in literature, art, philosophy and the biologic sciences: a focus for varied activities: a coordinating nucleus.[14]

This passage is uncannily prophetic in describing both the major concerns of Mumford's subsequent literary career and the interdisciplinary, organic outlook that informs his work, early and late. Mumford records that *Cities in Evolution*, which he read the following year, also "profoundly altered" his "habits of thought and ways of living."[15]

The immediate effect of both books was to inspire Mumford to explore "the streets and neighborhoods of New York ... with a new sense of both personal direction and public purpose" (SL 155). Mumford writes that Geddes "taught the reader, in simple terms, how to look at cities and how to evaluate their development." Responding to Geddes' emphasis on the "regional survey," Mumford roamed the streets of New York City "with a new purpose: looking into its past, understanding its present, replanning its future became indissoluble parts of a single process."[16] Geddes' influence also prompted Mumford to leave New York to take a temporary job in Pittsburgh for a few weeks in 1917; he wanted to study that city "as an example of [a] paleotechnic industrial centre" (a term he acquired from Geddes) and "to commune with Pittsburgh in every aspect."[17] Mumford's discovery of Geddes' writings on the city thus had direct, practical effects as well as continuing, permanent consequences. The young Mumford experienced such a complete conversion to the Geddesian philosophy that he wrote in an imaginary, but not completely facetious, dialogue with his future wife: "I am a disciple of Patrick Geddes, and I am an abject admirer of everything he has said and done."[18] In short, Mumford's discovery of Geddes' work while a student at City College encouraged him to pursue his native interests in the city; it helped him to define the content and focus of his studies; it provided a comprehensive, "organic" method of analysis which he took as a model for his own approach; it gave him a sense of ideological and practical purpose as a writer.

Upon discovering Geddes' work, Mumford began to dream of studying with him – a dream that was never realized. In late 1916 Mumford wrote to Frank Mears (Geddes' son-in-law in Edinburgh) proposing to study at the Outlook Tower under the tutelage of Geddes, but he was then engaged in teaching and city planning activities in India, where the first letter from Mumford reached him in 1917. So impatient was the young man to study with his intellectual hero that, as he later confessed, he "had the effrontery to regard the World War as an atrocious nuisance."[19] In one of his earliest letters to Geddes (probably written in late 1918), Mumford reiterates his desire to study with the master in Edinburgh, "even in advance of getting my bachelor's degree in America" (LM undated [1918]). By this time, of course, Mumford had little intention of finishing his undergraduate degree at City College. However, Geddes had been working steadily in India since 1914; and Mumford did not realize that back in Edinburgh there was "no one left to study with, and that, in fact, the tower, sans Geddes, was but a hollow shell" (SL 152), as it would remain throughout the duration of Geddes' life.

In the early letters, Mumford also expressed a willingness to help promote Geddes' work. He initially offered to try to interest New York publishers in Geddes' books, most of which he had found to be unavailable in America (LM 6 January 1916). More importantly, in what is probably his first letter to Geddes, Mumford describes how he hopes to build upon the master's work: "it is my own hope to work over the fields you have plowed up and develop a political philosophy of cities, to complement that civic interpretation of cities on which your first thoughts will long stand as the last word" (LM 1917 n.d.).

Thus, long before their first meeting in 1923, this "distant teacher" (SL 144) had profoundly influenced Mumford's thinking, education, and activities. Mumford writes that because of this influence Geddes evoked in him "a personal loyalty ... more intimate than with any other teacher" (GG 371). This enduring "loyalty" was inspired not only by Geddes' writings on the city but also by his prompt and generous letters to his young admirer.

INVITATIONS TO COLLABORATION

Geddes issued the first invitation for Mumford to collaborate with him in the letter written on Christmas Day 1919. Writing from India, Geddes proposes that Mumford "run my Town Planning and Civic Exhibition in America after a term of it here" and that he help to turn the "unusually disproportionate file of materials" Geddes had accumulated into "books beyond the small shelf of printed ones." Geddes laments that he has continually but unsuccessfully sought effective collaborators, and he mentions the tragic loss of his elder son Alasdair in the Great War as well as the fact that his other son, Arthur, had been "semi-invalided" – thus suggesting a relationship in which Mumford would be much more than a working assistant or literary collaborator (PG 25 December 1919). Mumford was astonished and delighted by this invitation. He recorded in his personal notes that "my imagination has been unlimbering itself by traversing unbelievable distances and fulfilling unexpected dreams" and that the prospect of working with Geddes, despite the obstacle of having to travel vast distances to exotic lands, has "bludgeoned my petty fears into annihilation."[20] Mumford begins his reply to this first invitation with the confession, "Your letter of December 25 has bowled me over"; and he responds with a "prompt and enthusiastic Yes!" (LM 2 February 1920).

This early letter marks the zenith of Mumford's enthusiasm for collaboration and his confidence that he would be able to work effectively with Geddes, despite the obvious differences of age and experience. He writes: "I could give my whole heart to your work, and head and hand would promptly follow in unison" (LM 2 February 1920). Within a few years Mumford would realize what an extraordinarily callow, naively optimistic statement this was. Yet even Mumford's reply to Geddes' first invitation and his simultaneous journal entries betrayed, perhaps unwittingly, his latent reluctance and doubts concerning both his ability and his willingness to devote himself to a "collaboration" in which he would play a subordinate role. Despite his apparent enthusiasm, he experienced deeply seated reservations that could not be suppressed. Mumford's response represents an effort to convince himself, as much as Geddes, that they could work together effectively and that he could bend his own temperament and interests to assist a much older, grief-stricken man who, quite obviously, was seeking a replacement for his slain first-born son. He also astutely sensed that Geddes was seeking an utterly devoted, wonder-working disciple who could redeem a lifetime's accumulation of unfinished tasks and unfulfilled dreams. But even in 1920 the young Mumford had grand, ambitious dreams of his own.

Political uncertainties prevented Mumford from joining Geddes in Jerusalem (where he was planning the new Hebrew University) in the spring of 1920 and accompanying him from there to Bombay as they had intended. He chose the alternative of going to London to work with Victor Branford, Geddes' colleague, as editor of the

Sociological Review, and perhaps later to join Geddes in Jerusalem or India. Mumford remained in London for six months, May–October 1920. While Mumford was there, Geddes issued a second invitation to join him in Bombay as his "secretary and editor [and] as collaborator too" (PG 2 August 1920). However, Mumford's experience as editor of the *Sociological Review* had dampened the appeal of collaboration with Geddes. As much as he liked the man personally, Mumford was beginning to find Branford an inflexible, dogmatic proponent of the Geddesian "system"; this was not encouraging to the young man who had been attracted to Geddes, the liberating iconoclast. Coincidentally, he was hearing disturbing reports from certain of Geddes' associates in the Sociological Society–Le Play House circle (two organizations concerned with sociology and the regional survey) who had found it difficult to work with the founder, indeed who were sometimes terrorized by him. Consequently, Mumford was beginning to doubt whether his own temperament would permit him to be an effective collaborator (as his unposted letter of 12 July 1920 reveals). Moreover, at the time, he had been reading Van Wyck Brooks' essays in which he heard a stirring call to return to America to establish his literary career (SL 356). He had ample leisure for such reading since his duties as "editor" were vague and unchallenging; nor was the income adequate, and he had not been able to find other employment in London. Most importantly, Mumford realized that he was in love with Sophia Wittenberg whom he had met while working for *The Dial* in New York the previous spring. This combination of factors eroded his resolve to remain in London or to join Geddes in Jerusalem or India.

After Mumford returned to New York in October 1920, Geddes continued to issue invitations. In January 1921, for example, Geddes laments the fact the he and Branford lack "younger partners" to carry on their work, and he identifies Mumford as their primary, perhaps even sole heir (PG 10 January 1921). Such a plea probably evoked in the young Mumford confused and troubling feelings of pride and guilt, responsibility and entrapment. The increasingly independent and congenitally reserved young Mumford must have recoiled from the inordinate pressure of the expectations imposed by the revered master whom he had never met in person.

Nor was Mumford encouraged by Geddes' failure to define clearly the role and duties he expected his disciple to assume or to describe the nature of their prospective collaboration in terms that were especially appealing. Mumford was certainly interested in an opportunity that would enable him to unite "the studious and the literary life with the practical"; but he was not attracted by the prospect of becoming "the selling partner" for Geddes' books, the Cities Exhibition, and other projects (PG 8 December 1922). Even had he accepted the literary-editorial role for which he was more suited, Mumford could hardly respond enthusiastically to Geddes' invitation "to tackle … many thousands of pages of notes and fragments in all sorts of disorderly accumulations in stacks of boxes, here [Bombay] and at Tower [Edinburgh]" (25 March 1922).

Geddes' hopes of enlisting the services of his young admirer were highest just prior to his New York visit in the spring of 1923. Shortly before his arrival, Geddes writes: "It is my dream that I shall find my long sought collaborator in you" (PG 9 March 1923). After that summer together in New York, which was disappointing to both as far as the prospects for collaboration were concerned, nearly two years elapsed before

Geddes renewed his invitations. These post-1923 invitations were considerably altered in tone and described much different kinds of tasks Mumford might undertake; they no longer offered to employ him as mere "secretary" or "selling partner." Instead, the invitations became urgent entreaties as Geddes began to realize that Mumford was his most worthy and sympathetic disciple, the person who could most effectively articulate his ideas and continue his work after his death. In 1925 Geddes writes from Montpellier: "So here I am, and as an old man making his will, to what legatees, and with what executors? ... Hence I ask you! Come next winter and see what legacies you can take over from my mingled heaps – and what executry here or there appeals to you" (PG 2 April 1925). Geddes was soon referring to Mumford as "my essential heir" and promising that "I shall not be excessive in my demands of collaboration" (PG 25 May 1925). Geddes' insistent pleas that Mumford join him at Montpellier as his designated heir continued until his death in 1932.

Unfortunately, Geddes never fully understood Mumford's personality – perhaps he had never really made an effort to do so – nor was he sympathetic with Mumford's single-minded dedication to the literary life. At one point, Geddes became infatuated with the misguided notion that Mumford might assume "a University headship," and that such a position would enable him to realize his destiny as Geddes' "heir," the "leading legatee" (PG 20 November 1927). In 1931 Geddes proposes that they "put our long-dreamed collaboration on a business basis," and he lists several schemes that might be mutually profitable financially (PG 27 April 1931). As Geddes' health deteriorated during the final years of their correspondence, his proposals and pleas became more urgent, more desperate, and more unrealistic. By 1930 the elderly Geddes did not realize that the possibility of fruitful collaboration had long vanished; if the younger Mumford had been hesitant a decade earlier, the mature man was certainly not willing to assume the subordinate role the master would inevitably impose. For by that time, Mumford had published three well-received books on American culture and was completing a fourth; he was beginning to plan and research the Renewal of Life series, the massive, four-volume project that would occupy him for the next twenty years. Rummaging through the chaos of a desperate old man's "Teufelsdröckian paper bags" (PG December 1924) of notes and manuscripts had little appeal for the increasingly successful and purposeful Mumford, who at age thirty-five was embarking on the major phase of his own career.

In *Sketches from Life*, Mumford scathingly depicts the type of collaboration Geddes sought from him: "What Geddes urgently demanded of me was an impossible lifetime of devotion as Collaborator (read docile filing clerk!), as Editor (read literary secretary!), or as Secretary (read handy drudge!). In short, the perfect disciple: his alter ego!" (SL 156). Their two meetings in 1923 and 1925 convinced Mumford that an attempted collaboration "would be so full of frustrations and humiliations that it might come abruptly to an end before anything was accomplished" (SL 402). Despite his obvious reluctance and published statements to the contrary, however, Mumford never completely abandoned the hope of establishing some sort of working association with Geddes. As late as 1929 he proposes to arrange a "conference" with Geddes and Branford "to discuss plans and future projects."[21] And in late 1930 Mumford lists among his "uncompleted tasks and failures" the fact that he has not taken on "various possible collaborations" with Geddes.[22] Even though he realistically anticipated the sort of frustrations and difficulties working with Geddes would inevitably entail, Mumford admits that only with Geddes' death did "the dream of faithful discipleship and helpful collaboration which had never come down to earth in either of our lives" finally come to an end (SL 407).

The reasons behind the young Mumford's failure to establish a collaboration with his "intellectual parent" reveal much about the development of his personality and habits of mind as well as the nature of their relationship. Although Mumford became a devoted disciple of Geddes from the moment he discovered his writings on the city in 1915, insurmountable differences in their temperaments, methods, and goals existed from the beginning of the friendship. Mumford was jubilant when he received Geddes' first invitation to join him and was convinced that he could devote "heart ... head, and hand" to his master's service (LM 2 February 1920). However, within a year he was beginning to question certain aspects of Geddes' intellectual system and to doubt whether he was capable of working with a man who was notorious for his volatile impatience, his frenetic and erratic work habits, and his sometimes capricious, even insensitive treatment of colleagues.[23] Mumford was also suspicious of Geddes' apparent obsession with his graphic method, the terms and diagrams he had developed to demonstrate his ideas and theories. Mumford recalls that that during his months in London, although he had not yet found what his "essential life work" would be, "Some inner conviction, however, already told me it would not be through any direct alliance with Geddes, still less through a permanent addiction to his diagrammatic ideological 'synthesis'" (SL 276). In May 1921 he writes that he is no longer "the earnest inquisitive student" he was in 1919, one who was willing to devote himself completely to Geddes' service: "I don't feel that I wish to give up the plan of active sociological collaboration with Geddes, ... but I feel that this is a minor and not a major theme in the development of my life."[24] Both Geddes' imperious personality and his dogmatic preoccupation with his "thinking machines" engendered considerable trepidation in the young man, who at the time was struggling to discover his own intellectual identity and literary vocation.

Before meeting Geddes in person, therefore, Mumford doubted whether he would be able to collaborate effectively with the older man. Geddes was a rapid, impetuous thinker; his copious "morning meditations" produced a spate of swiftly noted ideas and possibilities that he had accumulated in those "midden heaps" of "disorderly accumulations" (PG 9 March 1923). Mumford, on the other hand, was a slower, more methodical worker. He kept meticulously organized files containing research notes, outlines of books and articles he planned to write, correspondence, and other material. He first expressed reservations concerning his ability to work productively with Geddes while in London in 1920; Mumford tried to explain his inveterate deliberation in absorbing and utilizing the ideas of others:

> I have very great doubts whether my equipment is sufficiently complete to enable me to keep pace with you and to work profitably with you.... I cannot write at all until I have thoroughly assimilated all the material that is mine to work with, and to transcribe notes which are a result of your

whole lifetime of experience is a task which I could not attempt until I had taken sufficient time to appropriate them. The task seems ... merely one of mechanical journalism. I can't pretend to have the requisite equipment for this.

(LM 12 July 1920)

He reiterated these concerns about his suitability as a potential collaborator throughout their subsequent correspondence.[25] As a cautious, careful thinker and meticulous literary craftsman, Mumford recognized that he could not deftly produce either the various books and articles or the comprehensive "opus syntheticum" Geddes believed could be hewn from his massive piles of unorganized manuscripts. But Geddes apparently never understood his disciple's concerns along these lines, much less was he able to allay them.

Another reason for Mumford's early skepticism about the possibility of effective collaboration with Geddes was his observation of how Geddes' associates, particularly Victor Branford, insisted upon a rigid and unquestioning application of certain tenets of Geddesian thought, rather than building upon those original and liberating aspects of the outlook Mumford had found so stimulating in 1915. Instead of adapting Geddes' philosophy to "particular problems of social reconstruction" (LM 3 January 1921), or exploring its wider implications for social theory and cultural criticism, Branford and others reduced it to what Mumford viewed as an arbitrary and ossified system, to something purportedly complete, authoritative, even sacred. In 1920 Mumford did not realize the extent to which Geddes himself encouraged and even insisted upon such reductive dogmatism, unquestioning reverence, and sedulous imitation of his system and its "thinking machines" – as Mumford was painfully to learn when he at last met his master in 1923. When Geddes invited Mumford to criticize his thought and circle of followers, the disciple's response was at once diplomatically flattering and severely critical:

The weakness of the Edinburgh school so far has been the weakness of the Aristotelian school after Aristotle: the work of the founder has been so comprehensive and magnificent and inspiring that it has in appearance left nothing for scholars to do except to go over and annotate and dilute the master's work.

(9 May 1921)

However, this was exactly what Geddes had encouraged; and, tragically, such servile annotation and repetition were what Geddes obtusely demanded of Mumford, the most gifted of all his disciples.

As early as 1920, then, Mumford began to have serious doubts as to whether he could ever effectively collaborate with Geddes, doubts which were confirmed during their first meeting in 1923. Nevertheless, Mumford's deep sense of intellectual and emotional allegiance to his acknowledged master never wavered, then or during the subsequent six decades. Despite his recognition that a genuinely productive association would be impossible, Mumford was invariably tempted to assent, often falling prey to feelings of guilt and emotional turmoil, each time Geddes invited or, later, pleaded with him to journey to Montpellier to undertake collaboration. Mumford had difficulty bringing himself to reply to Geddes' proposals and invitations with definite, negative responses, even though after 1925 he had become absorbed in his own, independent literary career. In fact, contrary to what he claims in his autobiography, Mumford's letters to Geddes and various unpublished papers reveal that, after 1925, he was still contemplating some sort of collaboration (as editor or biographer), that he was willing to arrange a third personal meeting with Geddes at Montpellier, and that he did not decisively abandon either of these possibilities until Geddes' death in 1932.

TOGETHER IN NEW YORK, 1923

Geddes' ostensible purpose for coming to New York in the summer of 1923 was to deliver a series of twelve lectures on city and regional planning at the New School for Social Research, to meet with members of the Regional Planning Association of America, and to present a few additional lectures elsewhere – a schedule Mumford had arranged.[26] However, the unofficial but primary agenda was to initiate, at last, what each hoped would be a mutually beneficial and productive collaboration. Geddes expected to find in Mumford the "long sought collaborator" who would transform his unorganized piles of manuscripts into publishable form and who would assume responsibility for other of his projects. He even sent the Mumford a stipend of $250 so the young man could devote much of his time that summer to their work together, "instead of deviling at reviews and articles" for various periodicals (DR 345).[27] Although he had reservations about his suitability to undertake such tasks as Geddes would demand, Mumford keenly anticipated his first encounter with the man who had exerted such a profound influence on his life and thought, the man whom he idolized as the most important thinker of the day.

While Mumford hoped that the visit might initiate a fruitful and enduring working relationship, shortly before Geddes' arrival Mumford sent several warning signals expressing his concern that the mutually anticipated meeting might not proceed as smoothly or develop as productively as both may have assumed. As they were initiating plans for the visit, Mumford wrote that he had recently begun to discover his metier in the sort of "literary work" with which he was then engaged; he explained that after leaving London he had not renewed attempts to join Geddes in India because "my present vocation gives me a power of control over my time and energy which any other work would not permit"; accordingly, "for the present it is better that I should confine my energies to a province where I am quite effective than that I should disperse them over activities which would shortly bring me grief" (LM 7 January 1923). The "province" in which Mumford was then experiencing considerable success comprised working with the Regional Planning Association of America (of which he was a founding member), reviewing books for the *Freeman* and the *New Republic*, and beginning the study of American culture that would result in his next four books. The less congenial "other work" and "activities which would shortly bring me grief" may refer to certain secretarial tasks or other mundane duties that collaboration with Geddes would likely entail. A month before Geddes' arrival, Mumford mentioned another factor that made him "dubious about genuine *collaboration*": the great difference in age, experience, and outlook that separated them. He wrote:

> My life ... has followed a different trajectory as a result of being born in another part of the world, being almost a generation and a half apart in time, being affected by other circumstances.... So my relationship to you can never be fully that of a collaborator: it is rather of pupilship in which I absorb what I can of your thought and make it over and revamp it to suit the particular life experience I have encountered.
>
> (25 March 1923)

As one well acquainted with Geddes' personality and work habits, Dorothy "Delilah" Loch (who had befriended Mumford during his London sojourn) attempted to prepare him to cope with the man he would be hosting in New York:

> So I lay a serious charge on [you] – Geddes must be accepted (as good Catholics accept grief) with an open heart and no reserves, *if* he is to benefit those whom his presence scourges. He will brook no reserves.... Don't forget he is an old man ... and lonely – the very-most-vicious-cave-barbarian when sad, angered, or thwarted.[28]

Of course, Mumford had already heard other associates of Geddes describe the difficulties of working with him. For his part, Geddes several times warned Mumford that he lacked the "popular gifts" of effective public speakers and that his voice did not carry in large lecture halls and auditoriums (PG 25 November 1922). Given the frequency and candor of these warnings, it seems that Geddes may have been trying to prepare Mumford for the prolonged, beard-muffled monologues that he was to find so exasperating.

Two months before he arrived in New York, Geddes created a difficulty for Mumford that also boded ill for their anticipated collaboration. He had a thousand copies of his book *The Masque of Learning* (alternatively entitled *Dramatisations of History*) inexpensively reprinted in India under the imprint of Boni & Liveright, the New York publishers of Mumford's first book, *The Story of Utopias*. He did this without requesting permission from Boni & Liveright, and he intended to ship the copies to New York to coincide with his visit. Always the scheming opportunist, Geddes speculated that his presence there would help to promote sales. He put Mumford in an awkward position by asking him to deliver his letter of explanation and to plead his case with Boni and Liveright (PG to Boni & Liveright, 1 March 1923). In a diplomatic reply to Geddes, Mumford confessed that the letter and request caused him "considerable perplexity" and explained why he did not deliver it (LM 3 April 1923). This episode was not a propitious prelude to their impending first meeting or future collaboration: it revealed the sort of imperious high-handedness with which Geddes conducted so many of his enterprises; it asked Mumford to perform a service that would have embarrassed the young writer and that may have jeopardized his relationship with his publishers. Mumford must have wondered if this was typical of the duties he would be expected to carry out in "collaboration" with Geddes.

Mumford tells the story of their meeting in his autobiographical essay "The Disciple's Rebellion" (Appendix 2).[29] He describes this account as one which aims to depict "the immediate interplay of our two personalities, in the archetypal roles of master and disciple: a relationship whose seamier side was long ago revealed in Aristotle's constant sniping at Plato, after spending ten years under his tutelage."[30] While Mumford disavows any intention of denigrating Geddes' "great mind," the memoir recounts an experience that was preponderately unpleasant, in some respects bitterly disappointing, and in at least one instance even humiliating for Mumford. Yet the account does acknowledge important, positive results of the meeting. Mumford describes how he was deeply affected and transformed by his first encounter with Geddes in person: he discovered a completely new dimension of the man, an intense and powerful personality that neither his published writings nor his letters disclosed.

The meeting began inauspiciously the moment Geddes set foot on the pier in New York. He immediately became irritated with Mumford because he had not obtained a ticket to enter the area where passengers were disembarking. For his part, Mumford was aghast that his exalted master was not wearing a necktie, which Geddes had neglected to put on in the haste of packing before going ashore. Yet a scene far more distressing for Mumford occurred the next day when Geddes grasped him by the shoulders and exclaimed: "You are the image of my poor dead lad, . . . and almost the same age he was when he was killed in France. You must be another son to me Lewis, and we will get on with our work together" (DR 345). Geddes must have been anticipating, at least subconsciously, such a role and relationship for Mumford ever since mentioning the loss of Alasdair in the letter written on Christmas Day, 1919; in 1922 Geddes had similarly written: "your qualities and outlooks not a little correspond to those of our lost Alasdair" (15 April 1922). However much he had idolized Geddes, the reserved and introverted young man was stunned and embarrassed by this nakedly emotional overture. He lacked both the maturity and presence of mind to respond tactfully or sympathetically. Mumford writes that "the grief and desperation in this appeal" were "too violent, too urgent, for me to handle. The abruptness of it, the sudden overflow, almost unmanned me, and my response to it was altogether inadequate, not so much from shallowness of feeling as from honesty" (DR 345). In addition to being repelled by the shock of such a blunt and intimate request, Mumford knew that he bore little resemblance to Alasdair, physically or otherwise. Mumford writes that Geddes' "need to falsify our relations and warp them in accordance with his own subjective demands" provided "a clue to a certain blind willfulness" that had "undermined" his own work and that would continue to plague their personal relationship (DR 346).

Despite such an unsettling commencement to their first meeting, Mumford was immediately dazzled by the scintillating mind and personality of his intellectual hero. Mumford found Geddes in person to possess an exuberance and vitality that words could not communicate. Within a week, however, Mumford began to detect an imperious urgency and dogmatism that belied the astute sensitivity and bold originality that he had found so inspiring and liberating in his master's philosophy. Soon after Geddes' arrival, Mumford reports to Delilah Loch:

> "He" came last week. I speak of Him in capital letters; for now that I have seen a little of him I am more convinced than ever that he is one of the Olympians. Of course that is the difficulty. Jove never walked among the sons of men without the sons of men getting the worst of it, and I find that all the warnings and reservations I have put into my letters have had precisely no effect upon P.G.; for he is a terrible and determined man, and now that he is ready to set down his philosophy, he wants to make use of me to the full. Which would be good and proper if I were a secretary, with a capacity for writing shorthand at 200 words per minute; as it is, I am not a secretary and I cannot write shorthand; what is more, I am as much of a visual as Geddes is, and I simply can't stand listening to anybody for more than an hour at a time, whereas he has held me, as the Ancient Mariner held the wedding guest, for five or six hours at a time; with no result whatever upon my intellect, except to make it dazed and stupid. He calls this collaboration; I call it physical torture. It is not that he lacks a shrewd eye to the state of things between us; but he is determined to carry his work through, and he has the conviction that he knows what is best; whereas I know that I am of no use whatever as amanuensis, and I have an equally firm conviction that although Geddes may know what is best for people at large, I, with infinite pains and difficulty, have a certain limited knowledge as to what is best for myself, in which knowledge is the conviction that I have not either the capacity or the will to become Geddes Minor. I see now why he has lacked collaborators and has, intellectually speaking, always been surrounded by servants. . . . He gives one no opportunity for reflection in the intellectual sense; he demands reflection in the physical sense; and that is the sort of thing which even a poor second rate mind like my own finds it impossible to give.[31]

During the first two months of the visit, their relationship was not a tranquil or happy one. Geddes may have been hurt and perplexed by Mumford's chilly response to his plea that the young man become as a son to him. He must have quickly perceived that his disciple was an ambitious, self-absorbed young man who was reluctant to immerse himself in the work of collaboration. For his part, Mumford found the old man to be egotistical and insensitive, usually preoccupied with other people, often complete strangers, rather than the faithful disciple who had arranged the visit. During those infrequent occasions he did devote to Mumford alone, Geddes typically subjected him to protractive and tedious monologues and, on at least one occasion, demanded that he reproduce from memory charts and diagrams of the graphic system – an experience that angered and humiliated Mumford. In early July, Mumford wrote in his personal notes that "it came to me that we were not merely not getting on with our work together but seemed actually from some obscure misunderstanding to be driving apart."[32] However, certain causes of their "misunderstanding" could hardly be termed "obscure": their great difference in age and experience, the strong and independent ego each possessed, and the unrealistically high expectations each must have placed on this long-anticipated meeting.

On 6 July 1923 the disciple rebelled. He wrote Geddes a long letter in which he analyzed the differences that separated them, described his own personality and interests, criticized Geddes' methods and work habits, and pleaded with his master to give him a clearly defined role and direction in their collaboration. The fact that he chose to convey all this in a letter is significant: it suggests, on one hand, that Mumford lacked the courage to confront Geddes directly and, on the other, that Geddes was so oblivious to Mumford's feelings and so preoccupied elsewhere that there was no other way for the disciple to express his concerns. In the letter, Mumford discusses the profound difference in outlook between his cynical, modern generation and Geddes' more optimistic, Victorian one. He protests that Geddes appeared to view him as a "quack journalist," instead of appreciating his devotion to serious writing. He charges Geddes with attempting "to make me over a little into the idealized portrait, whose aims and interests and actions were more congruent with your own" – an attempt which Mumford "instinctively" resisted. Finally, Mumford pleads with Geddes to concentrate on the essentials of his thought and to shape it into some concise and coherent form: to "take all the sub-theses, which obscure the main problems, and cast them aside; and devote your attention to the architectonic whole, and to the problem of how that whole is to be presented." He criticizes his master's habit of producing a "morning's pile of paper" that "will be so big presently that it will defy almost anyone's patience and organizing power" (LM 6 July 1923).

The letter had an effect, but only temporarily. In a reply written the same day, Geddes responded with warmth and understanding. Mumford recorded the scene that occurred when he reported for their morning confabulation several days later: "He looked at me kindly and reached out his hand: 'That was a good letter!,' and we had such a day together as we had never had before."[33] In his reply, Geddes admits that "yes, we have yet to meet," but he denies having stereotyped his young admirer as a "quack journalist": "There is no caricature of you in my mind, whatever be sometimes said in half-jest . . . nor yet any *excessive* ideal either. I claim *some* little understanding of the real creature!" Beyond that, however, Geddes does not acknowledge Mumford's criticism of his work habits, nor does he outline any definite program with which his young disciple might assist him; he only assures that he does not intend "mere secretaryship" or that Mumford become "a mere exponent of mine." As for his apparent unwillingness to organize and to clarify "the architectonic whole" of his thought, Geddes' justification is that "fresh thoughts and theoretic visions come with and from each task"; this implies that he had no intention to cease adding to the "morning's pile of paper" or to condense the essentials of his thought into concise and coherent form. His only excuse for his inability to devote himself to collaboration with Mumford is that he finds himself in New York "on holiday and free" for the first time "in a decade" (PG 6 July 1923).

The summer 1923 meeting in New York must be accounted a failure insofar as it did not establish a continuing, productive collaboration. As Mumford had suspected, their temperaments, habits of mind, and routines of work were incompatible. Moreover, Mumford discovered that the great Geddes possessed some significant flaws and weaknesses, and he found the older man's treatment of him frequently perplexing, sometimes painful. He experienced contradictory feelings for Geddes: "Personally he seemed a mixture of great obtuseness . . . and swift, keen insight."[34] Mumford's reaction to Geddes must have resembled that of his fictional persona, Bernard Martin, in the presence of his "master," James McMaster: Bernard finds himself "shrinking into a chaos of complicated resistances, half paralyzed by worship, weariness, and fear . . . aghast at that great pride and energy of mind which takes life for its province and falters at nothing."[35]

However, despite the problems and disillusionments, Mumford writes in "The Disciple's Rebellion" that "my sense of Patrick Geddes's 'greatness' survived the whole summer I spent in his company" (DR 346). Although they did share "some good moments, brief but memorable" which remained with Mumford throughout his career, the aspect of Geddes' visit that most influenced him was not contained within these rare "moments" (DR 349). Rather it was his firsthand encounter with the living person that had the truly powerful, enduring impact on Mumford. In fact, he claimed to have been more profoundly influenced by Geddes' personality, which he first encountered during this otherwise "abortive" (DR 346) collaborative venture, than he had been by the discovery of the master's books on the city eight years earlier:

> [N]othing that he had written . . . had an influence on my thinking nearly as profound as he in his own person had on my life. . . . [T]he impact of his person shook my life to the core. . . . The most impressive thing about Geddes, even at sixty-nine, was the sense he conveyed . . . of what it is to be fully alive. . . . [I]t was by this magnificent aliveness that Geddes towered above those around him.
>
> (DR 349)

Only after meeting Geddes in person could Mumford appreciate Delilah Loch's cautions and understand why such an original and "prodigious" thinker had not been able to enlist and retain capable disciples: "the very intensity of his vitality, its exorbitance, made impossible demands upon those about him" (DR 350). He saw in the person certain features Geddes' writing could not convey: "A delight in every manifestation of life, a sense of its wonder and mystery, was what issued forth from Geddes' personality, a kind of radiant emanation, with a halo of visible and usable wisdom, not essentially different from what emanated from William Blake" (GG 372).

The 1923 summer meeting had a crucial impact on the young disciple, because of both its positive and negative aspects. In "The Disciple's Rebellion," Mumford writes: "though we met briefly in 1925 and corresponded at irregular intervals up to the month of his death in 1932, this parting was really our final one" (DR 351). However, this first encounter did not end their association; rather it defined both the nature of their subsequent relationship and the sort of influence Geddes would continue to exert. The experience allowed Mumford to discover key aspects of the man's "genius" and his "extraordinary gifts" which were embodied in the person and and which his writings did not disclose. Delilah Loch wrote him a letter of encouragement and commiseration during Geddes' visit:

You suffer, I fancy, from being the hope of his old age.... [Yet] you will never regret these months of travail. Geddes' chance words, Geddes' casual reflections will come back to you for untold years afterwards with the force of inspiration and illumination to thought, life, and conduct. In your present turmoil you may hardly know you have heard many things which will stick by you for life afterwards.[36]

The prediction was accurate. Geddes is cited in nearly all of Mumford's books, and he figures prominently in Mumford's last – the autobiography published sixty years after their their first meeting. Without the 1923 meeting, however disappointing, it is unlikely that Geddes would have assumed such significance in Mumford's later writing as an ideal, symbolic personality – as one who embodied the purposeful, intense, and balanced life.

Therefore, although a complete failure in achieving its anticipated, primary aim of initiating a working partnership, the 1923 meeting must be accounted an enormous success as far as its long-term effects on Mumford's thought and writing are concerned. Far more was gained than lost. Rather than promoting collaboration, the impact of Geddes' personality actually had the effect of inspiring and shaping Mumford's independent career. Instead of a definite collaborative role or an agenda for future tasks, what Geddes gave Mumford that summer was "an insight into the possibilities of intense activity and intense thought."[37] The encounter represented their "final parting" only in ending Mumford's youthful and naive dream of remaking the world in association with his master.

In a letter to his friend Josephine MacLeod written on the return journey to Bombay from New York, Geddes provides a surprisingly clinical evaluation of Mumford's potential as a collaborator: "Lewis Mumford is a very bright and keen writer, ... but not yet ready for the scientific grind, though not incapable of that; as I found by putting his nose to the grindstone repeatedly. Still he is not yet the positive collaborator I had hoped."[38] Geddes apparently experienced some measure of disillusionment in his gifted young disciple and prospective collaborator, but this is the only recorded indication of it. Despite whatever disappointment he may have felt in the summer of 1923, throughout the next nine years Geddes reiterated and intensified his pleas for Mumford to join him.

After learning of his unhappy experience with Geddes that summer, Gladys Mayer – one of Geddes' most devoted associates – also wrote Mumford a letter of consolation:

So he's devil-ridden and God inspired in a breath, and we don't know whether to bless or curse him most. And the only solution is to love him as I did and do. This is how I see the origin of your torment and P.G. [is] the most tormenting and lovable spirit that ever harried the earth.[39]

While Mumford must have cursed Geddes often during the summer of 1923, his affection and respect for the man – both for what he had achieved and for what he represented – endured and flourished. In their correspondence during the ensuing years, both Geddes and Mumford express regret that the 1923 meeting was not more productive. In fact, each ascribes a certain amount of blame to himself: Geddes admits that he was foolishly preoccupied with other matters and had not been "cleared up" sufficiently to exploit the situation (PG 17 May 1931), and Mumford confesses that he had lacked the maturity and sagacity to make the most of the opportunity (3 May 1931). Mumford writes in "The Disciple's Rebellion" that their failure to cooperate resulted, at least in part, from the fact that "there were two demanding, self-absorbed egos to reckon with" (DR 348).

Mumford began to address Geddes as "master" in the salutations of the letters written after their 1923 meeting.[40] This reveals that Mumford's esteem for Geddes had not been diminished by the disappointments of the visit and, moreover, that encountering the man in person had irrevocably established his revered and idealized status. Although Mumford from time to time criticized aspects of Geddes' thought and method in the correspondence of the following two years, both continued to discuss the possibility of arranging another meeting and initiating some sort of collaborative work. They hoped that a second meeting in Edinburgh in 1925 would be more productive than the first in New York. Shortly before the Edinburgh meeting, Mumford expressed his hope to accomplish "much ... more" by way of collaboration than he had been able to do in 1923 (LM 20 June 1925). Mumford's willingness to undertake a second attempt at collaboration, despite the failure of the first, also indicates the depth of his esteem for the man he now formally acknowledged as his "master."

However, as far as Mumford was concerned, their second meeting only confirmed his sense that productive collaboration with Geddes would be impossible. Mumford visited Edinburgh in September 1925 for five days on his homeward-bound journey after lecturing at Alfred Zimmern's International Summer School at Geneva. Even though he was able to fulfill his youthful dream of residing in the Outlook Tower, Mumford found Geddes in his native habitat to be even more insensitive and willful than he had been "on holiday" in New York two years earlier. Again preoccupied with other people and tasks, Geddes paid relatively little attention to his most brilliant, if not most devoted, disciple. But Mumford accepted the reality of the situation with philosophical resignation rather than with bitterness, with enduring respect for Geddes rather than anger: "And yet I love him; I respect him; I admire him; he is still for me the most prodigious thinker in the modern world."[41] As Mumford evaluated their relationship half a century later, this final meeting was decisive: "all possibility of any closer relation with Geddes was over" (SL 403). Nevertheless, between 1925 and Geddes' death in 1932, Mumford's enduring devotion to his master would occasionally rekindle the dream that they might somehow manage to meet and to accomplish great things together.

Even after the Edinburgh fiasco of 1925, Mumford never bluntly told Geddes that he had no intention of joining his master at Montpellier, that he had completely abandoned the dream of collaboration, or that he did not wish to become Geddes' literary executor and designated intellectual heir. In fact, shortly before Geddes' death in April 1932, Mumford was trying to arrange his itinerary to include a visit to Montpellier during the study trip in Europe he had planned for that summer. Mumford's reluctance to be candid with Geddes was certainly in part motivated by a wish to spare the feelings of an elderly, seriously ill, increasingly lonely and desperate man. But there was also enduring loyalty and affection for Geddes which made it impossible for Mumford to disappoint his master or to abandon completely his youthful dream of collaboration.

THE TRAGEDY OF THE RELATIONSHIP

The relationship between Geddes and Mumford never became what each hoped it might become. When he discovered Geddes' work, Mumford's initial reaction was one of exuberant intellectual excitement, and he immediately became an ardent disciple who deeply revered his master. Unfortunately, during their two personal encounters, Mumford experienced Geddes' "radiant" personality in a mere handful of memorable but fleeting moments they shared. The extent to which Mumford had idealized his master before their first meeting, making him into an exemplary figure of profound originality and intellectual authority, created expectations that it was impossible for Geddes – or perhaps any person – to fulfill. In retrospect, it is clear that the disappointment and disillusionment the disciple experienced were inevitable.

The relationship was unavoidably complicated by their great difference in age (forty-one years). The Geddes whom Mumford encountered in New York and Edinburgh was an old man who could be rigidly dogmatic in outlook, unpredictable in behavior, and insensitive to the needs of his young admirer. Mumford writes that he had initially been drawn to the younger Geddes of the 1880s: "the man of vision, the liberated and insurgent spirit. . . . But at the time we met, Geddes' graphic system, with its desiccated ideology, had taken possession of him and displaced the person I had been drawn to" (GG 371).

The relationship was also impaired by Mumford's reluctance, or inability, to express his true feelings for Geddes. Compared with what he wrote about Geddes elsewhere, Mumford's letters rarely and inadequately acknowledge the extent to which he felt indebted to his master; nor do they provide a candid expression of his avowed "personal loyalty... more intimate than with any other teacher" that he repeatedly emphasized both in his personal notes and in writings published after Geddes' death – including the autobiography completed over half a century later (GG 371). Had he openly expressed such feelings of loyalty and affection, perhaps they could have established some sort of collaboration; or at least, the relationship might have assumed a more honest and personal character in which the sense of frustration and loss was not so acute.

In the mid-1920s, Geddes began to besiege Mumford with a succession of invitations, proposals, and pleas which became increasingly urgent. Although he appeared to be as active as ever, the aging Geddes (who turned seventy in 1924) realized that the prospects of writing and publishing his "opus," not to mention the completion of his other projects, were rapidly diminishing, that "for the most of it, it may be too late" (PG 20 March 1926). Thus Geddes looked to Mumford to help him finish uncompleted tasks and to assume responsibility for his work after his death. In March 1929, for example, he importunes: "try to come – Look you I'm old ... We need your cooperation" (PG 20 March 1929); and in April he reiterates: "Now's the time! Neither of us [himself and Branford] has anyone of your calibre to look to for collaboration – and we are seeking for heirs and executors – as yet without success" (PG 10 April 1929). Re-invoking the kind of relationship he had unsuccessfully tried to establish with Mumford when they first met nearly a decade earlier, he began his penultimate letter to Mumford with the greeting: "Dear Mumford – (no – Lewis my Son!)" (PG 5 January 1932). This salutation represents a desperate, final effort to establish a level of personal intimacy and a sense of obligation in Mumford that would persuade him to assume the role of literary executor and intellectual heir; it also reflects Geddes' recognition that Mumford was the most innately sympathetic and loyal of all his followers, that Mumford was his true "son" at least in an intellectual sense.

Therefore, despite the deep esteem each man felt for the other, the letters sometimes fail to communicate what might have been more honestly expressed or what would have been more relevant to their primary concerns. Reading the letters, one occasionally senses a restraint, a suppression of feeling and candor, even a melancholic tone of regret and loss. This is particularly apparent in certain letters written during the later stages of their relationship. Mumford detected such a tone in the letters Geddes wrote him after their 1925 meeting in Edinburgh: "I am saddened by two things: the repetitiousness of their ideas and suggestions and, even worse, their irrelevance, their failure to reach the plane of easy personal intercourse" (SL 406). Geddes was never able to articulate a timely and coherent statement of his need for Mumford's assistance, nor did he recognize that, after 1925, collaboration and literary "executorship" were losing their appeal for the increasingly successful and independent young writer. For his part, Mumford's letters also failed to "reach the plane of easy personal intercourse" because of his reluctance, or inability, to express his deep sense of gratitude, loyalty, and affection.

In "The Disciple's Rebellion" Mumford makes clear that, as far as he was concerned, the possibility of collaboration with Geddes decisively ended in 1923 and that he indulgently endured, but firmly declined, Geddes' repeated, subsequent invitations to collaboration. However, the letters and unpublished papers show that he actually had ambivalent feelings about the matter, feelings which continued to trouble him until Geddes' death in 1932. Mumford writes in 1925 that his "internal processes" had somehow been "paralyzed" during their New York meeting but that he hoped to do "much more" by way of collaboration in the future (LM 20 June 1925). In 1928 he confesses to Branford that the invitation to join Geddes at Montpellier "tore at my heartstrings" and made him "wish keenly" to arrange a visit: "I would drop everything and join him for a while had I any feeling within me that our being together could bring anything more than a deeper sense of friendliness and encouragement. Still, if I can possibly manage it this spring, I will."[42] It was not until 1931 that Mumford forthrightly told Geddes that productive collaboration on the long-dreamed "opus" would be impossible and declined the offer to become Geddes' designated literary executor and intellectual heir, even though he was still apologetically accepting much of the responsibility for their failure to establish a working relationship in 1923; nor did he have any interest in assisting Geddes in other ways, such as promoting the Montpellier college project or pitching his articles to American journals – because they paid more than European ones (LM 3 May 1931). Nevertheless, a deep emotional bond and sense of allegiance remained. Mumford was clearly torn between the practical impossibility of working with a difficult old man and an enduring sense of obligation to his intellectual father.

But by 1930 it was too late to initiate the sort of personal relationship both had attempted to establish in 1920. A decade earlier Mumford would have relished the role as Geddes' personal assistant and designated heir. In 1930 Mumford had achieved considerable success and recognition on his own, and he had little interest in undertaking any work that would divert him from his flourishing literary career, no matter how desperate the pleas of his master. Mumford must have taken Geddes' requests to sell his articles and to recruit students for the Montpellier college as impractical, if not demeaning. Geddes' last letter to Mumford ends with the familiar, sad refrain: "So you must take over much of my further Sociology and Education, etc." (PG 2 April 1932; he died on 17 April).

Soon after Geddes' visit to New York in 1923, Mumford began to view that disappointing experience as one which boded considerable tragic loss for both of them. On the one hand, he revered Geddes as his master, mentor, and role model who had been his most important source of ideas and inspiration. On the other, there existed significant impediments to their collaborating as equals, not the least of which was the unavoidable conflict between two powerful and ambitious personalities. There was also the inevitable conflict of a father–son relationship: deep, instinctive loyalty and love existed on both sides, to be sure; yet the father's compulsion to dominate and control conflicted with the son's struggle to establish his freedom and independence. In 1924 Mumford writes Delilah Loch that news of Geddes' recent illness "gives me another wrench after the failure of our meeting last summer." In a passage of striking poetic eloquence and psychological prescience, he describes the tragic loss that looms ominously before them:

> O, Delilah: this damned flaw in life! This cracked gem; this vase too beautiful to be broken, and too broken to be wholly beautiful. The tantalizing nearness of everything we want most; were it not for some fatal, stubborn grain in both of us, Geddes and I, linked together, intellectual and emotional, might still conquer the world. For lack of this he will be imperfectly articulate, and I, perhaps, will have nothing worth articulating.[43]

In a subsequent letter, Mumford describes his "temperamental inability" to work effectively with Geddes as something that "happens in all discipleships: it is the real tragedy of the relation."[44] He also discusses the tragic dimensions of the relationship in his correspondence with Victor Branford. After receiving one of the many invitations to join Geddes at Montpellier, he writes Branford of his reluctance to assume any of the roles Geddes wanted him to fill – as "amanuensis or super-secretary," as "an active teacher or even as college principal," or "taking up" the master's work in some other capacity. He describes his relationship with Geddes as "a tragic dilemma":

Tragic, because no one realizes better than I the importance of Geddes' thought, and the necessity of making it available; tragic because Geddes has waited so long, and has gotten, as it were, into a habit of thinking which is almost as destructive to publication as apathy and laziness; tragic because, given us all as we are now situated, nothing as far as I can see can be done.[45]

Mumford felt the tragic element of the relationship most acutely following their second, and final, meeting in Edinburgh in 1925. At the time, he recorded his recognition that his youthful dreams of comradeship and grand achievement in collaboration with Geddes would never become realities. During this last encounter, Mumford had observed contradictions and flaws in Geddes which he could neither reconcile nor tolerate; he realized that he could not suppress his feelings of rejection, humiliation, and bitterness which the willfulness and egotism of his master evoked. Upon departing Edinburgh he described the aging Geddes as a man of perplexing paradox: "the weakness and the strength, the steadfastness and the impatience, the effacing humility and the ruthless arrogance of this great man." He despairingly asks:

[W]hat am I to do with a pathetic man who asks for a collaborator and wants a supersecretary, who mourns the apathy and neglect of a world that he flouts by his failure to emerge from his own preoccupations and to take account of other people's: this man who preaches activity and demands acquiescence. . . . What an affectionate, loyal relation he could have, if he would permit it to exist.

However, Mumford refused to allow the old man's behavior to tarnish the image of the magnificent "master" who had so influenced him: "I shall retain the memory of the comrade I found too late, the great companion who, he and I, might have made over the world, gaily and exuberantly, in our double image!"[46] The great dream of Mumford's youth had been to remake the world in collaboration with Geddes. But after their final meeting, he realized that he would have to undertake the task alone.

Although not involving a direct conflict of the two personalities, there is another, coincidentally tragic dimension of their relationship: each experienced the loss of his young firstborn son in war. There are related, rather eerie similarities of experience involving both their respective roles as biological fathers and Geddes' paternal relationship with Mumford. The depth of Mumford's esteem for his master is indicated by the name he gave his son, Geddes Mumford, born on 5 July 1925. Even though Mumford later wrote that this was a name "perhaps too piously bestowed," the young Geddes – like the man whose name he bore – was an astute observer of nature and a strong-willed individualist (SL 389). Geddes' brilliant eldest son Alasdair was killed on the Western Front in 1917; Geddes Mumford was killed during the invasion of Italy in September 1944. Alasdair's death left Geddes permanently embittered; and, as previously discussed, his expectation that the young Mumford would replace the son he had lost undermined their relationship from the beginning. The loss of his own son was a similarly devastating blow to Mumford. The amount of writing he produced was substantially reduced during the late 1940s and early 1950s, and he succumbed to periods of severe depression during that period. However, unlike Geddes, he was eventually able to recover and to become even more productive than he had been prior to 1945. Mumford took particular care not to repeat Geddes' mistake of attempting to reincarnate his lost son in an admiring follower: "when I was faced with a similar grief I was careful not to seek from my students . . . the response I would no longer get from my dead son" (DR 346).

GEDDES' INFLUENCE ON MUMFORD

There are three important aspects of Geddes' thought and example that powerfully influenced the young Mumford upon discovering the master's work. First of all, Geddes' two books on the city vividly disclosed the complexity and richness of urban life, taking into account both the city's historic past and its future potential. As "a child of the city," Mumford had been fascinated by the urban scene ever since his childhood walks through the streets of New York with his grandfather. In his 1982 autobiography Mumford writes that it is difficult to appreciate "the exciting effect" that *Cities in Evolution* had on him when he first read it in 1916, for the "new approach" that Geddes advocated there was unprecedented in the existing literature on the city (SL 151). A second aspect that profoundly influenced the young Mumford was Geddes' view of the intellectual as activist: he taught that the scholar's studies should be joined with practical activity, and he emphasized the individual citizen's responsibility to improve urban conditions. Mumford writes that "Geddes was almost alone in translating this new urban knowledge into action, or at least concrete proposals for action" (SL 151). Third, Geddes served as a model of the intellectual as generalist who did not confine his work to that of a conventional literary, scholarly, or professional specialization. Mumford felt an instinctive affinity for Geddes' interdisciplinary approach that sought to create an "organic unity" out of the disparate facets of urban life. Although he was trained as a biologist, Geddes' outlook encompassed multiple fields, which included philosophy, history, education, city planning, and sociology. Mumford described him as a "professor of things in general" who "refused to recognize the no-trespass signs that smaller minds had erected around their chosen fields of specialization" (SL 220) – and Mumford liked to characterize his own work in identical terms. Acknowledging Geddes' important role in shaping his thought and outlook, Mumford later described him as "the distant teacher who helped to bring all the diverse parts of my education and my environment together and transform them into an increasingly intelligible and workable . . . whole" (SL 144).

In addition to these three particular aspects of influence, the young Mumford identified in his master's writings a characteristic, general approach to life and thought that he later described as one of "audacious insurgency" (SL 158). He recognized in Geddes "the audacity of an original mind, never content blindly to follow established conventions, still less the fashions of the moment" (SL 145). Geddes' example as a model of "audacious insurgency" – his demonstration of how to pursue the intellectual life with courage, purpose, and passion – profoundly influenced and inspired Mumford during the crucial decade of his formative development, 1915 to 1925: "Geddes' greatest gift to me was to deepen and reinforce the foundation that other minds had already made, while he gave me the courage to build an original structure with new materials in a different style: radically different, necessarily, from his own" (SL 158). Geddes thus enabled Mumford to realize his own unconventional, wide-ranging literary vocation, the calling to which he would devote the rest of his life:

> Patrick Geddes' philosophy helped save me from becoming a monocular specialist.... [I]t gave me the confidence to become a generalist – one who sought to bring together in a more intelligible pattern the knowledge that the specialist had, by over-strenuous concentration, sealed into separate compartments.[47]

Had it not been for the influence of Geddes, Mumford writes, "I could easily have drifted into a purely bookish life, short in tether, prudently bridled. That might have been the smooth road to conventional academic achievement" (SL 409). Not only did Geddes save Mumford "from becoming just another specialist," but he also provided the young man who was just beginning his career with "a vision to live by" that would shape and inspire his work for the next sixty years (SL 168). Mumford had certainly been influenced by those to whom he referred as his "American elders" – figures such as Thorstein Veblen, John Dewey, and Van Wyck Brooks. But, as he wrote Geddes, they could give him no more than "valuable fragments"; whereas Geddes' example and teachings were crucial to his creation of a coherent intellectual outlook and a literary role that were uniquely his own (10 December 1925).

What Geddes taught Mumford about how to study cities remained an important influence throughout his long career. The "method and outlook" Geddes advocated provided a model of "how to look at cities, how to interpret their origins, their life, their cumulative history, their potentialities."[48] Mumford's later writings on the city extend and deepen Geddes' approach. This approach recognizes the dense complexity of urban life and the importance of understanding the city's historic roots; it emphasizes the city's organic relationship with the region; and it advocates "conservative surgery" as the means by which the city "could be kept alive and retain its original character."[49]

However, beyond this emphasis on the city, Geddes' demonstration of how certain biological principles could inform the study of human culture ultimately had a more critical, far-reaching influence on Mumford's thought and writing. Trained as a biologist in the laboratory of Thomas Huxley, Geddes became interested in the relationships existing throughout the natural environment – plant, animal, and human. Geddes' notion of a "human ecology" was important in shaping both Mumford's method of historical analysis and the scope of his interests. In fact, Mumford claims that Geddes went further than any other philosopher "in laying the ground for a systematic ecology of human culture" (GG 367). In an early letter to his disciple, Geddes uses the term "historic filiation" to describe the evolution of human culture as an ecological process involving "the accumulation and transmission, generation by generation, and day by day, of socially acquired characters" (PG 20 December 1922). And Mumford later described his master using the same terms: "the patient investigator of historic

filiations and dynamic biological and social interrelationships."[50] He identified Geddes'
primary, original contribution as that of introducing "organic methods of thought and
action into aspects of life hitherto severed, amputated, discrete."[51] Mumford writes that
Geddes "saw both cities and human beings as wholes; and he saw the processes of
repair, renewal, and rebirth as natural phenomena of development."[52] Mumford
adopted this organicist, ecological view of human development; it became important
not only in shaping his interpretation of the city but also as a methodological approach
in his historical surveys of human development.[53]

Geddes' scientific studies convinced him that organisms are not necessarily in the
grip of mechanistic, deterministic processes of development, and he also applied this
biological principle to his analysis of human and cultural development. This was
another key, biologically related insight that Mumford acquired from his master.
Mumford writes that "Geddes made an important contribution in restoring the
Aristotelian concept of potentiality and purpose, as necessary categories in the
interpretation of life-processes" (GG 359). For Geddes, such "potentiality and
purpose" are represented in humankind's capacity for "insurgence." This, Mumford
writes, was "the quality of life that seemed most essential" for Geddes: "Man for him
was not just an adaptive organism ... but increasingly the shaper and molder of his own
world."[54] A recognition of humankind's "insurgent" potential underlay Geddes'
"conception of life as essentially an enacted drama" (GG 360).

These Geddesian principles were particularly important to Mumford as he began
to realize his own literary identity as a cultural ecologist in the Renewal of Life series.
Initially, Geddes directed him to the study of cities and taught him how to interpret the
city as "an important human artifact."[55] But as his career evolved, Mumford did not
confine his interests and writing to urban studies. Mumford's mature and most
important work is characterized not so much by its focus on the city as by its concern
with "historic filiation" on a broader canvas: the evolution of the human mind and
human expression in language, ritual, and symbolic activities. The method and scope
of such works as *The City in History* and *The Myth of the Machine* reflect Geddes'
concept of a rich and complex human ecology: "this constant interplay of many factors
in space and time was an integral part of his thought; the habits of specialization, of
dealing with a single factor at a time, was to him anathema."[56] Mumford generously
acknowledged these various, biologically grounded dimensions of influence, but he
described one in particular as Geddes' single, most significant contribution to his
thought:

> My greatest debt I owe to him is one that underlies my whole work, and it
> has little to do with his leading me to the study of cities. What he gave me
> above all was ... 'a sense of the wonder of life' – of life as the so far ultimate
> manifestation of cosmic evolution.[57]

Mumford liked to say that in being faithful to his master the disciple must eventually
reject him. In pursuing his own independent literary career, Mumford revised, transformed, extended, and discarded various aspects of Geddes' thought. Indeed, he exhibited
the Geddesian spirit of "audacious insurgency" in the way he adapted Geddes' method

and ideas for his own purposes: "it was Geddes himself, in person, who taught me to beware of all self-enclosed systems – not least his own" (SL 158). Accordingly, Mumford claims that his "obligations to Geddes . . . are both closer and remoter than most people who have compared our ideas have discovered: closer in personal indebtedness, remoter in intellectual adhesion" (SL 408). In *Sketches from Life*, published when he was eighty-seven and which he knew would be his last book, Mumford writes: "Some touch of Geddes' aboriginal energy and vitality, then, remains present, I would hope, even in my work today" (SL 409). Their correspondence is significant because it documents the young writer's "obligations to Geddes" and conveys something of the master's "aboriginal energy and vitality" as Mumford first experienced it.

MUMFORD'S CRITICISM OF GEDDES

During the last decade of his life, Geddes became preoccupied with his "thinking machines," the graphic method he had developed to present his ideas. Mumford viewed this emphasis as an obsession that both prevented Geddes from putting his philosophy into coherent, written form and undermined their effective collaboration. However, Mumford never published a comprehensive criticism of his master. In fact, with the exception of "The Disciple's Rebellion," Mumford's published writings extol Geddes' thought and achievements and depict him as an ideal personality type.

Privately, however, Mumford was harshly critical of the sort of work and methods that preoccupied Geddes during the 1920s. He presents this criticism in a chapter originally intended for his autobiography entitled "The Geddesian Gambit" (Appendix 3). Although he wrote several drafts of the chapter, he was never satisfied that it met his standards of literary excellence.[58] More importantly, perhaps, even in his eighties Mumford was still reluctant to criticize the master of his youth. In 1977 he confessed that when he attempted to write "a systematic criticism" of Geddes' graphic method, "I find that my younger self scowls reproachfully at me."[59] Mumford explained that "the example of his life and work was so important that my first duty was to make this fruitful part of his achievement better known: so except in brief personal letters to him, I never published my criticism."[60] Thus, whatever the literary problems of "The Geddesian Gambit," and they are certainly not glaring ones, Mumford eventually abandoned the piece, deciding never to publish such a negative assessment.

"The Geddesian Gambit" contains Mumford's sharpest and most comprehensive criticism of his master. Focusing on Geddes' thinking machines, the essay argues that his obsession with such "intellectual shortcuts" undermined both his capability for more productive work and the possibility of meaningful collaboration with his young disciple. Mumford describes Geddes' absorption in his graphic method as a "fatal addiction" (GG 369). And in a vivid metaphor, he depicts how this "addiction" affected their prospects for collaboration:

From the outset, Geddes' over-valuation of his graphic method was what made anything that could be called collaboration impossible. And the closer

Geddes advanced toward me, like a Roman gladiator with his trident and net, ready to snare me in his cunning ideological net, the more warily I was inclined to retreat and preserve my freedom of movement: indeed my very life!

(GG 358)

Mumford argues that Geddes' emphasis on the graphic system betrayed those aspects of his thought and method that were most original and liberating: his comprehensive approach, his recognition of life's organic interconnectedness, and his emphasis on humankind's capacity for insurgence. He cites a fundamental tenet of Geddes' own philosophy to critize the master: "Any attempt to produce a single synthesis good for all times, all places, all cultures, all persons is to reject the very nature of organic existence" (GG 362). Mumford had initially been drawn to the thinker who refused to work within the confines of a single philosophy or academic discipline and who revered the wondrous beauty, variety, and complexity of life. However, Geddes' dogmatic obsession with his thinking machines belied such an approach: "though he was a merciless foe of all closed systems he had, in fact, put together a tightly closed system of his own, which left no openings for time, chance, experience, feedback, or future emergents" (GG 363). In this regard, Mumford's hallmark essay, "The Fallacy of Systems," is both a celebration of the Geddesian spirit that the young Mumford had found so invigorating and a denunciation of the method that preoccupied Geddes during his last decade.[61]

Mumford likened Geddes' graphic method, particularly the "square of 36," to a "game of intellectual chess."[62] In Mumford's view, this "game" became a form of solipsistic self indulgence as well as a means of evading the more important work he should have been completing – namely a coherent, written exposition of his life's work and philosophy, the "opus" often mentioned in the correspondence. Mumford first used the chess analogy in a 1921 letter to Delilah Loch:

[the] square of 36 ... bears the same relation to one's daily reactions and investigations as a chessboard does to the movements on a battlefield. To make the chessboard serve as a complete battlefield, or to pretend that the battlefield is as simple as the chessboard is to court defeat. Those who have followed Geddes have tended to substitute the doctrine for life.[63]

Of course, after their 1923 meeting, Mumford realized that Geddes himself encouraged and even practiced such a "substitution." Fifty years later Mumford writes that "the game of intellectual chess Geddes had invented could not be played except according to the fixed rules he had laid down.... In fact, it had become a form of intellectual solitaire" (GG 361). Geddes had left no open spaces on the board, so "the opponent was checked before he could make the first move," thus "the Geddesian Gambit" (GG 362). According to Mumford, during his last decade of life, Geddes would be absorbed for several hours each morning in this "intellectual solitaire," "a kind of private intellectual gymnastics," which had become a "devotional exercise, a daily ritual, which admitted no deviation or alteration" (GG 364). In Mumford's view, Geddes became "trapped" by the system he had devised – both by his daily ritual of working through it and by his efforts to make everything conform to its patterns and laws. Mumford sadly observes that "nothing of course could have been a greater betrayal of Geddes' own insurgent mind than the lifeless finality of this final chart" (GG 364). In the correspondence, Geddes often explains or solicits reactions to certain aspects of his chart, references and questions which Mumford usually ignores in his replies.

In Geddes' letters to Mumford as well as other writings, a favorite target of criticism is what he terms the "verbalistic empaperment" of contemporary life – in journalism, business, and the academy. He argued that such "empaperment" suffocated significant thinking and meaningful activity (PG 15 April 1922, 17 May 1931). Yet after 1923 Mumford came to believe that, ironically, the more valuable aspects of Geddes' thought "had been smothered in [the] *graphic* empaperment [of] his own mountainous middens of graphs" (GG 364). Mumford alleges that Geddes had become so enthralled by the "intellectual shorthand" of the graphic system that he failed to commit the most valuable aspects of his thought to writing. He believed that Geddes' ideas and philosophy would be widely influential only if they were expressed in articulate prose, not in the arcane, idiosyncratic charts and graphs that accumulated in the "morning's pile of paper" which he daily produced (LM 6 July 1923). In an unpublished note entitled "Last Word on P.G." written in 1976, Mumford cites Geddes' claim (in his letter of 1 March 1923) that his "brain still teems daily & thus interferes with writing, though always promising to make it better," as proof of his "intellectual narcissism." Such a claim, Mumford counters, was "a delusion: he was in love with his own intellectual image, which he revisited daily by gazing into [a] graphic counterpart."[64]

In theory, of course, Geddes never intended for his graphic method to represent a final, complete synthesis or to become a rigid system. Responding to Mumford's criticism late in their correspondence, he protests that "my squares are not to confine the world into my categories"; rather, "they are so many windowpanes for looking out into the world movement" (PG 6 May 1930). Yet Mumford saw only the mirror of "intellectual narcissism," not a portal to wider vistas. Of course, Mumford had originally been attracted to a different, earlier Geddes – "before he had manufactured his graphic straightjacket," when "his thought was free, vigorous, original."[65] In Mumford's view, the "sterile and static" method Geddes developed in his latter years betrayed his real genius: "His mind was so capacious, and his command of the scientific evidence in biology was so extensive, that he had no actual need of such a 'thinking machine': he had a far better instrument in his own infinitely richer brain."[66]

Mumford viewed Geddes' efforts to establish the Collège des Ecossais at Montpellier as another self-delusion, another waste of energies and distraction from more important work.[67] Mumford described the Montpellier college project as "a white elephant" that "produced in Geddes' fertile mind a whole herd of little white elephants" (SL 404). The impractical scheme, Mumford believed, allowed Geddes to evade the sterner, far more significant task of committing his philosophy to writing and thus giving it the complete and clear explanation that his numerous articles and idiosyncratic graphs and diagrams could not convey. Mumford thus described the buildings at Montpellier as "the empty substitutes for the books he had never disciplined himself to write" (GG 366). Yet representing more than the escapist fantasy of an old man who had lost the vigor and effectiveness of youth, the Montpellier project, Mumford believed, was symptomatic of a problem that had plagued Geddes throughout his career. As he asks in a 1926 letter to Victor Branford: "Why the terrible and incessant activity which has been so often just the *evasion* of some other task that demanded completion? I am puzzled by these psychological anomalies in P.G.'s life."[68] Two years later Mumford attempts to answer this question: he suspects that Geddes undertook such projects "out of a restlessness and out of a desire to bring a quick realization to his problems"; he theorizes that Geddes used these schemes as "a way of running from" the books he should have written.[69] Mumford confesses his own "tendency" to do "just the opposite" (that is, to eschew action for the self-absorbed literary life); but he contends that "one can run away into action just as well as one can escape into dreams and fantasies."[70]

Although Mumford had strong objections to the habits of mind and the educational scheme that preoccupied Geddes during the 1920s, he expresses his concerns only mildly and infrequently in the letters. The correspondence does not support Mumford's claim in 1976 that "I more than ever brashly challenged him."[71] Despite what he may have thought privately as well as what he inaccurately recalls writing to Geddes, Mumford could never bring himself to complete, much less to publish, what he felt would be an adequate "exhaustive critique" without immoderately "belittling" his master, or sounding "willfully derogatory," or "renouncing earlier assessments."[72]

THE COMMEMORATED GEDDES

Throughout their friendship Mumford frequently expressed an interest in writing a biography of Geddes. In 1921 he began to give the project serious consideration, and in 1930 the biography remained on the list of the books he planned to write. Mumford envisioned the biography as a panoramic study of Geddes – of his extraordinary mind and personality and of his many achievements and wide influence in the context of the times in which he lived. In 1921 he wrote Delilah Loch to inquire about the possibility of becoming Geddes' "official biographer": "the biography attracts me as a piece of creative work, in which one might by good fortune sum up and crystallize all that was good and permanent in Geddes' philosophy [which is] Geddes' life itself." Describing himself as a "literary man" rather than "a simon-pure sociologist," even then Mumford was more interested in writing the biography than collaborating on the "magnum opus."[73] In 1922 he declared that "I am the ideal biographer for Professor Geddes" and hoped that he might have the opportunity to live with his master for a time while working on the project.[74] Shortly before the New York visit, Mumford wrote Geddes: "I still stick to my notion of a biography as the biggest personal contribution I can make towards spreading your thought"; such a work would provide "the essential explanation and raison-d'être of your philosophy and plans, seen as things which have had a life and evolution" (LM 25 March 1923). In late 1923 (not long after the New York visit), he wrote to the publisher Horace B. Liveright proposing a book that would present "the vivid philosophy and work of Geddes" but "would never lose the human touch."[75] In 1927 he told Branford of his ambition to write a combined biography and cultural analysis that would treat "Geddes' life and times – and what they point to."[76] And in the same year, he notified Geddes of his "hopes of writing a connected 'Life'" (LM 13 December 1927). In his personal notes reviewing his accomplishments during 1928, Mumford wrote that he would make a definite decision during the spring of 1929 about undertaking the biography; a note dated 17 January 1930 is appended: "Decision still suspended."[77]

Geddes was delighted by his brilliant young disciple's interest in becoming his biographer. In 1922 he offered financial support that would enable Mumford to write a book on his work as a town planner (PG 18 November 1922), and throughout the ensuing years he continued to urge Mumford to write his biography. Why then did Mumford never undertake a book-length biography of Geddes? *Herman Melville* (1929), a work which is still often cited in Melville studies, demonstrated his talent for producing a probing biographical study that explored the psychological and spiritual dimensions of the person. And he was quite interested in the art and theory of biography as his 1934 essay "The Task of Modern Biography" reveals.[78] Mumford did publish several short biographical sketches of Geddes, the most notable of which is in *The Condition of Man*.[79] However, these sketches are predominantly idealized portraits that emphasize the features of Geddes' personality and work that Mumford considered most significant. They may depict what he considered to be the "essential" life, but they do not represent balanced or factually objective biography; nor do they provide the sweeping portrait of Geddes' "life and times" that Mumford had hoped to write. They are more akin to hagiography than disinterested, critical biography. In fact, Mumford confessed that these sketches of Geddes "tend to be idealizations, not in the sense that they falsify his achievements, but that, as in old-fashioned Victorian biography, they deliberately overlook his faults and intellectual defects" (GG 354).

In "The Geddesian Gambit" Mumford claimed that he "never heeded Geddes' testamentary injunction to be his biographer" because of his aversion to the graphic method (GG 354). However, this is not a wholly convincing explanation. Rather it appears more likely that Mumford simply could not bring himself to publish a dispassionately honest account of the master of his youth – an account that would necessarily include criticism of Geddes' graphic system as well as his personal foibles, his "evasion" of tasks that "demanded completion," and other shortcomings. At the same time, Mumford felt a deep sense of obligation to undertake the project, especially since Geddes had began to look to him as his most worthy intellectual heir and no other capable biographer had come forward. This nagging sense of obligation, and the contradictory feelings it evoked, ended with Geddes' death in 1932; yet even then Mumford did not completely abandon the possibility of writing a life of his master.

Mumford was also very interested in writing a fictionalized account of Geddes' life; in fact, he probably came closer to undertaking this project than a conventional biography. In 1928 he published "The Little Testament of Bernard Martin," a transparently autobiographical short novel in which Mumford is "Martin" and "James McMaster" is modeled after Geddes. Mumford's personal notes of the late 1920s reveal his intention to write a companion work entitled "The Great Testament of James

McMaster." Mumford envisioned this as a work of epic scope; it would be something of a fictional "life and times" of Geddes that would "deal with a much larger figure and a longer course of time [than that depicted in 'The Little Testament'] – from the heyday of Victorianism out into the future."[80] He projected "The Great Testament" as "a long philosophic excursus, with an apocryphal Geddes, a *quite* apocryphal Geddes, as the radiating center for all sorts of thoughts and experiences."[81] This fictional account, he writes, would be "a transcription not of the Geddes I knew, but of the Geddes I sought," another "self" who "would have rounded out my imperfections instead of being blessed with a different set of them."[82] Mumford believed that such a work would do greater service to Geddes than a conventional biography that sedulously recorded the facts of his life and his teachings. Indeed, the fictitious "McMaster" would beg "the one disciple he has had who has fought against him and evaded him to write his intellectual biography since he had rather be supplanted by his follower, as Socrates was supplanted by Plato, and be raised into a myth, than be entombed as an effigy by those who passively 'accept' him."[83] It appears "The Great Testament" would have celebrated more than one powerful ego.

Although he may have contemplated it more seriously than a standard biography, "The Great Testament" was also a work Mumford never began in earnest. Because of his continuing devotion to his master and, perhaps, a lingering sense of guilt for never assuming the role of collaborator or literary executor, Mumford kept the project on his literary agenda as late as 1938.[84] But by then, of course, he was well established in his true metier, which was that of neither biographer nor novelist.

Although he never wrote a full-length biography, Mumford did remain faithful to his commitment to commemorate his master as "the most prodigious thinker in the modern world." Descriptions of Geddes in various essays, reviews, prefaces, and segments of books published throughout Mumford's career together form a composite portrait of the man. This portrait reflects what Mumford considered to be Geddes' distinctive characteristics and most important contributions. It does not depict the dogmatic and willful person whom Mumford had encountered in New York and Edinburgh. Nor does it describe the desperate old man who besieged Mumford with such misguided and pathetic pleas during the final years of their correspondence. Instead, Mumford commemorated Geddes as an idealized, symbolic personality, as the living example of the ideas and outlook that had so influenced the young disciple. This idealized portrait presents a complete, balanced, and fully realized personality. Mumford makes bold, sometimes grandiose claims for Geddes' intellectual stature and potential influence. He describes him as "one of the truly seminal minds" of the nineteenth century, as "a philosopher whose knowledge and wisdom put him on the level of an Aristotle or a Leibniz."[85] Mumford's greatest claims for Geddes' contemporary significance occur in *The Condition of Man*; here he depicts Geddes as a person whose life example offers hope and serves as a model of purposeful living in a world devastated and disordered by the Second World War. Mumford describes him as "the Bacon and the Leonardo, perhaps the Galileo" of the modern age; he argues that "by [Geddes'] example and practice the path of redevelopment and renewal for modern man becomes clearer."[86]

No other writer – certainly no one of Mumford's stature – has offered such unqualified praise of Geddes' life and teachings or such an estimation of his eminence. Mumford claims that he was reluctant to publish a comprehensive criticism of Geddes' thought because of his deep sense of indebtedness to and esteem for the master of his youth. Yet there are other factors that must have influenced what he chose to write, and not to write, about Geddes. Very early in their association, Mumford made the decision not only to adopt Geddes' general outlook and method but also to link his reputation with that of the master, and he never renounced this commitment. Mumford disseminated Geddes' ideas and approach more effectively than any other disciple. Geddes' reputation has been enhanced and continues to endure, to a significant degree, because of what Mumford has written about him. And without question, Mumford used his association with Geddes to his own advantage. To extol one's master is also, at least indirectly, to extol oneself. Thus, by celebrating Geddes' originality and greatness, Mumford was adding luster to his own reputation. Moreover, the person Mumford commemorated is a man he rarely, if ever, encountered in the flesh: it is an idealized figure that Mumford had begun to create from Geddes' letters and published writings several years before they first met; it is as much the fictitious "James McMaster" as the real Patrick Geddes; it is a figure who embodies Mumford's personal vision of the good life. In other words, Mumford's several biographical portraits of Geddes more accurately and completely represent the essence of Mumford's philosophy and ideals, his "utopia" in microcosm, than the lineaments of the actual man. Consequently, had Mumford criticized Geddes or depicted him in anything less than ideal terms, he would have risked diminishing his own reputation as well as that of the master whom he revered deeply.

Geddes' initial and decisive influence on Mumford came through his published writings, particularly *City Development* and *Cities in Evolution*, and through the six-year correspondence that preceded their first meeting in 1923. However, the man whom Mumford idealized and commemorated was Geddes the Socratic teacher and incandescent personality, the man who taught most effectively and brilliantly through his spoken words and personal example rather than his writing. Mumford describes this Geddes as one who is "vigorous, incisive, systematic, satirical, who, like the teachers of ancient Greece, communicates more fully in his conversations and diagrams and brilliant impromptus" than in his writing.[87] He praises Geddes' "insights, his gift for swift and penetrating observation, and the life-wisdom he brought to each fresh situation"; in such instances, "Geddes the teacher properly takes precedence over Geddes the systematic thinker."[88] Writing half a century after Geddes' death, Mumford claims that it was these qualities manifest in "the living man and his formative example . . . that had drawn me to him and that still attract me" (GG 358). Yet, obviously, Mumford had not been able to observe "the living man" in the flesh when his impact and influence on his disciple were greatest (from 1915 to 1923), and he had scarce opportunities to experience Geddes in person – and on his good behavior. Both personal meetings were bitter disappointments for Mumford; yet Geddes' personality at its best, glimpsed only in rare and brief moments, was so powerful and impressive that it could redeem for Mumford their otherwise frustrating encounters. Although Geddes was not a disciplined or gifted writer, his letters to Mumford also, from time to time, reveal the great teacher whom the disciple so admired. However, the documentary evidence shows that

Mumford must have encountered the idealized "living man" more often, more consistently, and more meaningfully in his imagination than anywhere else.

The portrait of Geddes that emerges from Mumford's various sketches contains several prominent features. Not surprisingly, these are the features of Geddes' thought and personality, as outlined above, that had most influenced Mumford and that became important aspects of his philosophy and writing. First of all, the thought of this idealized Geddes is characterized by its broad comprehensiveness, its interdisciplinary fluency, and its capacity for synthesis: he "took all knowledge as his province" and sought "to make possible a continuous interchange between the isolated provinces of thought."[89] This comprehensive, synthetic outlook "shows a constant interpenetration of the general and the particular, the philosophical outlook and the scientific outlook, the universal and the regional."[90]

Second, Mumford's portrait emphasizes Geddes as a man of action who exemplified the relationship between thinking and living. The commemorated Geddes stressed the intellectual's moral obligation to improve the human condition through purposeful activity. Mumford writes that for Geddes, "thought ... does not imply divorce from activity or responsibility; into it has gone not merely the studies of the scholar and scientist, but the feelings of the lover, the husband, the father, the friend, and the experience of the artist, technician, planner."[91] This was a person who embodied his motto: "Vivendo discimus: we learn by living."

Mumford emphasizes a third important characteristic in the portrait of his master: "Geddes added to the usual list of organic traits a quality he himself exhibited to the utmost degree: insurgence – a capacity to overcome, by power or cunning, by plan or dream, the forces that threaten the organism."[92] While what Geddes taught and exemplified has special significance for the intellectual, it is relevant for anyone who aspires to the balanced and purposeful life: "he saw that over-specialization produces arrests, regressions, failures of inventiveness; and he tirelessly sought to encourage the processes of insurgence, of self-transcendence, of creativity."[93] Insurgence in thought and action enables one to realize life's "high destiny": "that of revamping out of nature's original materials, with the help of nature's original patterns, a more perfectly harmonized, a more finely attuned, a more complexly balanced expression of both personality and community."[94]

Fourth, the most salient characteristic of the commemorated Geddes, infusing both his life and thought, is his profound "reverence for life" in all its manifestations (SL 155). As discussed above, this was the Geddesian trait that made "the deepest impression" on Mumford and most powerfully influenced him: "the essence of Geddes was his reverence for life, and his passionate commitment to thinking itself, along with sex and art and love, as the highest manifestations of life." Mumford sees this quality represented best in the person, embodied in the way the master lived: "No man in our time has shown a higher degree of intensity: an intellectual energy ... a practical grasp ... a sexual vitality."[95]

In *The Condition of Man* Mumford asserts that "the Geddesian doctrine of life" offers a "new sense of the organic," a vision of "growth, reproduction, renewal, and insurgence" that might renew and redeem human life and culture after the horrors of the Holocaust and Hiroshima.[96] In the essay "What I Believe," Mumford describes Geddes as the embodiment of and the inspiration for his own philosophy of "organic unity":

> Geddes showed that a conception of life, unified at the center and ramifying
> in many inter-relations and comprehensions at the periphery, could be
> rationally lived; that it had not been outmoded by the age of specialization
> but was actually a mode that might, through its superior vitality and
> efficiency, supplant this age; that one could practice in one's own person, in
> the germ, a type of feeling and acting which might ultimately be embodied,
> with fuller, deeper effect, in the whole community; that, even on the crude
> test of survival, a life that was organically grounded and pursued with a little
> courage and audacity had perhaps a better chance than the narrow goals and
> diminished possibilities of our dominant civilization. My utopia is such a
> life, writ large.[97]

The Geddes thus depicted symbolized for Mumford the ideal character and highest
goals of life, both contemplative and active.

The scope of Geddes' influence is not confined to the ideas and views that
Mumford borrowed from him, for it also includes Mumford's revisions and rejections
of those aspects of Geddes' thought that he considered incomplete or erroneous. He
writes that it is important to "realize how stimulating [Geddes'] negative contributions
were; for some of my most original work has come through supplying certain essential
components left out of Geddes' systematic thought" (GG 355). Mumford believed that
Geddes' other followers, most notably Victor Branford, merely "repeated his for-
mulas," yet he claims that "my distinction was to apply them. . . . I filled out their skinny
theoretical shapes, tightened up some of their weak places, departed from schematical
exactness when it did not fit reality."[98] Sophia Mumford succinctly formulated the
essential difference between Geddes and Mumford as she saw it: "Geddes spent the
latter part of his life codifying the insights he had between 1880 and 1900. Lewis spent
the latter part of his life developing and enlarging the intuitions he had in his youth."[99]
Mumford's youthful "intuitions" were, of course, in large measure inspired and
nurtured by Geddes. Under the spell of Geddes' influence, in late 1920 Mumford wrote
a brief note proposing a "Book on cities: their origin, their function, their future."[100]
This early idea became the nucleus of his life's work, for he would years later publish
such a book in *The Culture of Cities* (1938) and even more expansively in *The City in
History* (1961). Mumford recalls that Geddes "awakened" him to undertake the
Renewal of Life series in 1930 when the cumulative effect of the master's influence
ignited "a sudden explosion of energy on a scale I had never experienced before."[101]
The first two books in the series, on technics and the city, applied and developed many
of Geddes' ideas; the third volume, tracing the history of the human personality,
culminates with a portrait of Geddes as the ideal personality; the fourth, Mumford's
philosophical and moral treatise, emphasizes the Geddesian concepts of insurgence,
balance, and unity. Mumford concludes this last volume in the series with a tribute to
his master: "While in the present book I have, I trust, pushed beyond the natural
limitations of Geddes's period and culture, I should never be surprised to find the blaze
of his ax on any trail I thought to have opened alone."[102]

NOTES

Philip Boardman provides a detailed, narrative life of Geddes, with substantial excerpts from his writings, in *The Worlds of Patrick Geddes: Biologist, Town planner, Re-educator, Peace-warrior* (London: Routledge & Kegan Paul, 1978). Boardman includes a discussion of Geddes' "thinking machines" as well as a useful bibliography of works by and about Geddes. Helen Meller provides a scholarly, intellectual biography in *Patrick Geddes: Social Evolutionist and City Planner* (London: Routledge, 1990). Meller presents an illuminating discussion of the sources and development of Geddes' thought; her bibliography is also very helpful.

    The authoritative life of Mumford is Donald Miller's *Lewis Mumford: A Life* (New York: Weidenfeld & Nicolson, 1989). Miller provides a comprehensive discussion of both Mumford's life and his writing. Also important, of course, is Mumford's autobiography, *Sketches from Life: The Autobiography of Lewis Mumford: The Early Years* (New York: Dial, 1982). Elmer S. Newman compiled the helpful *Lewis Mumford: A Bibliography, 1914–1970* (New York: Harcourt Brace Jovanovich, 1971).

Sources frequently cited in the text are noted parenthetically and identified by the following abbreviations:

DR: "The Disciple's Rebellion," first published in *Encounter*, 27 (September 1966): 11–21; included with minor changes in *Sketches from Life* as Chapter Twenty-Three, reprinted below as Appendix 2, to which page references are given.
GG: "The Geddesian Gambit," unpublished manuscript by Lewis Mumford, 50 pp., UP f. 6906, reprinted as Appendix 3 below, to which page references are given.
LM: Letter from Mumford to Geddes.
PG: Letter from Geddes to Mumford.
SL: *Sketches from Life*.
WD: *My Works and Days: A Personal Chronicle* (New York: Harcourt Brace Jovanovich, 1979).

1. Published editions of Mumford's letters include: Robert E. Spiller, ed., *The Van Wyck Brooks–Lewis Mumford Letters: The Record of a Literary Friendship, 1921–1963* (New York: E.P. Dutton, 1970); Michael Hughes, ed., *The Letters of Lewis Mumford and Frederic J. Osborn: A Transatlantic Dialogue, 1938–70* (New York: Praeger, 1971); Bettina Liebowitz Knapp, ed., *The Lewis Mumford/David Liebovitz Letters 1923–1968* (Troy, NY: Whitston, 1983). No editions of Geddes' letters have been published.
2. Undated note, UP f. 7969.
3. Note dated 31 July 1946, UP f. 7963.
4. Mumford considered *Green Memories* to be one of his most accessible and stylistically effective books. *The Van Wyck Brooks–Lewis Mumford Letters*, 371; *The Letters of Lewis Mumford and Frederic J. Osborn*, 161.
5. Geddes describes his "long-dreamed *Opus syntheticum*" in "A Note on Graphic Methods, Ancient and Modern," *Sociological Review*, 15 (July 1923): 228.
6. Lewis Mumford, "Introduction" to Philip Boardman, *Patrick Geddes: Maker of the Future* (Chapel Hill: University of North Carolina Press, 1944), viii–ix.
7. Mumford to Philip Boardman, 12 January 1979, Oslo.
8. See Miller, *Lewis Mumford: A Life*, 46–90, 219–232; Rosalind Williams, "Lewis Mumford as a Historian of Technology in *Technics and Civilization*," in T.P. and A.C. Hughes, *Lewis Mumford: Public Intellectual* (New York: Oxford University Press), 43–65; Casey Nelson Blake, *Beloved Community: The Cultural Criticism of Randolph Bourne, Van Wyck Brooks, Waldo Frank, and Lewis Mumford* (Chapel Hill: University of North Carolina Press, 1990), 191–201.
9. Mumford writes elsewhere: "I first, at eighteen, came upon Geddes' writings, when a student of biology at the City College of New York in 1914" ("Author's Note" on "The Disciple's Rebellion," UP f. 7716). However, in other documents, Mumford recalls first reading Geddes in the fall of 1915, at the age of nineteen (Mumford to Victor Branford, 5 November 1919, NLS ms. 10557; and Mumford to Josephine MacLeod, 7 May 1922, Oslo).
10. For a description of the Outlook Tower, see: Meller, *Patrick Geddes*, 101–113; Charles Zueblin, "The World's First Sociological Laboratory," *American Journal of Sociology*, 4 (March 1899): 577–592.
11. Mumford to Josephine MacLeod, 7 May 1922, Oslo.
12. "Biographical Note," UP f. 8135.

13. Mumford, "Introduction" to Boardman, *Patrick Geddes: Maker of the Future*, x.

14. Note dated 9 January 1916, UP f. 6866.

15. Note dated 15 December 1916, UP f. 7937.

16 "Mumford on Geddes," *Architectural Review*, 108 (August 1950): 84.

17. Note dated December 1917, UP f. 8223.

18. "Personalia," 7 July 1920, UP f. 8233.

19. Mumford to Victor Branford, 13 February 1920, UP f. 5676.

20. "Personalia," 4 February 1920, UP f. 8233.

21. Mumford to Victor Branford, 21 April 1929, NLS ms. 10575.

22. "Random Note," 25 October 1930, UP f. 7966.

23. Shortly prior to Geddes' arrival at the Le Play House in London, Dorothy "Delilah" Loch writes to Mumford: "And WHAT SHALL WE DO WITH Geddes here – so soon – so dreadfully soon. Even dear Davies says 'it makes me tired to think of it'! Geddes torrenting about, making or breaking his own and everybody else's engagements.... Come over and help us 'ere we die!! Take Geddes away... spare our overworked bodies and wearied brains this new demand!" (26 February 1921, UP f. 2919).

24. Mumford to Delilah Loch, 13 May 1921, UP f. 2920.

25. See Mumford's letters of 25 March 1923, 5 January 1924, and 22 May 1926.

26. Geddes' New School lectures, as recorded by a stenographer and edited by Paul Kellogg and Geddes, were published in *Survey Graphic*, 53–54 (February–September 1925); they are reprinted in Marshall Stalley, ed., *Patrick Geddes: Spokesman for Man and the Environment* (New Brunswick: Rutgers University Press, 1972), 289–385.

27. Norah Mears to Mumford, 6 March 1923, UP f. 3224.

28. Delilah Loch to Sophia Mumford, 8 April 1923, UP f. 2923.

29. In reviewing that significant first meeting of master and disciple, it should be noted that most of the recorded facts and evaluations are Mumford's; there is little record of Geddes' impressions.

30. "Note on Disciple's Rebellion," UP f. 7716.

31. Mumford to Delilah Loch, 17 May 1923, UP f. 5980.

32. "Geddesiana," 10 July 1923, UP f. 6863.

33. Ibid.

34. Ibid.

35. "The Little Testament of Bernard Martin Aet. 30," in *Findings and Keepings: Analects for an Autobiography* (New York: Harcourt Brace Jovanovich, 1975), 130.

36. Delilah Loch to Mumford, 8 July 1923, UP f. 2923.

37. "Geddesiana," 10 July 1923, UP f. 6863.

38. Geddes to Josephine MacLeod, undated letter, NLS ms. 19995.

39. Gladys Mayer to Mumford, 24 November 1923, UP f. 3202.

40. Mumford had earlier referred to Geddes as "dear Master" in the close of a letter (LM 10 October 1922); and he closed another with the phrase "Yours in discipleship" (LM 26 December 1922).

41. Note dated 19 September 1925, UP f. 8181. Reprinted as Appendix 1 below.

42. Mumford to Victor Branford, 20 January 1928, UP f. 5677.

43. Mumford to Delilah Loch, 1 April 1924, UP f. 5980.

44. Mumford to Delilah Loch, 20 June 1925, UP f. 5980.

45. Mumford to Victor Branford, 27 April 1926, UP f. 5677.

46. Note dated 19 September 1925, UP f. 8181.

47. *Findings and Keepings*, 101.

48. "Mumford on Geddes," 87.

49. Mumford to Philip Boardman, 29 December 1968, UP f. 5669.

50. *My Works and Days: A Personal Chronicle* (New York: Harcourt Brace Jovanovich, 1979), 115.

51. "Mumford on Geddes," 82.

52. "Introduction" to Jacqueline Tyrwhitt, ed., *Patrick Geddes in India* (London: Lund Humphries, 1947), 11.

53. In early 1921 Mumford proposed to write "an ecological study of three communities" (LM 3 January 1921). For a discussion of Mumford as an "ecological" historian, see Frank G. Novak, Jr., "Lewis Mumford and the Reclamation of Human History," *Clio*, 16 (Winter 1987): 159–181.

54. "My Debt to P.G.," 13 June 1967, UP f. 8181.

55. "Introduction" to Tyrwhitt, *Patrick Geddes in India*, 8.

56. "My Debt to P.G," 13 June 1967, UP f. 8181.

57. Ibid.

58. The UP typescript of "The Geddesian Gambit" is dated June 1973. In a note dated 3 March 1976, Mumford writes: "I am still not satisfied with this for many reasons. The tone is wrong, to begin with, and it is over-written – clumsily over-written at that." In a second note, dated 12 March 1976, he writes: "Another reading has not yet satisfied me. This has everything but inspiration." In a letter to Philip Boardman, Mumford writes: "Your devotion to Geddes himself is admirable; but since some of my best thinking has been done through my breaking away from his limitations and his arbitrary categories, your advocacy of Geddes only made me more conscious of how far I have gone beyond the Geddes of the final quarter century, and how much I nevertheless owe to his much freer and more brilliant thinking in the eighties and early nineties. I have tried three or four times to sum up these differences in a single chapter; but I am not yet ready to publish any of the versions; for a superficial reader might think that I was trying to reduce Geddes' contribution, whereas my only effort has been to show how his aversion to the written word and to ideas which can't be imprisoned in one of his graphs, ossified his thinking in the concluding years of his life: though the essential Geddes from time to time would spring up in his letters. But if I published what I have so far written it might seem I was trying to play down his influence on me: so none of my drafts seems any more acceptable to me than they would to you" (12 January 1979, Oslo).

59. Mumford to Philip Boardman, 25 January 1977, Oslo.

60. Mumford to Philip Boardman (unposted), 24 January 1977, UP f. 5669.

61. *The Conduct of Life* (New York: Harcourt, Brace, and Co., 1951), 175–180.

62. Philip Boardman reproduces and explains the "square of 36," also referred to as the "Chart of Life," and the "IX–9" chart, often mentioned in the correspondence, in *The Worlds of Patrick Geddes: Biologist, Town planner, Re-educator, Peace-warrior* (London: Routledge & Kegan Paul, 1978), 465–484.

63. Mumford to Delilah Loch, 29 January 1921, UP f. 5980.

64. "Last word on P.G.," 1976, UP f. 8181.

65. Mumford to Philip Boardman, 2 December 1972, Oslo.

66. Mumford to Philip Boardman, 24 January 1977, UP f. 5669.

67. Geddes describes the setting and philosophy of the Collège des Ecossais in letters of 14 November and 3 December 1924 and in "Ways of Transition – Towards Constructive Peace," *Sociological Review*, 22 (January 1930): 10–31.

68. Mumford to Victor Branford, 16 October 1926, UP f. 533.

69. Mumford to Victor Branford, 22 August 1928, UP f. 5678.

70. Ibid.

71. "Last word on P.G.," 1976, UP f. 8181.

72. Mumford to Philip Boardman, 24 January 1977, UP; and 28 November 1979, Oslo.

73. Mumford to Delilah Loch, 13 May 1921, UP f. 5980.

74. Mumford to Josephine Macleod, 7 May 1922, UP f. 5994.

75. Mumford to Horace B. Liveright, 25 December 1923, UP f. 5672.

76. Mumford to Victor Branford, 6 February 1927, UP f. 5677.

77. Note dated 16 December 1928, UP f. 7959. Shortly after her father's death, Norah Geddes Mears wrote Mumford: "We [Norah and Arthur Geddes] infer he wanted you to undertake his biography.... [W]e would like very much if you were to be his biographer" (23 September 1932, UP f. 3224).

78. "The Task of Modern Biography," *English Journal*, 23 (January 1934): 1–9.

79. *The Condition of Man* (New York: Harcourt, Brace, and Co., 1944), 381–390. See also: "Who Is Patrick Geddes," *Survey Graphic*, 53 (1 February 1925): 523–524; "Patrick Geddes, Insurgent," *New Republic*, 60 (30 October 1929): 295–296; "Mumford on Geddes"; "Introduction" to Boardman, *Patrick Geddes: Maker of the Future*; "Introduction" to Tyrwhitt, *Patrick Geddes in India*.

80. "List of projected books," 1928, UP f. 8234.

81. Mumford to Van Wyck Brooks, 30 April 1924, UP f. 7963.

82. "Random Note," 1926, UP f. 8229.

83. Mumford to Van Wyck Brooks, 30 April 1924, UP f. 7963.

84. "Random Note," 22 December 1938, UP f. 8235.

85. "Introduction" to Boardman, *Patrick Geddes: Maker of the Future*, vii.

86. *The Condition of Man*, 389.

87. "Vivendo Discimus," Review of *Life: Outlines of General Biology*, by Geddes and J. Arthur Thomson, *New Republic*, 68 (16 September 1931): 130.

88. "Introduction" to Tyrwhitt, *Patrick Geddes in India*, 10.

89. *The Condition of Man*, 286.

90. "Introduction" to Boardman, *Patrick Geddes: Maker of the Future*, xi.
91. "Patrick Geddes, Insurgent," 296.
92. "Mumford on Geddes," 83.
93. "Mumford on Geddes," 87.
94. *The Condition of Man*, 384.
95. *The Condition of Man*, 383.
96. *The Condition of Man*, 389–390.
97. *Findings and Keepings*, 322–323. "What I Believe" was first published in 1930.
98. "Random Note," 27 November 1936, UP f. 7960.
99. "Random Note," 3 March 1976, UP f. 8181.
100. Note dated 24 December 1920, UP f. 7963.
101. "Random Note," 24 April 1980, UP f. 7963.
102. *The Conduct of Life*, 318.

Lewis Mumford AND *Patrick Geddes*

THE CORRESPONDENCE

1915–1919

Dear Sir,

Sometime during the next two years, circumstances and the War permitting, I purpose to follow my studies in Europe. To Edinburgh I have been attracted by the sociological work of Professor Geddes. May I now ask you, therefore, for information – by catalog, bulletin, or otherwise – concerning the courses and conditions of study, and so forth. Any expenses attached to this I shall be happy to pay.

With anticipatory thanks I am, Yours faithfully,

Lewis Mumford

100 West 94 Street
New York City

15 November 1915

THE SECRETARY
OUTLOOK TOWER
EDINBURGH
SCOTLAND

[Ts NLS]

Dear Sir,

I enclose some pamphlets, etc., on the work of the above – they will give you some idea of what goes on there. There are no advanced classes, etc., but a good many people work there on lines suggested by Professor Geddes. There are no fees except membership subscription, in the ordinary way; though, occasionally special courses are arranged. People get the best results for themselves by settling down to work out some problem in which they are interested, and through that they get into touch with the general scheme of the Tower.

Our working membership has been hard-hit by the war – but we are carrying on so far as possible. By the time you mention for your visit things should be moving more satisfactorily again.

You will note a list of other pamphlets – also larger books – on back page of syllabus.

I shall be glad to hear further from you when you have looked over the enclosed.

Yours faithfully

F.C. Mears[1]

Outlook Tower,
University Hall
Edinburgh

13 December 1915

[Ms UP]

1. Frank Mears (1880–1953) was an architect who assisted Geddes with many of his projects, including urban renewal in Edinburgh, the Cities Exhibition, and the planning of the Hebrew University in Jerusalem. He married Geddes' daughter, Norah, in 1915.

Dear Sir,

Let me thank you, at this rather late date for the Outlook Tower Report sent me last October. In the hope that the allied offensive in the fall would prove efficacious I had put off this little duty, thinking that it would be possible for me to take up my studies at Edinburgh in a not too distant future and to be on hand to witness the huge transformations which will follow the war's cessation. This amiable hope is now indefinitely put off: so the present letter must mend the breach in courtesy. While reading manuscripts for Mitchell Kennerley I had the good fortune to come upon Professor Geddes's amplification of *Wardom and Peacedom*, and with regret I learnt it was waiting Mr.

100 West 94 Street
New York City

6 January 1916

THE SECRETARY
OUTLOOK TOWER
EDINBURGH
SCOTLAND

[Ts NLS]

Branford's request for return.[1] The necessity for importing Professor Geddes's books seriously hinders the accretion of a body of readers proportionate in size to his own genius, capacity and sociological initiative: and for that reason it is doubly deplorable that neither *Cities in Evolution* nor *Wardom and Peacedom* have appeared in an American edition. At present the surveys of the Sage Foundation, the Cleveland Foundation, and others, while thorough in method and exhaustive in immediate presentation, have the effect of being simply clinical studies of palpable ills; and while every community will be found ailing in some function and thus in need of examination and diagnosis by means of the *social* survey, the necessity for a continual, fundamental survey, regionally devised, has not yet been established and acknowledged and acted upon by American workers in the field: and their surveys have thus a tendency to one-sided emphasis on administrative changes, technical innovations and the like. Here Professor Geddes's plea for regional surveys, with a bio-geographic background and geotechnic application is especially called for; and that in his own words, rather than at second hand through his collaborators and students. If I can be of any use in broaching the matter of publication to New York publishers please command me: such service would be small repayment for the vast debt I owe the founder of the Tower. With apologies for this presumptuously long letter and renewed thanks, I am,

Yours faithfully,

Lewis C. Mumford

1.   Mitchell Kennerley (1878–1950) edited *The Forum*, which published some of Mumford's early essays. "Wardom and Peacedom: Suggestions Towards an Interpretation," *Sociological Review*, 8 (January 1915): 15–25.

---

100 West 94 Street
New York City

9 January 1916

MR. F.C. MEARS
OUTLOOK TOWER
EDINBURGH
SCOTLAND

[Ts NLS]

Dear Sir,

I thank you for your letter of December thirteenth, which accompanied the pamphlets you then sent me. My interest in the work of the Outlook Tower has been further deepened; and I am determined, if possible, to pursue my studies there next fall.

Enclosed is a money order for fourteen shillings to cover the cost of the pamphlets received (26d. with postage); and Professor Geddes' *Evolution of Cities* which, since I know of no American edition, I must ask you to send me at your convenience. Out of the balance of cash, please add as many of the following as may be covered by it:

A Great Geographer; *City Surveys for Town-planning*; *Two Steps in Civics*; *Classification of Statistics*; *Principles of Economics*; *Books of the Masques of Learning.*[1]

With sympathy for you in the difficult situation created by the present holocaust, and with hopes that its speedy resolution will release new energies for the task of construction, I am,

Yours faithfully,

Lewis C. Mumford

1. Works by Geddes: *Cities in Evolution: An Introduction to the Town Planning Movement and to the Study of Civics* (1915); "A Great Geographer: Elisée Reclus, 1830–1905" (1905); "Civics: As Applied Sociology," *Sociological Papers*, 1 (1905): 103–138, "Civics: As Concrete and Applied Sociology, Part II," *Sociological Papers*, 2 (1906): 57–119; *The Classification of Statistics and Its Results* (1881); *An Analysis of the Principles of Economics* (1884); *The Masque of Ancient Learning and Its Many Meanings: A Pageant of Education from Primitive to Celtic Times* (1912).

Dear Sir,

The book and pamphlets are being sent you by this post. I enclose one more, which the caretaker says was omitted.

It is cheering to find that many citizens, even some who are much occupied with war problems, are keeping various civic organizations and projects alive and active.

We are quite within the extended Zeppelin zone here and may have a visit any time.[1]

Yours faithfully,

F.C. Mears

Outlook Tower
University Hall
Edinburgh

3 February 1916

[Ms UP]

1. A reference to the German airships used in raids in World War I, 1914–1918.

My dear Mr. Mears,

In a press of study and work I have neglected to acknowledge the receipt of the books sent me from the Outlook Tower on February third. They arrived last week in good condition; and I thank you for attending to them. There must be a touch of ghastly frivolity about sending books to America while one is expecting bombs from Germany. From the insulated United States the war seems like a morbid dream; I daresay that in the island which is no longer 'tight' it is a gripping nightmare. Let's hope that the present German offensive will flatly collapse and leave open the way to an heroic peace.

With thanks again, I am, Yours faithfully,

Lewis Mumford

100 West 94 Street
New York City

25 February 1916

F.C. MEARS, ESQ.
OUTLOOK TOWER
EDINBURGH
SCOTLAND

[Ts NLS]

My dear Professor Geddes,

Now and again during the past year I have corresponded with the secretary of the Outlook Tower, but the thoughts which I am trying to put together in the present letter are in large measure of personal application, and I have directed them to you in the hopes that you will pardon the inevitable officiousness of an uninvited communication from an altogether unknown friend.

[8 January 1917][1]

[Ts copy UP]

How well able you are to keep in touch with current developments in America, with slow and badly disrupted mail service, and (I suppose) endless local anxieties and difficulties, it is hard to guess, but I assume that the stream of current events reaches Edinburgh with most of its burden deposited as it empties on the American side, into the Atlantic, and that at least some of the following news will really be new to you. Communicating about civics while the present war is raging gives the sensation of trying to converse at night during a thunderstorm – perpetual rumbles make connected discourse impossible whilst occasional flashes of lightning only throw the speakers into a more impenetrable gloom when they have vanished: and up to April foreign correspondence had this further disability, that those whom we knew in Europe were standing in the open battered by wind and rain, and we in America were talking leisurely in the security of snug houses, and thus there was a further gap that made understanding doubly difficult. Now that breach is over, however, and one writes now with a sympathy that the most willing imagination could not induce hitherto.

Prime among the achievements of Regional Civics is the establishment of a system of university extension throughout the Connecticut Valley. A department of University Extensions was created by the Massachusetts bureau of education in 1915: and the work now includes Class Instruction, Correspondence study Groups and Co-operative extension lectures *throughout the Valley*. (It is notable, by the bye, that in the report the valley region is taken as the unit of control.) I quote from the Bulletin issued September 1916:

"... Coming now to Manhattan various new and significant initiatives may be noted. In city planning itself, a series of conferences between the municipal engineers and civic officers throughout the conurbation, held under the auspices of the Civic Club is perhaps most noteworthy for its promise, just as the Zoning and Districting act, for all its timidities, is most important for its achievement. In encouraging a popular appreciation of Civics the New York Public Library, thru the activity of the chief of its Public Documents Department, Miss Adelaide Hasse, has taken a lead."[2]

Three exhibitions are now being held: one consists of early prints and maps of the chief seaboard cities: another of the development of the present water supply system: and a third on the teaching of civics and citizenship. At a glance one sees that these exhibitions are complementary to each other, and the thought at once occurs; why may they not form the nucleus of a permanent synoptic survey of the city? There is good reason for hoping, I believe, that such a survey may be the outcome. In the exhibit on the methods of civics teaching (with its new school syllabuses, its numerous local maps, its magnificent engravings, its suggestive volumes of poetry) your own papers have a prominent place: and a card placed on the open pages of the *Sociological Papers* advises teachers especially to consult them. Finding that Miss Hasse was also one of your disciples I pressed upon her the wisdom of taking advantage of the present moment for utilizing the three separate exhibits as the basis for a better co-ordinated and more completely rounded out survey – treating one aspect at a time until sufficient material had been gathered to present a unified "synopsis" of the city in its manifold aspects. This suggestion she warmly seconded at the time, and is at present pondering.

Civic workers have been keen enough to turn the present vague and more or less blindly gregarious agitation for Americanization into an effective demand for citizenship, and this not in merely the sense of membership in the National State, but more realistically, as active citizenship in the community. As an indication of this re-orientation of the static civics of bureaucratic dispensation, comes one of the latest books on *Elementary Civics*, that of McCarthy, Swan and McMullin, with almost one-half of the vividly written little volume dealing with the intimate affair of the community.[3] Dr. McCarthy, as you probably know, is the leader of the co-operative movement in Wisconsin, and he lays his stress on co-operation and co-ordination of local professional groups rather than upon the relations between a discrete individual and an all-embracing state. Both individualism and staticism (as it might well be called) are doomed in political theory: and the younger political scientists, like Harold Laski, are going back, thru Maitland and Gierke, to a Theory of Corporations whose new developments may have far-reaching civic implications.[4] The idea that cities exist at the will of the state, and that formal boundaries laid down by the state are more important than functional lines created by regional developments of transportation, light, poetry, education, and so forth, is no longer ascendant, even if it be still dominant; and we are recovering (on a modern spiral as you would say) some of the essential ideas which were latent in the community life of the Medieval cities, and yet were never quite lucidly or definitively formulated by the medieval scholars. At present, however, this theory of corporate autonomy is confined to professional groups, and it is my own hope to work over the fields you have plowed up and develop a political philosophy of cities, to complement that civic interpretation of cities on which your first thoughts will long stand as the last word.

This, of course, Fors and Mars permitting. In the meantime I remain, in unavoidable anonymity,

Your respectful and affectionate pupil,

1. This undated, unsigned carbon copy at UP is apparently a draft version or copy of Mumford's first letter to Geddes. In the two replies below, Geddes refers to Mumford's letter of January and 8 January.   2. Adelaide Hasse (1868–1953) served as Chief of the Economics Division of the New York Public Library; she compiled the *Index of Economic Material in Documents of the States of the United States, 1849–1904*.   3. Charles McCarthy, Flora Swan, and Jennie McMullin, *Elementary Civics* (1916).   4. Harold Joseph Laski (1893–1950), political theorist, author of *The Problem of Administrative Areas* (1918) and *A Grammar of Politics* (1925); Frederic William Maitland (1850–1906), English law scholar, author of *Township and Borough* (1898); Otto Gierke (1841–1921), political and social theorist, author of *Political Theories of the Middle Age* (1900).

Dear Sir,

I was pleased to have your letter of last January from Outlook Tower & should have written sooner but the constant work, migrations, & repeated bereavements from war, etc.

But it is very encouraging to hear of your continued interest. The Regional Survey Association is growing & spreading. Drop a line (with a dollar!) to D. Morris c/o The Friends' School Saffron Walden Essex, England,

c/o Sir J.C. Bose[1]
93 Upper Circular
Road
Calcutta

8 August 1917

[Ms UP]

& as its Secretary he will send you particulars & documents & keep you in touch with the work on this side & Prof. Fleure of University College Aberystwyth, Wales,[2] is a very active spirit & brain in these matters, & others of whom Mr. Morris's budget will give indications.

As to books, herewith such particulars as I can send from here. You'll see we are at last struggling into print.

If you come to Europe with or after the war, let me know for I hope to return from India next spring. And I shall be glad to hear from you meanwhile.

Indeed I have also still have the idea of bringing the Cities Exhibition to U.S.A. after the war. (You may have noted in *Cities in Evolution* that it was nearly there four years ago.) Perhaps by the time the war is over, its line of appeal may be better understood than then, thanks to friendly co-operators in its outlooks like yourself.

Yours faithfully,

P. Geddes

1. Mumford's annotation of 20 March 1963: "His first letter to me." Sir Jagadis C. Bose (1858–1937), physicist and plant physiologist, was a close friend of Geddes in India. Geddes wrote *An Indian Pioneer: The Life and Work of J.C. Bose* (1920).   2. Herbert John Fleure (1877–1969) was a geographer and admirer of Geddes; author of *Human Geography of Western Europe* (1918).

Bose Research
Institute
93 Upper Circular Rd.
Calcutta

31 August 1918

[Ms UP]

Dear Mr. Mumford (yours of 8 January),

I have stayed on here in India working out my City Report for Indore – as I hope suggestive to others.[1] I am soon sending you this, bulky though it is, as you will be interested in many of its points – eg. definite treatment of Past–Present–Possible – break with conventional drainage of "all to the Sewer" – for all to the soil – with result on Garden city planning, & so on. Note mention of Sinha's wall-tile invention – making tile-walls so cheap that total cost of building here comes down 40% – Edison's problem solved I trust more simply.[2] If you feel interested, my friend and London colleague Victor Branford / 3 Chisholm Road / Richmond Hill, Surrey (whom you should in any case see when you come to England) will keep you in touch.[3]

See also re Regional Surveys, etc., Miss Barker c/o Assn. for Regional Surveys – 11 Tavistock Square, London WC.[4]

I hope you'll send me your essay here; as I shall not come home until peace – but then I hope to be active in reconstruction now imminent if energies last.

With all best wishes Yours cordially,

P. Geddes

1. *Town Planning Towards City-Development: A Report to the Durbar of Indore*, 2 vols. (1918).   2. The problem of providing inexpensive housing for workers. See Geddes' letter of 10 January 1920.   3. Victor V. Branford (1864–1930) was Geddes' close friend and devoted colleague; an accountant, he served as chairman of the Paraguay Railroad in South America before helping to found the British Sociological Society in 1903; he edited the

Sociological Review (1912–1930); he wrote *Interpretations and Forecasts* (1914) and *Science and Sanctity* (1923).   4. Mabel Barker was Geddes' god-daughter and worked intermittently as his personal assistant; author of *Education for a State of Peacedom through Regional Study* (1915).

Dear Professor Geddes,

[1918]

[Ts copy, undated and

unsigned, UP]

    Your letter of August reached me a little over a fortnight ago: but I send this to Europe because I infer that the rush of events during the last quarter will hasten your return to Great Britain; and this likely before the present note could reach India. Your reference to Mr. Branford and Miss Barker increases my eagerness to visit England: and I am quite resolved now to spend a post-graduate semester in Edinburgh and London even in advance of getting my bachelor's degree in America. The last six years I have spent intermittently (on account of health and purse!) at Columbia, New York University, and the College of the City of New York. Coordinately with this I have been cultivating a literary garden whose adolescent blossomings were at the beginning of this year just giving evidences of maturer fruit. My most fruitful scholastic work, however, has been done in the museums, the libraries and the city at large; and while I yet lack about a year's credits to obtain the bona fide certificate of "education," I have decided to continue on the present lines in the faith that my literary work will give me the position that my aversion to the conventional academic life robs me of, until the eventual, *incidental* possession of the proper diploma gives me access to the tutorial staff of a college. In this connection the projected School of Social Science, under Professors Veblen, Dewey, and Beard, with its scheme of academic apprenticeship and of research unembarrassed by the "point and credit accountancy" gives hope of finding a place even in America.[1] My term in naval service will be slightly lengthened because of my being in the radio branch, which will be one of the last to get discharged; but at the longest (unless new hostilities rise out of the peace conference out of the conflict of statecraftiness with statesmanship) not more than half a year on the seas lies ahead of me.

    I am very grateful for your courtesy and thoughtfulness in sending me the city planning report of Indore: when it reaches me I shall try to review it for *The Public*, the Fels Foundation weekly. Your business in India seems to be in its larger implications to regionalize British imperialism; in other words to give it a larger raison d'être than the sanguine map-painters of the City or Downing Street have any conception of. This is such an admirable adventure, in addition to its contributions to the cities movement itself, that I am itching for an opportunity to talk about it and explain it to others. Have you had an opportunity to glance at the *Journal of the Institute of Architects* during the past year? The pause in building activity during the war has given the architect opportunity, it would seem, for survey, criticism, and introspection, and the spectacle of sacrifice has stimulated the impulse for service, with the result that architecture renascent has identified its purposes with those of labor triumphant, with a regard for the facts of industrialism which the planners of the Chicago Exposition (of the elder heyday of "cultured appreciation") might well marvel at – timidly and dubiously.

Hoping that the even greater activities and interests of the coming year will heal in some degree the wounds that the war has, I am sad to learn, inflicted upon you,[2] I am with renewed gratitude,

Yours faithfully,

1. Thorstein Veblen (1857–1929), economist and sociologist; John Dewey (1859–1952), philosopher; Charles A. Beard (1874–1948), historian and political scientist. These prominent American intellectuals were personally known to Mumford during his early literary career in New York and were fellow contributors to such journals as *The Dial*, *The Freeman*, and *The New Republic*. Beard helped to establish the New School for Social Research; Mumford took a course there taught by Veblen.   2. Geddes' brilliant elder son, Major Alasdair Geddes, was killed in action in France in May 1917.

---

c/o John Ross, Esq.
C.A.[1]
5 Victoria Street
Westminster,
London S.W.

11 August 1919

[Ms UP]

Dear Mumford,

Glad to have yours of 8/6. Very interesting! First as to your enquiring regarding my plans. I am now done with St. Andrews–Dundee, though my retirement only takes effect in 1920 – & have accepted a chair of Sociology & Civics founded for me at University of Bombay – but as duties are only from 20 November to 20 March yearly I shall not be quite isolated from Europe, etc., while I have also my town planning practice in India. I was asked to go to Palestine, for Zionists, & may still do so, though now not whole time job, which may not suit them. I am just off to Ireland to see if any constructive ideas can still be put before wartime politicians – but there is little chance of this. Then off to France to see what may be going on regarding Reconstruction – & so out to East, one way or other. Thus we cannot meet this year.

However, if you come over, see all you can of Victor Branford – 3 Chisolm Road, Richmond – put up at Roebuck Hotel or the like on Richmond Hill, for a week or so, so as to be near him – you will find him a most congenial spirit, & he will introduce you to others. In Edinburgh look up my daughter, Mrs. Geddes Mears, 14 Ramsay Garden, near Outlook Tower, who will introduce you to my friends & her husband, though very silent, is a good fellow, & knows Edinburgh well.[2] It *may* be revived & anyway I keep paying rent; though in sad plight & closed, they will show you what they can of it. Don't expect much alas! (Write her beforehand.)

Now a general question. My Sociology chair, in conjunction with Branford here, Tower, etc., affords a means of raising some general discussion with other Sociology Departments as to treatment, methodology, etc. – & if you will be so very kind as to help with this, by letting me – who have read nothing for 5 years in India & little for years before that – so busy is life – know what live Schools of Sociology, & men, you have in Unistates, I'll be greatly obliged. What books, Surveys, etc. should I own for my Dept. as I have a little money to spend?

The numbers of *Dial* on museums do not appear to have arrived – yet *may* be snowed under in Outlook Tower, which I had no time when in Scotland even to enter! Better send till further notice, either to c/o Branford instead of above address or after October to Sociology Department, University of Bombay. (You may think me deserting Edinburgh & so in a way I am as now too full of sorrows, since I lost wife & son – I may return of course some day, when scars are less fresh.)

I am asking Branford to make publishers send our book to *Dial* for your kind offer of review – also his lively *Papers for the Present* – But all we do remains unknown practically. The concrete outlooks are submerged for most minds by the political or financial – which are as remote from these as could ever be the religious!

Yours faithfully,

P. Geddes

1.   John Ross was a chartered accountant and friend of Geddes whom he had met while a student at the University of Edinburgh.   2.   Norah Geddes Mears (1887–1967) frequently collaborated with her father.

Dear Professor Geddes,

I have been delinquent in following up my last letter. This has been in part due to the unduly large amount of time I was forced to give up to editorial work, and partly because in the interval of leisure I undertook for the first time a hurried reconaissance of Washington. The other factor of slowness was the list of books whose titles I was getting together and verifying. Your inquiry about live American schools of sociology, and men, compelled me to verify by strict inquiry an impression I had for a long time labored under; namely, that sociology had drifted into an academic backwater in America, with the result that the only work of importance under this head was done in the detached sciences, while the really vital schools for me, as an American, were in London, Edinburgh, and Paris. I have discovered nothing to make me alter this impression; much, rather, to confirm it. Except for work done in the agricultural department of the University of Wisconsin, under Professor Galpin, the major part of American sociology has risen out of the environment of books, largely those of Simmel and Bücher, et al., and not out of the demands of American cities and regions.[1] Hence the emphasis upon historical sociology, with its processional sweep and its glib obscurity about details, and with this the corresponding neglect of regional evolution and local problems. Sumner, Ward, and Giddings, with their lesser devotees and imitators have each applied to the (largely) dead body of Spencerian sociology the elixir of a comprehensive abstraction, such as Giddings' consciousness of kind and Ross's social control; and, to change the metaphor, they have sought to illumine the whole field with a single light only powerful enough to illuminate a tiny sector of it.[2] Even those who have limited their conception of sociology to that of social psychology, like Ellwood and Ross and Cooley,[3] have remained persistently within the purview of generalities, no nearer to observational facts than historic record will permit. I have an idea that the failure of American sociology to develop on other than Teutonic lines was due to a breach in personal and intellectual continuity occasioned by the Civil War. Before that period America had developed two men whose importance we are now only faintly beginning to recognize. One was Rae, the originator

100 West 94 Street
New York City, N.Y.

17 November 1919

[Ts copy, unsigned, UP]

of a sociological theory of Capital, and the other was H.C. Carey, who is known to the present generation, if at all, not as a sociologist but as a strong proponent of the principle of the protective tariff. Carey was a contemporary of Comte: his *Principles of Social Science* was published in 1858 in three volumes. In spite of extravagances in metaphor and a certain hardihood in framing exact laws – as in the attempt to state the attraction between the members of a social group as varying inversely as the square of the distance – Carey remains more nearly at the center of present day sociological interests than any of his successors. The war alone can explain the fact that he was so completely forgotten and neglected. Writing in a pre-Darwinian dialect Carey nevertheless anticipated MacIver in conceiving the science of sociology to be concerned with the conditions and means of human association, and in an early part of the volume he deals with associations in cities and regional groups in contrast to difficulties and weaknesses of association in great states and empires. The acuteness of his observation comes out in his criticism of Ricardo's theory of rent; where he showed that in spite of the logical beauty of Ricardo's doctrine the actual practice in America was to occupy the lighter, shallower soils first, before undertaking the drainage and more arduous labor on the richer valley bottoms.[4] Carey, alas! didn't found a sociological school: he died in the seventies, and just about that time the Ph.D. came into fashion and everyone made a great go of studying in Germany, and specialization set in; and the science of human association was lost. Within the last generation the most hopeful events in sociology have taken place outside the sociological departments. The civic survey came out of the Chicago exposition; the social survey issued from the schools of philanthropy and the settlements; the soil survey (essentially a geologic survey with agronomic applications) out of the Department of Agriculture. On the library cards which I am sending you in a separate packet the best of these surveys are listed. Unfortunately interpretation has not proceeded hand in hand with investigation, and the labrynth of inquiry is still, except for the threads we are able to pick up from Europe, and especially from you – a labrynth. In the preliminary sciences we have been more fortunate: Davis in geology, Veblen in economics, Huntington and Ward in climate, Boas and Lowie in anthropology have all opened up ground which their recent retrospective schools are already working over fruitfully.[5] I have listed the books that seem to bear immediately on the inquiry at hand, and if time allows me I shall try at some later date to append a periodical bibliography. Of practical initiatives those that seem to me most worthy of note are: the social unit experiment, the New School for Social Research, and the Play School experiment. The first was started in Cincinnati in 1918. A fund was provided by a national organization, and a group set out to explore the social resources of a defined neighborhood, for the purpose of reinvigorating the local life and evolving, out of materials at hand, chiefly by voluntary service, a more adequate set of social agencies. On the basis of the results obtained in Cincinnati the social unit group hopes to build up a new constellation of active, socialized communities, within the limits of the present mass cities. It is the reflection of the need for reconstructing and vivifying the life of non-occupational groups which Miss Follett writes about in the New State. This initiative has not yet been adequately appraised: it has been lauded by Mr. Lane, our enlightened secretary of the Interior, as a return to democracy, and denounced by the frightened boss of political Cincinnati as a relapse into "bolshevism."[6] Neither estimate is accurate. What is lacking in the social unit plan up to the present is the vision of the townplanner; what it does indeed contain is the immediate resourcefulness of the social worker who sees that the problem of poverty is not limited to districts altogether depauperate and depraved. The New School for Social Research I mentioned briefly in my last letter. There is little to add. The effort is brave, but only chaotically sociological; the atmosphere is intimate, but the thinking still too departmentalized, and the participation of the student body too remote; the interests are academic, but the materials with which they deal remain somewhat too abstract. Miss Caroline Pratt's Play School (in Manhattan) is small in dimensions but big in importance.[7] She is working with children from four to eight years of age; teaching and experimenting at once. Independently, it appears, she has utilized the method of working with the complete environment which you and Miss Barker initiated at Edinburgh before the war: a large part of the day is spent in the open and the children explore the waterfront and the harbor and the markets and come back to the school to discuss and rationalize their experiences, and express them in maps, models, composition, and play. Miss Pratt uses the natural impulse to play and make believe as the mainspring of the educational tasks provided; she finds that this impulse runs into different channels after the age of eight and a somewhat more mature technique must then be provided – a task which she has not yet essayed. The play school has as yet published no detail[ed] reports of the experiment; but a large quantity of data has been gathered, and as soon as it is done into form I shall send you copies – likewise of course Mr. Branford.

During the last week the *Dial* underwent a change which will leave me with a new set of problems to face, and as yet I am a little at sea. The same fate has overtaken us as overtook the Reconstruction *Athenaeum*. The group headed by Mr. Thorstein Veblen, in the effort to build up a free, non-political review, had won an audience of readers only to find that financial support failed them in the days of transition, and as a result control has passed out of their hands and at the moment I write the *Dial* is on its way to becoming a purely literary magazine. I relinquish my associate editorship accordingly, in company with the rest of my editorial colleagues. What I shall now turn my hand to I am not quite certain. As a free lance I shall not be able to give my mother and nurse more than a modicum of support: hence I shall either tie myself to another journal, such as the *Journal of the Insititute of Architects*, or I shall tide over the interim in some other occupation than journalism. My real problem is to continue my education and to support my family at the same time. Unless I make a satisfactory compromise along these lines during the the next couple of months I shall probably ship to sea as radio operator in the spring, for at least a six months cruise. Your chapter in *Our Social Inheritance* on the possibilities and prospects of the wandering student has stirred me more than ever (for I had already been stirred) to a realization of the educational possibilities of such a cruise: perhaps for the first time in my life I am equipt to get the the full educational advantages out of such a *wanderjahr*.[8] The very charm of my present position was not without its menace: for the temptations of a safe and restricted and more or less sessile life seem to lurk in the background of all urban existence: and perhaps the present jolt out of a condition of economic security and physical ease will prove the best thing that could have happened to and for my educational program. I want very much to sit at your feet for a while, and perhaps if I am footloose for a period this, too, will be possible. . . .

The new Making of the Future books which Mr. Branford had sent to me arrived only a few weeks ago, and because of a protracted printer's strike I have not been able to write my review. If the *Dial*'s policy does not change too suddenly I hope to do a long essay review, embracing the whole series; but if I do not succeed in doing this I shall suggest a somewhat shorter review to the editor of the *Nation*. A duplicate set which was sent me by mistake I turned over to Mr. C.H. Whitaker, of the *Journal of the American Institute of Architects*, on account of his interest in your work and in the subject, and his access to an audience which it is of first importance to re-orient sociologically.[9]

I would be grateful if you would continue to inform me of your itinerary, in so far as you are able to forecast it. A friend of mine in London wrote me of having met you at a dinner to Brandeis.[10] I felt a twinge of envy . . .

Please pardon the delay of this letter and believe me,

Faithfully and gratefully yours,

1. Charles Josiah Galpin (1864–1947), author of *Social Anatomy of an Agricultural Community* (1915) and *Rural Life* (1918); Georg Simmel (1858–1918), German sociologist, author of numerous papers which appeared (in translation) in American journals; Karl Bücher (1847–1930), German sociologist, author of *Industrial Evolution* (1901).   2. A list of important early American sociologists: William Graham Sumner (1840–1910), author of *What Social Classes Owe to Each Other* (1883) and *Folkways* (1907); Lester F. Ward (1841–1913), author of *Dynamic Sociology* (1883); Franklin H. Giddings (1855–1931), author of *The Elements of Sociology* (1898).   3. Other pioneer American sociologists: Charles A. Ellwood (1873–1946), author of *Methods of Sociology: A Critical Study* (1933); Edward Allsworth Ross (1866–1951), author of *Foundations of Sociology* (1905); Charles Horton Cooley (1864–1929), author of *Human Nature and the Social Order* (1902).   4. John Rae (1796–1872), author of *Statement of Some New Principles on the Subject of Political Economy* (1834); in fact, H.C. Carey (1793–1879) published *Principles of Political Economy* in three volumes 1837–1840; Robert M. MacIver (1882–1870), author of *Community* (1917) and *Society: Its Structure and Changes* (1931).   5. William M. Davis (1850–1934), Harvard geologist; Ellsworth Huntington (1876–1947), author of *Civilization and Climate* (1915) and *The Human Habitat* (1927); Robert DeCourcy Ward (1867–1931), author of *Climate, Considered Especially in Relation to Man* (1918); Franz Boas (1858–1942), author of *The Mind of Primitive Man* (1911); Robert H. Lowie (1883–1957), author of *Primitive Society* (1920).   6. Mary Parker Follett (1868–1933), author of *The New State: Group Organization: The Solution of Popular Government* (1920); Franklin K. Lane (1864–1921), a conservationist interested in reclamation projects who served as Secretary of the Interior under Woodrow Wilson (1913–1920).   7. Caroline Pratt (1867–1954) established the City and Country School along lines suggested by Dewey; play and excursions were incorporated into the curriculum.   8. *Wanderjahr*: Period of travel or apprenticeship.   9. Charles Harris Whitaker (1872–1938), editor of the *Journal of the American Institute of Architects* and later Secretary of the Regional Planning Association of America.   10. Louis D. Brandeis (1856–1941), U.S. Supreme Court justice.

Dear Mumford,

    Your letter is the most gratifying of all my Christmas morning budget, & so I answer it first.

    I note – indeed with great regret – that the Veblen group is losing *The Dial* – but I hope they will be able to keep the school afloat – & before long perhaps start another and smaller organ – perhaps a monthly in which their essential thought can still be addressed? (If so, please enroll me as a subscriber to it, as I was first going to have become for *The Dial* but now shall not.)

    Your outline sketch of the settling down of the American schools of sociology too fully confirms my impressions of it – but with live critical minds like Veblen, & constructive ones coming on, like yours, this is only the too common phase in the rhythm of life. And this sub-Germanic phase of university life on both sides of the water is surely ending.

    Has not Veblen a constructive purpose – & faculty too, if he would give it scope? Is Charles Ferguson able to make anything of his often excellent ideals and aims?[1]

    I am much obliged – as also I am sure is Branford – by your proposed reviewing help to our series. I hope this last couple – *Our Social Inheritance* & *Provinces of England* – may help readers to see what we are after – an escape from the politics of the metaphysical – lawyers, etc. (essentially "*Barristeria*") – into more real forms of study not merely of "Association," but of the actual human *associations*, i.e. the 15 million villages, towns & cities which dot this earth, & with reconstructive *action*, in their life & lives accordingly is Sociology, & Civics.

    You ask for my plans. Here until end of term, 10 March; sailing if possible 13th (but only too likely not immediately to get passage). Then if all's well, back to Jerusalem for the summer; & due here once more for 10 November thus leaving Palestine by 15 October.

    Of course the new Jerusalem *may* fall through, as too many of my fine schemes have! But for the first time I have had clients in the Zionists, who were ready for all I could offer, & not reluctant to leave their old & easy going habits, as with all towns I have had to do with elsewhere. The Zionist Commission in Jerusalem were delighted – the Christians too! – with our version of Revelation XXI-2 (I think it is) (despite incompleteness of given measurements!) for the university on Mount *Scopas* – well named as greatest of world outlooks – both historic westwards, & cosmic eastwards.[2] So they kept my assistant [my old Town Planning Exhibition Assistant & collaborator in various enterprises till carried off to war (also my son-in-law) Frank Mears] to go on with its perspectives & model presentments, etc., & then accompany them to London & Geneva to the general gatherings of Zionists there next month, at which they hope permission to proceed actively. We'll soon hear (end January) of course as above said. There *may* be disappointments & delays, but I am sufficiently hopeful to have given up much remunerative work after term

University
Department
of Sociology &
Civics Library,
College of Science
Bombay

25 December 1919

[Ms UP]

– e.g. Colombo City, etc., which I am handing over to my admirable friend H.V. Lanchester, who is on his way here at present.[3]

Next as to your own plans. I am much pleased that you still have the idea of coming to me for a time: and in view both of this new & ambitious Department here & its associated civic aims, as also of this larger scale & more thorough going ones for Jerusalem and Palestine, I think you might do worse.

Alas, your difficulty is the unending one – which I too have still – for though now between this chair & my practice, my income is swallowed by the deficits & overdrafts of past life's adventures, plus the peculiar disasters of the war years; & this absorption will go on for years to come. Still, you might do some potboiling, e.g. as correspondent in India & in Palestine, & will write illustrated magazine articles, etc., if not books. At your age, & very long after (indeed till lately, as above), I was obliged to overdraw on future; & I believe if a man in our lines, without means, wishes to be effective later, he must often do this – if not always!

Of course with me, you might take hold toward potboiling work of the best qualities – i.e. towards the larger civic constructiveness which must soon come to the front beyond your present civic contributions, etc. Have you any planning aptitude? And what architectural feeling & imagination, even if not technical skill? What would you say to run my Town Plannning & Civic Exhibition in America after a term in it here?[4] Or to help me (& so, more or less directly, Branford too), ageing as I am (& he also) towards overtaking the unusually disproportionate file of materials for books beyond the small shelf of printed ones? I have always tried to find collaborators. Thomson & Branford are the especial cases of this;[5] and now my admirable assistant professor here (lately University Librarian & much else before that); but as you know, I lost my son, while the other is (& may too long be) semi-invalided: so I must find others!

The best I can afford to suggest is (a) a sum of £100 towards expenses, if you care to come for me for a year (this payable on your arrival in London or Paris); plus (b) another £100 towards the same in course of year, and (c) more, if you can help me to earn it! But this last (c) alas, is unlikely; for all my writing, throughout life, despite occasional trifling pay, has but in aggregate added to the above-mentioned deficit; & even this latest series substantially so! Of course this is not to be understood as any proposed exploitation of you for a sum so far below your ordinary earning power: but as a small scholarship for research and such collaboration as you may give me in work mostly unremunerative, in any conditions, etc.

Yours cordially in any issue,

P. Geddes

Better send a weekend deferred cable to Geddes, University Bombay, *if you decide to come* to Palestine this summer, & wait reply, lest my visit be delayed after all so give your cable address. PG

Example of possible collaboration (the first which occurs to me, offhand)

Consider this theme – of which I am thinking & gathering jottings – of Neo-Politics (in *Civics*), i.e. of collective action, but no longer of the Bar –

which practically invented this, into the modern game of Ins and Outs called politics, by making the Jury into the constituency; and then marking time, and postponing detection of the real nature of their game, by various devices: e.g. (a) Legislation-cobbling (in the interest of *Ins* – or professed interest of *Outs* sometimes, so as to "dish the Whigs" etc.). (b) Jury ("franchise") extending. (c) Popular "Education" for the three R's (Reading for Liberals, writing for Bureaucrats of "empire," and arithmetic for Financials) to exclusion of three H's, which lead to their defection, among other changes towards new order.[6] (d) Consequent for all these, the formation of the still active-minded of this barristerial culture into amateur barristers (Gladstone, Morley, Balfour, etc., & like in every country). Mrs. Pankhurst is but the last or one of latest.[7] (e) Side-issues – e.g. anti-clericalism, which effectively threw Radicals (Socialists & Anarchists too) off the real scent, and so on.

Whereas is not real politics each and every form of collective action: e.g. Militaristic & Bureaucratic – in various measures in the various empires / Financial – "City," *Haute Finance*, etc. / Artisans – Trades Unions, Syndicates, etc. / Peasants – Soviets, etc., & Agricultural organizations, cooperatives, etc., generally to Rustics / So why not increasingly e.g. Doctors to Public Health, Teachers to Education & so to Town Planners, Architects, Artists, etc., to Civics.

To map out this would be a real bit of contemporary sociology, & to plan it out further would be real civics.

---

1. Charles Ferguson, author of *The University Militant* (1911).   2. In 1919 the Zionist Commission appointed Geddes to plan the new Hebrew University in Jerusalem. Revelation 21:2: "And I John saw the holy city, new Jerusalem, coming down from God out of heaven, prepared as a bride adorned for her husband."   3. Henry Vaughan Lanchester (1863–1953), architect who worked with Geddes in India; author of *Town Planning in Madras* (1918) and *The Art of Town Planning* (1925).   4. Using materials he had collected in the Outlook Tower, Geddes first mounted his Cities and Town-Planning Exhibition in London in 1910 as part of the First International Conference of Town-Planning; it was then displayed in Edinburgh, Dublin, and Belfast; competing with much more expensively produced exhibits of other countries, the Cities Exhibition in 1913 won the Grand Prix at the World Congress of Cities, part of the International Exposition held at Ghent, Belgium. Helen Meller writes: "From 1910 onwards, the Cities and Town-Planning Exhibition was to travel more or less everywhere Geddes went and it brought him his greatest successes" (175). Much of the Exhibition was lost in transit to India when the ship carrying it was destroyed by a German raider in 1914, but it was soon reassembled.   5. John Arthur Thomson (1861–1931), loyal associate of Geddes; collaborated with him on several books, including *The Evolution of Sex* (1889), *Evolution* (1911), *Sex* (1914), and *Life: Outlines of General Biology* (1931); held the Chair of Natural History at Aberdeen University, 1900-1930.   6. The "three H's": head, hand, heart.   7. Personalities prominent in British politics: William Gladstone (1854–98), statesman; Arthur Balfour (1848–1930), statesman and philosopher; John Morley (1838–1923), statesman and man of letters; Emmeline Pankhurst (1857–1928), militant suffragette.

Lewis Mumford AND Patrick Geddes

THE CORRESPONDENCE

1920

Dear Mumford,

University
Department of
Sociology & Civics
College of Science
Bombay

10 January 1920

[Ms UP]

Pray read, and post for me, the enclosed to Secretary of Edison Laboratories: I know not where they are!

How can one obtain any real account of Edison's contributions to science & invention, which I know are many & great; but as yet know mainly through too impressionistic journalism. e.g. Have his concrete houses satisfactorily materialised?[1] (Also of Luther Burbank – whom we botanists, on this side at least, know of mainly in that way, & from his own 3 volumes not very convincing. Thus, e.g. Mrs. Besant once invited me to her Adyar Garden to see the famous "Spineless Cactus" (*Opuntia*) which she had imported from him at some expense.[2] But on the walk down from Madras City, I saw it often in the hedge-rows! – it is a common and natural variety, in India at least. She was not too well-pleased (*with me*) when I showed her them. I saw it wild on my last country excursion on this side also.)

Returning to our main common interest of Sociology and Civics – if it be not asking too much of you, pray indicate some books I ought to buy of the last ten years. I got one or two of Patten, Giddings, Ward & other serious writers in Library & bookshops here, but 0 recent. Giddings does not grip me in any way. Has Small ever written a general book? Patten and still more Veblen are the only two of whom I have felt impressed, as personal contributions and points of view – though there are no doubt good sensible compilations, like Hayes', etc., & these I might give students to read. What now of Zueblin? I knew him 20 years ago, also C. Henderson as of civic impulse? Have they effected something? I hope so; both moving fellows & full of ardour.[3]

What of the Surveys up & down America – Pittsburgh, etc. How far by the way do these derive from the "Survey of New York," published by Dr. Strong and Dr. Johnson for Institute of Social Service after Johnson's visit to me at Outlook Tower about 1898 or 9.[4] It would be curious if that had some result in that way?

You see in my past ten years active planning I have fallen behind with reading. Even my contacts with France, which through life I have most valued and profited by, are largely interrupted these ten years. You must be living in a different world of discussion now are you not? Or have people been largely marking time? What post-war writers can you recommend? Or have they not appeared?

Why have you no adequate American bookselling, even in Britain, much less any in India? Ford cars, etc., everywhere, but 0 books. Even Macmillan does not stock his American issues here! Is it worthwhile buying back numbers of *Small's Sociological Magazine*? How long since I saw it. (If you know a live bookseller, who would care to send me sociological publishers catalogues here – as also to c/o Zionist Commission Jerusalem – should hope to buy some in the next year or so. He would best send them by post, not in cases as merchandise!)

Another question. Has higher Education with you taken up the Cinema? Have you any precedents for putting them in new University of Jerusalem – which I keep on with planning. Did you see my *Masques of Learning* book (1) Ancient (2) Medieval & Modern? If not ask Branford to send you a couple. Tell me when people will be ready for such as these (and their attached *Repertory*) as Cinema productions? (In these matters too I have seen little: too old to be attracted!)

I hope America is not to withdraw from League of Nations? The present chaos is not a little from these delays. Who is to be your next President? Hoover? Would not he be a more concrete statesman? like Plunkett in Ireland – are such types coming on? Or have you also fresh infective swarms of Lawyers? – like Carson & Smith (now Lord Chancellor with us)?[5] Enough of such worrying of you. I'll be interested to have your answer to my last.

In any issue, cordially yours,

P. Geddes

1. From 1906 to 1908, Thomas Alva Edison worked on plans to mass-produce pre-fabricated concrete houses for workers; he hoped to sell each of the three-story, six-room houses for $1200, but the scheme was abandoned. He had earlier patented a type of composition brick and a machine to produce it. Geddes was interested in using Edison's construction techniques in India. 2. Luther Burbank (1849–1926), American horticulturist and plant breeder. Annie Besant (1847–1933), theosophist, Fabian, active in Indian politics and educational reform; founded Central Hindu College in 1904.   3. Simon N. Patten (1852–1922), American political economist, author of *The New Basis of Civilization* (1907). Albion W. Small (1854–1926), held the first American chair of sociology at the University of Chicago, founded and edited the *American Journal of Sociology*, author of *The Meaning of Social Science* (1910). Edward C. Hayes (1868–1928), author of *Introduction to the Study of Sociology*. Charles Zueblin (1866–1924), sociologist at the University of Chicago, attended Geddes' Edinburgh Summer Meetings and wrote an article on the Outlook Tower, "The World's First Sociological Laboratory," *American Journal of Sociology*, 4 (1899): 577–592, author of *American Municipal Progress* (1902). Charles Henderson (1848–1915), sociologist at the University of Chicago, author of *Introduction to the Study of the Dependent, Defective, and Delinquent Classes* (1901).   4. Probably Josiah Strong (1847–1916), social reformer, founder of the American Institute for Social Service. The references to "Dr. Johnson" and the "Survey of New York" are obscure.   5. Horace Curzon Plunkett (1854–1932), Irish statesman who founded the Land Organization Society; Edward Henry Carson (1854–1935), Ulster lawyer and politician, as member of Ulster Unionist Party opposed home rule; Frederick Edwin Smith (1872–1930), lawyer, Lord Chancellor, M.P., supported the Irish Treaty (1921).

New York City[1]

2 February 1920

[Ts copy, unsigned, UP]

Dear Professor Geddes,

Your letter of December 25 has bowled me over, and I have just enough wits left to run to a cable office and announce a prompt and enthusiastic Yes! Short of breaking up our household altogether my mother joins me in being prepared to make any sacrifice necessary to embark on the venture: and we have fortunately sufficient funds to make my year's absence at least possible. With these financial considerations properly subordinated and forgotten I have rolled over your proposal for the last thirty-six hours in a mood of humble introspection: for I wished to be sure that I had enough to offer to permit my taking advantage of your adventurous generosity without doing either of us injustice. The judgement I reached at last was favorable! And if in forming it the wish to work with you was father to the thought of my ability to, ultimately the wish itself rests on the fact that I have long marched mentally in your company and that despite deficiencies and immaturities in my own equipment the fact of my being in tune with you will enable me to keep in step. Architecturally my aptitudes are undeveloped, but not entirely absent. I am naturally a "visualizer," and I have not rambled around our eastern cities from Portland to Washington without getting an insight into the obverse side of the architects' business – into that which should be remodeled or strengthened or obliterated. Of technical facility I have nothing except the ordinary rudiments of mechanical drawing and a fair ability with freehand. (In the Cement Laboratory in Pittsburgh I used to make sketches of the terra cotta cornices and "ornaments" before they were broken up in the compression tests.) But except for the fact that I did try my hand in a National Housing contest last year, with drawings made while under restriction in barracks during an influenza epidemic, my answer with regard to planning must be that of the man who didn't know whether he could play the violin because he had never tried. When it comes to taking charge of the Town planning exhibition or helping with your materials I can however speak with greater confidence. My acquaintance with your field has been deepening steadily during the last five years and since most of my energies have been poured into the business of expression I think something might be done toward overtaking the disproportionate pile of which you speak. Here my very youth may be an asset and the dissimilarity of two generations may prove a common stimulus – although I realize that intellectually you are in the van and that my generation is panting – so far as it has been moved to exertion – to keep abreast of you! I realized that when I failed to put the "science of human association" in the plural number.... At any rate, I could give my whole heart to your work, and head and hand would follow promptly in unison.

Along with your cable I am sending one to Mr. Branford. When he heard of the *Dial's* demise he was moved to offer me a part-time post in London, on the assumption, for which I was no doubt responsible, that my primary interests were in journalism. I answered with a tentative acceptance and have spent the last month in disentangling myself from a host of interests, and promises and conditions which conspired to keep me on this side of the water. My letter informing him that I would leave sometime in April crossed the one in which he told me I had better come before Easter or postpone the trip to next September. There was of course nothing pressing in his offer: it was obviously made for much the same reason that prompted yours: and I think he will understand my reason for wishing first to take advantage of yours, although the difference in mailing time gave his the earlier start. I am cabling him however to say that I shall join you on his approval: and I trust also that I will have your approval in making this decision. My very warm respect for Mr. Branford, with whom I have corresponded frequently during the last half year, is of different quality from my attitude toward you, and any attempt to treat the difference quantitatively would prove needlessly disparaging to him. (Perhaps the figure of the source-stream and the reservoir would cover it, with all implications as to energy, difficulty of access, and so forth: but it is not fair to lose a good and trusted friend of yours in a metaphor and I beg pardon for the impudence!) I owe a debt of thanks to Mr. Branford for preparing me to take advantage of your offer: and I await your cable to find out whether and when and where I am to meet you. By the time you receive this all that should be settled, and I am sending this in duplicate to Bombay and Jerusalem in the hope that in one place or another it will anticipate my arrival. By that token the enclosed photograph will break the ice of outward unfamiliarity: I long ago acquainted myself with you in Alfred Gardiner's book.[2] I am trying to forestall passport difficulties by getting in touch with some prominent American Zionists: Irishmen and Jews are privileged at Washington and the mandarinate quickens its movements on occasion under their stimuli. . . .

I postpone all renewals of thanks and gratitude until I am able to voice them personally. With happy anticipations I am,

Yours faithfully,

1. Mumford's annotation of 31 March 1978: "Important!".   2. Alfred G. Gardiner, author of *Pillars of Society* (1914).

University of
Bombay

5 February 1920

[Ms UP]

Dear Mumford,

I have unexpectedly got offer of a passage as early as 19/3 & thus am cabling you – "*Britain April October, try come.*"

My plan is to see old friends 2–3 days at Montpellier, perhaps *sense* atmosphere of Geneva, & then spend a few days in Paris – perhaps a visit to war zone & to Brussels – & then to V.B. in London before end April. I'll need to stay there or thereabouts some time, for collaboration with him is a main motive of coming home of course. Another is with J. Arthur Thomson (read his *System of Animate Nature* – 2 vols. but very readable). I am asking both to consider whether they could not pass part of their long vacations at Edinburgh where accumulations of Outlook Tower may be utilised. Failing this, I may go to Aberdeen or district to suit Thomson. Both lines of interest would suit you, I think, would they not? And we old fellows must not longer delay, I feel.

I shall do my best towards meeting your expenses; though in view of having worked for 0 all last summer in Palestine, & of much spending & no earning this coming one, I can't offer all I'd wish. Let me hear as to this, of course.

This rapid decision involves a busy mail morning: but I'll write you again before long. (Do not of course write here, but c/o Le Play House.)[1] (If you cable – Geddes–University–Bombay is address.)

1. Le Play House: named after the French social theorist Frédéric Le Play (1806–1882), who was an important influence on Geddes' thought; the Le Play House (65 Belgrave Road, Westminster) served as the headquarters of the Sociological Society, the Le Play Society, and other organizations founded by Geddes. Mumford resided there during his 1920 sojourn in London.

Department of
Sociology and Civics
University of
Bombay

14 February 1920

[Ms UP]

Dear Mumford,

Your cable just in – (It reads will join Palestine. What date. Mumford hundred west ninety fourth New York.)

What date? Alas, the U.S. Senate is holding up Eastern settlement; that holds up Palestine Settlement & that delays Zionist Congress – which had to discuss my plan & appoint me (I hope) to proceed actively with them. With the Zionist Congress thus deferred, I can't go to Palestine myself, though I am down for passage in a month hence! much less can I ask you to come. Under these circumstances (unless you can buck up your Senate, & before you get

this!) your only course of cooperation with us is to take Branford's offer, &
stop in London until I can ask you to come further at least.

You see how the situation stands, and so I am sure will not blame us. I
may come home myself.

Pray write not here, but to c/o V. Branford who I keep informed of my
address.

Yours cordially,

P. Geddes

———————————

Dear Professor Geddes,   100 West 94 Street
                                                                              New York City

I have duly forwarded your letter to the Edison Laboratories and in a
sketchy way I shall try to reciprocate the rest of your communication of January   22 February 1920
tenth. What I am not able to answer offhand at the present moment I shall get
possession of before I leave: so that you may count upon my being able to fill up   [Ts copy, unsigned, UP]
the vacant spaces, either by memoranda or conversation, when we meet. (Which
last will have been arranged, I trust, by the time this reaches you.)

First as to Edison's inventions. His plan for single unit concrete houses
contemplated, I believe, the use of steel molds, and this part of the scheme has not
advanced beyond the stage of the original experiment. The idea of pouring the
concrete through the whole mold at one time, however, has been successfully
applied by Ingersoll, the watch manufacturer, and great ecomomies seem to have
been effected by this method, with the additional advantage that the house-molds
can be more widely varied than Edison's more expensive frames. Grosvenor
Atterbury has experimented for the Sage Foundation with small houses put
together out of large cast concrete blocks and seems to have achieved economy
without undue uniformity by this method.[1] I shall endeavor to get hold of the
literature on both methods, for what information I have on the subject has been
acquired casually. From the first Edison has been an eager advocate of the use of
motion pictures in teaching, and various large moving picture companies now
have special educational departments. Two kinds of film have been developed.
The first kind employs the conventional methods of photographing natural
objects: the second utilizes a somewhat different technique of photographing a
succession of line drawings and makes it possible to explain a complicated
process (in mechanics for example) by a moving diagram. The latter was
developed during the war and used at Columbia and elsewhere in the instruction
of engineering classes. Science lecture halls in American universities are usually
equipt with a screen and a picture projector: hence the introduction of motion
pictures encounters few difficulties in these institutions, whereas in the lower
schools it is generally necessary to make use of the school auditorium. Through
the cooperation of research groups like the Rockefeller Foundation for Medical
Research it is possible that the moving pictures that will soon be available will
surpass the textbooks. There is no lack of American precedent for the cinema
itself: before another decade is over there may even be precedent for a motion
picture library with special, small projection rooms!

Your *Masque of Learning* was one of the first things I imported from the
Outlook: and the pages of each book are now well-thumbed. The thing could

be done magnificently in motion pictures and I shall consult a friend of mine in one of the live companies to see what prospects it may have. The few great moving pictures I have seen have been woven around historical themes, some very ancient, and it was by reason of historical interest, rather than of drama that they held the spectators. The Bible itself has been filmed in a production longer than *Parsifal*:[2] but this was done under the auspices of a Christian sect, and it may be that the better sort of educational film will have to be financed cooperatively by colleges, etc., before the commercial producer is tempted to make a venture on the public's intelligence.

In America the chief change that has come over thought in sociological fields has resulted from the emergence, during the past ten years, of a class of socially conscientious architects and engineers. Following in the wake of the earlier conservationists, these men have discovered wastes of material and miscarriage of effort at every turn of productive enterprise. Where they have been given a free hand they have been able to achieve remarkable results through reorganization based upon an accurate industrial survey, and they are beginning to ask why the methods which have been found successful in reconstituting factory units should not be applied to other institutions. They do but lack the word "Regional Survey" to describe their characteristic method of approach. The other day one of these socialized engineers, now lecturing at the New School, was discussing with two members of a newly formed technical board of American Zionists the problem of repopulating Palestine. One of the Zionists, Horace Kallen, was adumbrating the plan for socializing the land and creating a suitable industrial regime when Marx broke in with a series of questions.[3] Did they know the water resources of Palestine? The extent to which established towns had adequate water supplies? The difficulty of obtaining fuel? Had they a resource engineer on the spot getting the lay of the land by first hand survey? He insisted upon establishing a concrete basis of fact before discussing political and social possibilities, and he indicated intense dubiety about the whole Zionist project until I assured him that the method he advocated had been persuasively urged and acted upon by one, Patrick Geddes, who had long ago claimed for sociology the instruments now jealously appropriated by the engineers. Men like Marx, Steinmetz of the General Electric Company, and others have not yet begun to occupy any of the higher offices of the government: but they are plentifully sprinkled in various government departments and in the great industrial corporations – the governments behind government – they are achieving a place which makes the reign of the pecuniarily crafty businessman more and more precarious. Hoover is the only technician of presidential calibre, and much might be expected of him did he not, at times, seem to think with an older generation rather than with his own.[4] Veblen tends to overemphasize the antagonism between the business man and the engineer. Outside the performance of the immediate job it is the business view of social relations rather than the energetic standard that tends to be dominant. The technique of engineering must be made consciously to affect the engineer's political outlooks: it does not operate automatically. The socialized engineer must become more of a social philosopher before he succeeds in being more of an engineer. The question with Hoover is whether, as a politician, the business man or the engineer would be in the saddle. I cannot answer confidently. It was easy for the engineer to be uppermost in Belgium, for the same reason that communism is readily established in a shipwreck: in a more normal situation vested interests are stronger and tend to obscure the necessity for facing the real business in hand.

Before leaving the Unistates I shall order a small budget of books in civics and sociology so that you may be sure to have them on hand in Bombay next winter; and in addition I shall make arrangements for obtaining either a long list from a bookseller or a series of catalogs from the publishers. There is one shop in the city that makes a specialty of getting together libraries for special purposes; but their service in filling some orders of mine was highly unsatisfactory and I shall make inquiries before recommending them. In how many languages can you stock your Bombay library? Monographs like Blanchard's study of Grenoble are excellent examples of the geographic–historic survey. And some of our government surveys are worth having merely as models of investigation. There are a few good studies of cities from the geographic standpoint in German: Kurt Haddert's *Die Stadt Geographisch Betrachtet* is one. I have lost most of my facility for reading in German during the past four years and I share the common revulsion against the moral cretinism of German civilization: but I note with a tinge of admiration that the publishers of the *Sammlung Goschen* did not seize the war as an excuse for patriotically doubling their prices. Perhaps a later generation may profit from a chastened Germany – Thorstein Veblen still believes however that the chastening has yet to be administered![5]

Last week I invented a system of note-taking and recording and filing which will multiply my effectiveness as a student many times. For many years I have been experimenting with paper, index boxes, and the like: and I had not up till now devised any method of keeping notes which applied equally well to all kinds of data – a small quotation as well as a long memorandum, a continuous series as well as a disconnected one. By selecting a standard size, six inches by ten, I obtain a size large enough for a long note, suitable for storage in a loose leaf folder. By dividing the paper into three sections I get two four by six slips, with a two inch marginal flap for further classification. Hence, for long notes as well as short I am able to make use of the regular (and convenient) four by six index without departing from a single, uniform size of paper. A six by ten sheet thus gives flexibility and uniformity, a combination I have long been in search of. I have already done a long essay, a sort of guidebook for students, on "The Tools of Study": fortunately this has not been published, for this new system will call for drastic revisions. In six months I hope to be able to report whether its charm still remains.[6]

At this moment I am still awaiting your cable, but of course by the time this reaches you I may have made my departure. How soon you can expect me will depend of course upon what has happened in the meanwhile. Branford cabled me his assent to my joining you. My ties at home have proved particularly hard to unbind and for that reason I am reluctant to leave a day before it is necessary. Hence probably I shall have to miss the fortnight with Mr. Branford he so kindly invited me to enjoy.

Always cordially yours,

1. Grosvenor Atterbury (1869–1956), architect and town planner; designed Forest Hills Gardens in Queens, New York, an early model town providing workers' housing; author of *The Economic Production of Workingmen's Houses* (1932). The Russell Sage Foundation was established in 1907 by Margaret Slocum Sage (1828–1918), the widow of the American financier Russell Sage (1815–1906), with an endowment of $15 million for "the improvement of social and living conditions" in the United States. The Foundation sponsored research in health, social welfare, education, government, and city planning.   2. *Parsifal*: opera by Richard Wagner, 1882.   3. Horace M. Kallen (1882–1974), philosopher and social theorist, author of *Why Religion* (1927).   4. Charles Proteus Steinmetz (1865–1923), mathematician and engineer employed by the General Electric Company; renowned for his research on the theory of alternating current and lightning. Herbert C. Hoover (1874–1964), mining engineer and thirty-first president; in 1920 he published several articles on social reconstruction in Europe.   5. Raoul Blanchard, *Grenoble: étude du géographie urbaine* (1912). *Sammlung Goschen*: the inexpensive series of books in German first published by the Leipzig firm of G.J. Goschen and taken over in 1919 by Walter de Gruyter & Co.   6. Mumford's note on the manuscript dated 27 June 1951: "This great invention I dropt almost immediately! I doubt if it lasted a week."

---

100 West 94 Street
New York City

14 April 1920

[Ts UP]

Dear Professor Geddes,

For the last three weeks I have been waiting for a serene moment to come along; so that I might not merely answer your letter of the 14 February but resume at the same time the sociological and civic topics that have engaged us in earlier letters. My activities however have been unwontedly disconnected and tumultuous and I fear I shall not regain normal mental tranquility until I have made my departure. Your letter of course brought disappointment but it was not altogether a surprise; for between my knowledge that the Zionist Congress had been postponed and the newspaper report of political difficulties between Great Britain and France in the Near East, with an incalculably active nationalism developing to boot, I was able to interpret your silence as a negative way of indicating your inability to plan for the immediate future as confidently as you had figured in December. Your own disappointment at the turn events have taken weighs much more heavily on me than my own, and I trust that by the time the Zionist Congress will have reached a favorable decision the political atmosphere will be sufficiently clear to enable you to proceed actively. As for my own cooperation, there is some doubt in my mind as to how far, speaking territorially, I shall be able to go. The British Military Control maintains its wartime vigilance, and when I applied for a visa last Saturday my passport was held up for a couple of hours until my history could be investigated; at the end of which time the polite young clerk at the consulate informed me with a faint air of dubiety that I had been an editor of the *Dial* and that the *Dial* was known as a radical paper! The visa to England was given me without further comment, and I shall leave on April 24 in due order: but whether my menacing record as an associate of Mr. Thorstein Veblen will handicap me from traveling to the fringe of the Empire, should the opportunity itself arise, is a matter which has yet to be proved. Were my main occupation other than journalism the difficulty would doubtless dwindle, but were that the case it is likewise quite obvious that the difficulty would never have arisen. A more secondary matter which also tends to limit my movements, and which makes London a desirable place to work in, is the additional limitation which has been placed upon my finances by the rise in rents and the exactions of illness in my household, both of which have increased the difficulties of departure to a point where I have been tempted to abandon all my plans for sociological collaboration in favor of a limited bread-and-butter program. The temptation has never been very imperious however and as long as I can hold my own individually while abroad I am prepared to meet the demands at home by a direct capital outlay if necessary. I have no doubts about the wisdom of this procedure, and your words of encouragement in your letter of December confirmed a principle which I had already adopted and acted upon. At any rate I am accepting Mr. Branford's invitation and after the first of May expect to be installed at 65 Belgrave Road, and whether I stay in England or am able at some later date to join you I shall do my best to contribute to our common work. Fiat labor, pereat homo![2] It seems rather idle to devote a whole letter to these personalia; and I hope during the next fortnight to rally my thoughts together for a less limited expedition. In line with the present mood however is the news that Walter Fuller is now one of the editors of a new weekly, *The Freeman*. I have spent hours in his company, and you were the only topic of conversation; Fuller boswellizing you in sputters of enthusiastic anecdote![3]

Sympathetically yours,

Lewis Mumford

1. Mumford's annotation: "Returned from Jerusalem."   2. "Fiat labor, pereat homo!": Let the work be done, let the man perish!   3. Walter Fuller subsequently became editor of the *Wesminster Weekly* in London. "Boswellize": to describe a life with the personal detail and in the admiring manner of James Boswell's *Life of Samuel Johnson* (1791).

Dear Mumford,   Jerusalem

At last – only today – I can give you my timetable. I leave this 22nd for   7 June 1920
England – only arriving London July for Zionist Conference where I have first
of all to explain & defend plans for City and University of Jerusalem This will   [Ms UP]
take perhaps a week, though we may meet briefly meanwhile. Then I hope to
be free for you (though there *may* be upsets!).

My idea is to get your help to arrange my tumbled chaos of papers – worse perhaps than Branford's – & in so doing to get your further help in writing various points – partly those urgent, partly those which may most appeal to you. I can't say for how long – for I ought to go to Ireland, & make a run to Edinburgh, etc., too. (These might be interesting for you also.)

If you can apply your mind & experience to problem of arranging papers so as to be portable also, I'll be grateful. I know lots of bibliography & methods, but in practice can't apply them in my migratory life, & so lose not only time & trouble, but valuable notes altogether (2–3 years work *lost* this way!).

For theory the more you know of our books & thought the better. Look over *Evolution* & *Sex* with J.A. Thomson as well as those with V.B. & get hold of what Regional Association is after.[1] Prepare any criticisms as well as suggestions, developments, etc., for I have had little time to read for many years past, & Branford is also too busy.

If as I hope you know & get on with W. Mann, & also with Alexander

Farquharson & George Sandeman, pray have some talk with them regarding these matters. Mrs. Davies & Miss Barker are not ignorant of them – & particularly good on regional work.[2]

Yours,

P.G.

1. Geddes and J.A. Thomson, *Evolution* (1911) and *Sex* (1914).   2. Members of the Sociological Society–Le Play House group of Geddes' associates in London.

---

12 July 1920[1]

[Ts copy, unsigned, UP]

Dear Professor Geddes,

This is a hard letter to write, and I write it with my mind not fully settled, with my mind still clinging to the hope that some unpredictable turn of fortune will annul the decision which I have reluctantly reached. I have not so much reached a decision as found a decision cast for me by the events that have occurred during the last six months, and perhaps the simplest way of coming to the point is that of relating what my new circumstances are. Last January I had only a financial difficulty to settle with before I accepted your offer, and I did not waste five minutes in reckoning with it. Today the financial difficulty has been increased by a long siege of illness in my household at home: but the major point is that a whole set of personal difficulties has been superimposed upon it. Two things happened during the spring: I got badly run down in health and – I fell in love! During the last few days, in discussing the matter with Branford and in looking forward to the conditions under which we should carry on work in India, the first disability has seemed to me the greater obstacle: but I do not think it would have bulked so hugely (it surely didn't last January) had not the second been there. Sophie is not yet so sure of herself or so willing to go onward to marriage in spite of the fact that Sophie was my secretary in the old days of the *Dial*: and my coming to London broke off our comradeship in the middle of that period of half-light that precedes the dawn, with the result that neither of us will be able to work or live satisfactorily until we can take up our relationship where we dropt it off.[2] I can't look forward with tranquility to a whole year's separation from her: and I know that even if I made the drastic decision to spend the rest of the year with you in India the conflict that would result would probably throw me out of effective mental adjustment, so that my work would be incomplete and second rate. I conceive that if Sophie and I were married I might throw financial prudence etc. to the winds and join you: but I don't see how anything like this is possible in our present relationship. I have perhaps overstressed this personal difficulty because it seems to me the clue to other things that stand in the way. It is the inner reason that makes all the outer reasons count so much the more. The outer ones are no less real but I wish properly to subordinate them. Apart from the increased expenditures at home and the fact that the expenses of travelling to India and back would leave me with hardly any capital in reserve, I have very great doubts, now that I have had an opportunity to touch part of your work which I never had access to before, whether my equipment is sufficiently complete to enable me to keep pace with you and work profitably with you during the first six months of our association. I cannot write at all until I have thoroughly assimilated all the material that is mine to work with, and to transcribe notes which are a result of your whole lifetime of experience is a task which I could not attempt until I had taken sufficient time to appropriate them – which of course is not transcription at all but recreation. Branford, in describing your method of work to me, seemed to think that it would be possible for me, after you had spent a day outlining and making notes upon the subject of the opus, to work straight ahead on the basis of these notes without much further excogitation and reflection on my own part. The task seems to him merely one of mechanical journalism, and it could accordingly be organized on the basis of large-scale production, with anywhere from two to six books as the result. His prospectus interested me and I see plainly that this would be the ideal method to get your thought put in final shape for publication – but the difficulty is to catch your mechanical journalist. I can't pretend to have the requisite equipment for this: my output would be lower and poorer in quality under these conditions than the meanest hack on Fleet or Grub Streets.[3] The very fact that I sympathize with every aspect of your work, and that I am deeply debted for the most fruitful results of my own thinking (and acting and living too: my attitude toward Sophie, perhaps, above all!) would keep me from serving effectively as amanuensis. With a reasonably long time together, a year let us say, I could probably catch up with you sufficiently to overcome that handicap of superficial knowledge and reflection which would probably handicap me from doing effective work in any shorter period.

All this, I hope, will not remove the possibility of our collaboration: at worst I trust that it only postpones it. If in spite of present circumstances I was sure I could be of genuine assistance to you I should not hesitate to accept your offer.

Always gratefully yours,

1. Mumford's annotations: "Not sent"; "Important!! 31 March 1978." 2. Mumford married Sophia Wittenberg in September 1921. 3. Fleet Street: the center of London journalism; Grub Street: former street in London inhabited by writers and associated with literary hack-work.

Dear Mumford,

Instead of answering your letter on personal lines, I begin with the general.

Here I'm busy with what is becoming more than the Town Planning of Jerusalem, Haifa & Jaffa (ports), Tiberias, etc., (inland towns), with garden suburbs, etc., etc. Through intensive development of cities on one hand, & of agricultural colonies, etc., on other, are increasingly, correlat[ing] Region & City.

But this Palestine, & Jerusalem are they not as of old, & in war lately – in future politics & policy too, strategic centres & in spiritual sense even more than temporal? Here in fact is the best possible object-area, & object-lesson, of Friedenspiel, for Jerusalem & Palestine are still names to conjure with, more than we of lapsed masses think, & with appeal to them also. We have to muster

c/o Zionist
Commission
Jerusalem

2 August 1920

[Ms UP]

the forces & think out the campaign; fight in it too if we can. You know Blake's verses ending

> "I will not cease from mental strife
> Nor shall my sword fall from my hand
> Till we have built Jerusalem
> Within this green & pleasant land!"[1]

Quite definitely this impulse exists in all regions of the West, all cities more or less. To realize something here, & idealize yet more, is what Jews are already after & not without real beginnings, so now will come turn of projecting this back on other pleasant lands, greener, too than this, e.g. Ireland, of which the old trouble is not merely of alien exploitation & tyranny & bungle not least of all, but of the concrete results of these in region & city. The agrarian is partly solved, between the cash down of John Bull & the constructive leadership of Plunkett, but the mingled neurasthenia & exasperation of homelessness in the miserable towns & cities & the starvation of soul which is all they offer, are hardly seen by any; & not yet generally conscious (though in morbid fashion all the more active, as the psychologists know, Freudians especially. In passing do you know that doctrine? – look into it! It has social bearing beyond what they see – *Empires* with *Sadisms*!).

Well, if I were not here, I'd have been in Ireland, studying & reporting on towns & cities, which I know fairly already. See too the correlation with areas of reconstruction – Belgian, French, etc., Poland, etc. Also the potential reaction in countries not demolished from without, Russians, & Germans too. (I leave you to consider Regionalism in America – too long since I was there – but civics is plainly stirring, & beyond externals I trust.) In short then here is a policy – (1) of making the best we can in Type-Region, (2) of projecting it on others, (of course *mutatis mutandis*.)[2] Details cannot here be gone into, but one developing towards clearness.

Now more general questions. In October, I hope middle, I go back to Bombay. Come! Help me to pull together my general theoretic *opus*, sociological, civic, evolutionary too, & in graphic terms but also written explanation. I need a sympathetic mind to elicit and elucidate my by best, & to help me as secretary & editor (as advocatus diaboli[3] too, for I trust you have the critical side – cannot but have). You will find live men among my students – all graduates, not boys.

Though I say secretary & editor, I look forward to you as collaborator too: that will rapidly develop.

Above then is a programme of Civics & Sociology. The Civics you may say is all very well, but you are not acquainted with Palestine, & are asked now to come direct to Bombay (else no chance of passage there!). But there is Civics too in Bombay & India, & I have much material as well as phenomena around. Here is your great American city in different forms, & suggestive to you accordingly. Besides we may come back here together in Spring. University, etc., may go on then – I can't say yet. Anyway I propose you a year of this adventure, if I live, or indeed in any case for I owe you guarantee.

So with this to personal matters. (1) As to your great good fortune, of finding your love, I have only congratulations & good wishes to you both. (2 & 3) *Money problems*, personal & family, quite reasonable. I guarantee them plus passages India: you for your part will I daresay reduce my liability by writing your impressions of India & the like, & help me to earn something from my manifold unfinished manuscripts. (4) *Health*: Voyage a great sanatorium – Bombay location good, ½ mile between sea & sea. We must try not to overwork each other! (5) *Return to America*: Of course, every man to his own city & region.

Now I don't think I am simply egotistic in these proposals. I think I can help to accelerate your development to mature work, if you'll absorb my material: so that your work will sooner take larger forms of action & fuller universe of discourse: and I especially won't forget that you have to be getting on towards your wedding. (If it be any use, I'll ask you to be recognized by University as Assistant or Lecturer – if you think it can be any use in academic record. Professing need not be the cloistered monkery some make it – but as correlation of thought with practice it helps.)

I don't know if there be much more to say – save that if you decide to come, see about journey at once! crowded ships! You'll need to take any line, any class 1st or 2d you can get – the 2d is decent enough. Try other lines before P. & O.; but all extortionate. Cook will give you list of lines and fares. Take return ticket, as there is reduction (& it would be too dear to go home by East, even if you wished to).

Yours cordially,

P. Geddes

Write of all matters you vaguely mention. Yes, Mann has a graphic mind, but his theosophy sports its working not a little. I'm glad you got on with Farquharson – give him my most cordial regards & ask him to write.[4] P.G.

I never heard if Mrs. F. is better – or gone – I fear the latter. If he wrote me, I did not get it. Tactfully ascertain, & reassure him lest he think me unfeeling.

1. From William Blake's *Milton* (1804–1808). The lines actually read: "I will not cease from Mental Fight, / Nor shall my Sword sleep in my hand, / Till we have built Jerusalem / In England's green & pleasant Land." 2. "*Mutatis mutandis*": with necessary changes made. 3. "Advocatus diaboli": devil's advocate. 4. Alexander Farquharson, associate of Geddes and proponent of the regional survey; author, with Sybella Branford, of *An Introduction to Regional Surveying* (1924).

Dear Mumford,

I daresay you are quite right in wishing to go home this winter, & that I was wrong in pressing you! Not that I go back on my suggestions; I shall gladly welcome you: but I feel as if I had written after all in the impatient egotism of "an old man in a hurry," and not with sufficient consideration for you.

Still I must be on guard that I am not again deceiving myself; it is possible that Bombay is not the best place for our beginning: but better London or somewhere in Britain – with (say) some stay in Ireland? – next summer? For I shall almost certainly *not* be needed here next vacation, for my planning of

Zionist Commission
to Palestine
Jerusalem

31 August 1920

[Ms UP]

towns will be advanced enough, & the University building is pretty sure not yet to be started, & even if it does, that will be more for Mears than me. So if you can return next summer, as I understand from Branford you think of doing, we can have more time together in England than in Bombay, perhaps 6 months instead of 4. Still, as said above, if you can come, welcome!

Yours faithfully,

P. Geddes

P.S. Branford speaks again of my writing for *Sociological Review.* I find this very difficult. I work at the problems before me. But it occurs to me that if you get from B. a few of my old letters, you may marginally blue-pencil passages which could be typed out as *Sociological Notes*, & printed (with such trifling omissions and emendations as may be required). They could go into small type – *Third Alternative*, etc. For I have been accustomed to write him freely of what was in my mind.

I see I have just done the like to Farquharson, so I ask (in p.s.) him to show you that letter (pages 3–4 anyway) as an indication of this though it was written without premeditation of this, like Branford's & other letters.

By the way, have you taken any stock in psycho-analyzing – to which I have only been awakened by Dr. Eder here, last summer & this. I read Ernest Jones' paper on it in *Sociological Review* 1915 with interest the other day.[1] Could not a group be formed to carry on such work in sociology? & perhaps with Le Play House as centre & *Review* as organ?

1. Dr. M. David Eder (1856–1939), Freudian psychoanalyst and British Zionist; persuaded the Zionist Commission to appoint Geddes as planner of the new Hebrew University in Jerusalem. Ernest Jones (1879–1958), British disciple and biographer of Freud; the article is "War and Individual Psychology," *Sociological Review*, 7 (July 1915): 167–180.

---

65 Belgrave Road
London, S.W.1.

8 September 1920

[Ts copy UP]

Dear Professor Geddes,

Branford's letter about our plans for the future is already in the mails, and so you know the course we've decided on. Without pausing to consider what my change in program would mean I inquired at Cook's and the steamship offices the very day your letter of 2 August arrived, and it was only when I found out that it would be impossible to engage passage to India in the early part of the autumn that I proceeded to busy myself with other alternatives. It would have been rather a wrench had I altered my itinerary and joined you during the winter in Bombay, and since that would have involved staying more than a year away from home it seemed to me just as well (all points considered) that I go back to New York for the winter and prepare myself to join you anywhere you designate in the spring. It is needless to tell you how much your program means to me, how urgently I wish to join you, and how grateful I am for your persistent generosity and kindness. The posponement will mean that I shall be able to collaborate with much greater effectiveness when we do meet. A word then about next Spring. If the general scheme which Branford has outlined in his letter seems acceptable to you, please let me know next winter as soon as your itinerary is outlined when and where I am to meet you. It is still necessary for an American to get permission to enter Palestine from the British Military authorities and if that is our destination it will be possible to save tedious delays if I know beforehand. (Of course I appreciate your difficulty in making an early forecast.) My New York address, you will remember, is 100 West 94 Street.

The week that I spent with Branford was the most profitable I've had in England: he cleared up a great many points I was going to bother you with, and while I can't pretend that I've digested the whole system, it is at any rate true that a good part of it has been ingested. In addition I am still feeding on the typescripts of your London lectures on Sociology, which I ransacked out of one of the bookcases in Le Play house. The lectures [...]

... broken down the barriers between the organic and inorganic, the vital and the mechanical – but I assured him that Bose's researches only extended the province of vitalism![1] I am not sure that he was comforted, for in the present state of his health he finds mental re-adjustments difficult, and is constantly tempted, I fancy, to mold the facts that confront him a little nearer to the heart's desire. That is a characteristic of what the Freudians would call the introvert type of mind and what I should prefer to call the introverted phase of activity: when one's energies become feeble in relation to the external situation the introvert reaction is to concentrate attention upon some carefully laid down plan of action or line of thought and resolutely ignore or flee away from anything in the environment which tends to divert you from the restricted situation in which O is equal to E. The extroverted phase of activity is normal when O is relatively greater than E, and when O is therefore prepared to face the situation, to appraise it for what it is worth and to embrace it in the general line of its activities. These phases of activity, it seems to me, become fixed only as abnormalities, producing the extreme introvert type (whose reactions are so irrelevant to E as to constitute insanity) and the extreme extrovert type (whose concentration upon E produces the mechanical automatism of the perfect soldier and the model administrator).[2] I am rather keenly aware of the psychology of these reactions because for the last few months I have been passing through, and I hope *out of*, an essentially introverted phase, which has been accompanied by debility and slackness and general good-for-nothingness. It was in that situation that I had to make my answer in July to your first letter, and it is probably for this reason that I have not written you in the meanwhile. This answers, at any rate, one of the questions in your letter, whether I know anything of the Freudians. I have not kept pace with their literature during the last two years, but I've studied them long and zealously, and have applied their technique again and again to my own affairs, and with good results. Your suggestion of its social applications recalls one or two attempts I have made in that direction; rather patent interpretations, e.g. pointing out America's interest in the "wrongs of Ireland" as a case of transferred reproach in order to forget the incapacity to check similar audacities on the part of the government against foreign "communists," negroes, and others at home. Empires and sadism is good: how about conscientious objectors and masochism? the proletariat revolution and the fixed introversion of workers in Mass Cities? repression and the fixed extroversion of the governing classes? I must try to work a little of this into the book I am going to write this fall on "A Sociological Interpretation of Economics" based upon a course of lectures on the "Principles of Reconstruction" that Farquharson and I gave at the Summer School of Civics in High Wycombe.

The more work I do in Sociology the more convinced am I that the study and pursuit of sociology is overwhelmingly difficult unless it be synoptically charted somewhat after the manner you have worked out. The sociologist must do what not one of the specialist workers is capable of undertaking by himself. Does it not therefore follow that the number of genuinely competent sociologists in any generation can't be very much greater than the number of mathematicians capable of handling every department of mathematics? About every two generations some great sociologist, on this hypothesis, should give the same sort of impulse to the social sciences that a Descartes, a Newton, or a Clerk-Maxwell gives to the physical sciences.[3] It would be fun, if one had the time and the scholarship, to work this out graphically. Henry Adams, the American historian, satisfied himself that the time interval between such fundamental propulsions of thought could be graphed and a regularity detected.[4] (Did you ever meet Adams on any of your wanderings, by the way? He had many points of contact with you; had the same longing for synthesis; had mastered many fields of knowledge, etc. He shares with you the honor of predicting in 1910 the apporoaching catastrophe of 1917. His sociological thought is concentrated in two books, *The Education of Henry Adams* and *The Degradation of the Democratic Dogma*. His method is warped by the materialist fallacy: but his work is nevertheless full of suggestion and interest. I shall send you one of the volumes when I get back to New York. Both of them should be in every sociological and historical library: they represent in some ways the highest product of American scholarship.)

Please command me if I can be of any special service to you in New York this winter. I shall be in contact with the American Zionists through Horace Kallen, with whom I believe you are already acquainted. These younger men are all for carrying on your work, but alas! they do not hold the purse and they are not always able to untie the purse strings. Another important thought occurs. Pray do not be alarmed at the prospect of my returning next year, burdened by a wife! Unless my intuitions and interpretations are all astray Sophie has found someone in New York who has interested her much more passionately than I have, and the likelihood of our getting married seems at present remote. (Incidentally, in making my second decision, the impulse, as far as she was concerned, worked *against* returning to New York!) Also, if she does marry me in the near future, she won't be a burden because (1) as sub-editor on the *Dial* she has been able to put aside a certain amount of money, and (2) as a capable stenographer and secretary she would enable me to get off an amount of literary work which I couldn't attempt to do alone. The last year has clarified a great many of my ideas on the ethics and the psychology of sex relations – the subject drifted out of mind when my first love married someone else three years ago and I had never troubled to analyze out the genuine requirements in sexual behavior and the formal demands of one's code – and I now feel much more sure of myself than I used to. An external adjustment of one's sexual life is easy enough: but it means nothing: formal compliance with the "high" standard of sexual abstinence is little better than a dodge for concealing the necessity for more fundamental adjustments in one's daily activities and one's inner life. Marriage provides that basis for a satisfactory adjustment: but I am not sure it is a permanent one (great peace and happiness, of course, when it is) and I wish the laws of England and New York more adequately recognized this fact. Now putting marriage aside, an effective adjustment of inner and outer sexual life has proved confoundedly difficult, especially during the period when I have been living alone and have had few social contacts. It is a curious fact (but some of my other shipmates noticed the same thing, and they were by no means given to sexual abstinence) that the only period during which continence in deed and mind came without difficulty was during my period of training and barracks life in the navy. It was a very strenuous, socialized existence, much like that of the Benedictines, and a harmony seemed to arise naturally out of it. It was an extroverted harmony however and the price of it was mental dullness and routineering. The reward came in the brisk thinking and the unprecedented amount of work I was able to do in 1919, and my relative failure in 1920 has been due to the maladjustment which has become increasingly evident, in part at any rate. I have been running lightly through the literature of sex, including the injunctions of Gautama[5] and Christ, without any great satisfaction. The problem isn't an individual one: it is social too; and the pitiable lack of success in the moralizers has been due to the fact that they have not sufficiently faced the necessity for definite civic readjustments – such as Jane Addams dreams about and you plan.[6] A readjustment of the individual alone, leaving the social situation what it is, is bought at too heavy a price, is it not? (Anxiety and other neuroses, etc.) In short, the current third alternatives between abstinence and genuine marriage leave most sensitive and socially minded men and women with some more or less positive sense of repugnance, and yet in our present-day communities abstinence is so difficult as to be frequently worse than the "third alternatives" are to people who are not unduly sensitive or socialized.

I've been pouring out my thoughts here in the hope that I might stimulate you to give me the benefit of your counsel. (Of course I've read the books you and Thomson have written: but they leave off where my problem begins!)

My best wishes and affection are with you always –

Lewis Mumford

1. Page 2 of the letter is missing.   2. Refers to the Geddesian formulation: OFE = organism, function, environment.   3. James Clerk-Maxwell (1831–1879), Scottish physicist known for his work in electricity and magnetism.   4. Henry Brooks Adams (1838–1918), American historian and prolific man of letters; Mumford was greatly influenced by Adams' essay "The Rule of Phase Applied to History" (1909).   5. "Gautama": family name of Siddhartha who became known as Buddha.   6. Jane Addams (1860–1935), social reformer and founder of Hull-House in Chicago; author of *Democracy and Social Ethics* (1902).

Dear Mumford,

Thanks for your interesting letter – as also for two budget of papers, *Nation, Daily News & Freeman* – first I had from anybody for an age.

Ghastly business in Ireland: full of evil omen. Empire can't stand long with Dyer & Carson & Churchill, etc., working their will, while we are all powerless to prevent them![1] And the folly of it! Is not *"delenda est Britannia!"*

Zionist Commission
to Palestine

25 September–
2 October 1920

[Ms UP]

getting written up in flame, in minds alike in America & Continent?[2] Are not even Canada, Australia too, etc., getting sick of us, let alone Egypt & India? I feel the Lord Mayor of Cork our Dreyfus case – or as a man said to me just now, our Lusitania torpedoed.[3]

Glad you got on with Branford! What of putting my lectures together. I don't remember ever seeing, much less correcting the typescripts you mention – what year? 1919? or earlier? – though I may have done so. I have little attraction, indeed none to authorship, since always chasing & noting new ideas too fast to use up the old. But if you will send me that manuscript to Bombay, posting late in October or early November with any notes or suggestions, I'll see what I can do with them. Do you think you could find a publisher in U.S.A., or is that hopeless? Did V.B. ever show my old syllabuses? Herewith such samples as I can lay hands on – not a full set – but probably he has the rest with his own, which you'll also find interesting.

Yes. I am encouraged by what you say of my having some audience in younger generation – but I suppose the failure to make any impression on most contemporaries, let alone seniors, had no doubt been repressive, though not really depressive, so far as working & thinking has been concerned.

I am sorry I can't quite give definite plans for next summer. I'd like to come home and work with you, including 2 months in Ireland; but one never knows! I have to earn, & largely, having many claims & cares; so may not have whole 6 months free; but I'll do what I can, you may depend on that, for it is time to be writing (66 today!).

I have a town planning Exhibition here with much exposition & lecturing, so cannot write as you wish (nor is it easy what you ask!). Moreover till I sail for Bombay there will be high pressure! Unless you write by return of post, I won't get it here: better write to University Bombay, posting by end of October or a little sooner, if this still finds you in London.

Yours cordially,

P. Geddes

Thanks for papers, very welcome.

---

1. Reginald E.H. Dyer (1864–1927), brigadier general in India; responsible for controversial use of force to quell a public protest at Amritsar. Geddes elsewhere described Winston Churchill (1874–1965) as a "poisonous firebrand" (Boardman, 458).   2. "delenda est Britannia": "Britain must be destroyed," an adaptation of Cato's remark about Carthage.   3. Alderman McCurtin, the Lord Mayor of Cork and a prominent member of the Sinn Fein movement, was murdered on 20 March 1920 by a band of masked men who broke into his home. The Coroner's jury found that he was killed by members of the Royal Irish Constabulary acting under orders of the British, and the Government was so charged. However, the evidence suggested that McCurtin, an I.R.A. Commandant, was actually assassinated by his own revolutionary organization because of internal disagreements.

Dear Mumford,

I trust you are now happy on your native heath – & have found matters both personal and social in more hopeful state than your last letter indicated. Pray let me hear.

And pray also reply as to a matter grieving me. In the last few months, I had begun corresponding with my old friend President Stanley Hall (Clark University, Worcester, Mass.), & have been expecting his reply to my last. But a Jewish acquaintance in Jerusalem from America lately told me with all but positive certainty that he was dead! Is that so? I earnestly hope that news is mistaken. Please tell me. But – If so, can you get for me any notices, giving account of his life and recent work – indeed since *Adolescence*? The Librarian of Clark University will I daresay do this for you, & for my sake as an old friend, though obviously I cannot write to ask him while I remain uncertain if that bad news be true. And what of Mrs. Stanley Hall, to whom I must in that case write? In 1900 & '01 I made the acquaintance of Professor Chamberlain there (writer on *The Child*) & of Professor Hodge, neurologist, & author of excellent book on *Nature Study*, also Professor Sanford whose subject I forget – but have heard nothing of them since. Can you tell me? I am sure you will forgive this trouble.[1]

I got a large posted ship catalogue from *Sunwise Turn* Bookshop, Mr. M.M. Clark, & lately sent an order, returning catalogue, registered. If you deal there, & see any more recent books (or translations) than in this, i.e. appearing this season, which you think *my Department ought to have*, say up to $50 or $100, pray have them sent in usual way. Or if you deal with another bookseller, let him do it, & University librarian will promptly remit price & postage on receipt; as he holds the funds for my purchases.

Thus if Veblen, Patten, Dewey, etc., etc., erupt, give me the benefit as soon as may be, or as new prophets appear! Of course within the (fairly broad) limits of my Department, but leaving economists, philosophers, psychologists (true, as distinguished from social) to other Departments. So much then for business worries! But with the stimulus of your change, & return, I am sure you can tell me something of sociological thought & progress both in U.S. & in Europe, & I shall be much interested in your reflections. And as you more or less know, I am so far repressed, & self centred, through the non-demand in the past for such ideas and methods as seem to me fundamental, & thus have written little to push them. But if you have any indications, I'll be glad.

I have had a fairly satisfactory season in Jerusalem, etc., though of course not accomplished all I hoped in planning, largely through absence of contour surveys of Jerusalem & other towns. But the University quarter & one or two others are broadly planned in Jerusalem, and I hope with swelling of the New Jerusalem effect required, as also Haifa, Tiberias & its baths (the oldest of *Spas*,

(Red Sea to Aden)
but address till 10 March:
University Department of Sociology & Civics College of Science Mayo Road Bombay

13 November 1920

[Ms UP]

& still one of the best). But the agricultural colony villages are still too much mere squatting, and not the germinal centres they should be: it is difficult to get hold of these – the Zionist & older groups being too middle-class in their ideas.

Have you in America as yet any literature of *Civilization Values*? (as per my *Indore* Book Vol. I). I know of nothing beyond literature of Le Play School – especially Demolins' journal – *La Science Sociale*, and books like his *Les Français d'aujourd'hui*, with its estimates of effect of *vine* & other cultures on the producers.[2] This doctrine is much needed for Palestine, with its peculiarly ambitious & would be cultural colonization; but no Zionist knows anything of it. He still thinks of the olive as for soap or salad, & never really understands that it *is* above all others the tree of peace & wisdom – & hence alike Jerusalem & Athens (which *understood* always more clearly, deeply though Jerusalem & Judea *felt*) had the wisdom of Solomon & of Pallas Athena. To apple culture your best east-coast civilization must be largely traceable, & not simply as the most brain-clearing of fruits ("tree of knowledge"). But the superior promise of California is not merely founded on its unusually numerous culture-institutions – these arise (however sub-consciously) from the finer fruit-cultures its climate & soil promote. So what literature have you as yet of that? I do not yet know of any! Yet surely you have some agriculturists *thinking* as well as technical? Is Bailey of Cornell still to the fore?[3] (He came to see me once at Dundee, & we got on well.)

Marsh, long ago U.S. Minister at Rome, etc., was geographically minded historian as well as English Philologist, wrote a book which impressed me. He changed its name, & I forget it – but I think the sub-title was *The Earth as Modified by Human Action*.[4] He was one of the first to expound the significance of deforesting in the decline & fall of Mediterranean Civilisation. I should think G. Pinchot & his Resources Commission must have been a good deal inspired by it. If your bookseller can find it pray have it sent me. Also Huntington's more recent and kindred books.[5]

I am sure you have taken back very unfavorable impressions as well as hopeful ones as for Irish affairs, etc., etc. If so, do not hesitate to write them – you will thus help the growing reaction; just as there was after Boer War and its conduct.

I can't yet form an idea whether I'll be in Palestine again next season, planning in India, or home. Of course I'll let you know as soon as I possibly can, but it will be difficult & delaying to arrange. If and when you know your own plans, & if these still admit of your working with me, let me have a cable address & I'll let you know my movements as soon as I can.

Always yours cordially,

P. Geddes

1. Granville Stanley Hall (1844–1924), psychologist and first president of Clark University, 1889–1919; author of *Adolescence* (1904). Alexander F. Chamberlain (1863–1914), anthropologist at Clark University; author of *The Child: A Study in the Evolution of Man* (1901). In his letter of 3 January 1921, Mumford identifies "Hodge" as Clifton F. Hodge, author of *Civic Biology: A Textbook of Problems* (1918). Edmund C. Sanford (1859–1924), Clark University psychologist; author of *A Course in Experimental Psychology* (1894).   2. Edmond Demolins (1852–1907), Frenchman who popularized the work of Frederick Le Play; friend of Geddes and lecturer at the Edinburgh Summer Meetings; author of *Anglo Saxon Superiority: To What is it Due?*

(1899).   3. Liberty Hyde Bailey (1858–1952), botanist and horticulturist; proponent of regionalism; author of *The Holy Earth* (1915) and *Universal Service: The Hope of Humanity* (1918).   4. George Perkins Marsh (1802–1882), American linguist and diplomat; author of *The Earth as Modified by Human Action* (1874). Mumford credits Geddes with introducing him to the work of Marsh, a figure whose work had an important influence on the twentieth-century conservation and ecology movement.   5. Gifford Pinchot (1865–1946), forester and conservationist; served in the United States Forest Service and as governor of Pennsylvania (1923–1927); author of *The Fight for Conservation* (1910).

Dear Professor Geddes,

  Your letter of the 25 September was forwarded from London, which I left on October 5, and I have been waiting for a favorable opportunity to answer it. Two short notes are all that I have had from Branford since coming back: he is still in miserable physical condition and will, I fancy, pospone his operation once more. One of my tasks in New York was to enlist co-operation, financial and scholarly, for the *Sociological Review*; but Branford has been unable to write to any of his friends here and for the present I am unable to proceed. I gather from the folk at Le Play House, however, that the Society is showing signs of re-animation and is gathering members and collaborators at a healthy pace. Farquharson, with his Civic Education League Group, is no doubt partly responsible for this, and I am afraid that at times Branford is a little alarmed lest the Soso be swallowed up entirely by the voracious young body to which it is at present host![1] I have been so occupied with literary work during the past seven weeks that I have hardly had time to see what was going on beyond my desk. The vote at the Presidential election may look to observers at a distance like an assertion of American provincialism; but it is at least equally a reaction against the fierce illiberalism that infected the Wilson administration during the last two years, and it is not so much a rejection of Wilson's international policies as a protest against a series of domestic abuses. At present there is a reaction of sympathy in favor of the pathetic shadow of a man that still lives in the White House, and there is a general disposition to condone his failures as a statesman and to remember his common humanity. The greatest cloud on the international horizon at the present moment is the Irish situation: your diagnosis of the reaction of the other English-speaking communities to the government's present policy is accurate and it is rather strange that the international aspect of the situation doesn't weigh a little more heavily in Whitehall. I am not afraid that America will come to blows with England over Ireland: but I am afraid that the mass of antipathetic sentiment that is being accumulated here will be employed to sanction recourse to arms in, let us say, a dispute over oil in Mesopotamia or Mexico. Doubtless in the end the British government will do the right thing, as it did with the Boers; unfortunately the end may come a little too late and all sorts of disasters may occur in the interval. The actual course of events during the next two generations may not be as harrowing as they seem in prospect; but I confess I fail to understand Branford's unshakable imperturbability and hopefulness. The chicane of the Great Powers at the League of Nations meeting in Geneva lives up to one's worst expectations.

  My work this winter consists of regular editorials, articles, notes, and

100 West 94 Street
New York City

3 December 1920

[Ts NLS]

reviews for the *Freeman*, and an intensive review of the latest work in sociology, psychology, biology, and geography. My stint for the *Freeman* provides me with no more than a bare financial minimum, but I would have to abandon the greater part of my study programme if I were to seek a steady position, and while there is any prospect of working with you next year it seems advisable to devote as much of my time as possible to study and research. Next spring I may in fact take a couple of courses at the university; but for the present I am studying by myself, with such contacts as I get, in the social sciences, by occasional talks with Horace Kallen and Veblen. Ferguson seems to have dropt out of existence; Bruno Lasker I am arranging to have lunch with next week.[2] My information as to how things are going in housing, city planning, civic surveys, and the like is lamentably deficient, and I shall have to put off giving you news on these subjects till my next letter, when I hope to have caught up. The *Athenaeum* accepted a paper of mine on "Sociology and its Status in Great Britain" last October and I trust that Branford or Farquharson will send you a copy when it appears.[3] In it I followed up a suggestion of one of your diagrams and sought to show that all the constituent sciences were properly speaking sociological to the extent that they were used for the purpose of controlling and directing the development of a community. Unfortunately the article didn't have the benefit of Branford's criticism and I am not sure how faithfully I interpreted your ideas. The paper was really a pièce d'occasion, for its main purpose was to call attention to the existence and revival of the Sociological Society! Please give me your criticism – if you have time for it.

I trust that your work in Bombay is proving this winter to be a little more pleasant and rewarding. Ghurye, in London, told us something of the difficulties under which you have labored.[4] Please let me know if I can be of any service to you here. I have exhausted personal news, I think, in this letter, and in my next will get back to our muttons!

Devotedly yours,

Lewis Mumford

1. "Soso": the Sociological Society of London.   2. Bruno Lasker (1880–1965), on the editorial staff of *The Survey*.   3. "Sociology and Its Prospects in Great Britain," *Athenaeum*, (10 December 1920): 815–816.   4. G.S. Ghurye, Indian sociologist; author of *Caste and Race in India* (1932).

Department of
Sociology and
College of Science
Esplanade Road
University of
Bombay

[Undated ms., page 1 only,
placed after Geddes'
letter of 13 November
1920 in UP file]

Dear Mumford,

Here I am back for the winter's term – in a vast room 200 ft. × 30+ verandah-gallery outside, and with plenty of tables to lay out papers towards order & bookcases, etc. But what disorder from last year's accumulations moved about once & again by unskilled & careless hands! No trained assistant though some willing & more or less capable students – but all deficient in practical comprehension of material & mental order as needing interrelation anew & so not readily seeing that tidy papers on desk, and new Jerusalem, are thoroughly parallel & corporate tasks, each complemental & integral to the other! And I too alas in practice falling away from this ideal, as well as short of it! Yet this idea of work is clear – towards an orderly presentment of sociological ideas in this gallery & a Town Planning Exhibition in the Museum across the street; this latter towards educatively aiding – by its expression of the progress of cities – the Bombay Development scheme now maturing under government, and the former towards better "making of the Future" more generally & not merely by books, etc., alone.

This then is an outline of what I should have hoped your help on had you been here: but after all why should we not be co-operating as it is, though materially far apart? We are in general agreement, so far, but I should also value and be helped by your criticism of *our* limitations as you have felt them also, during your stay in England: so pray let me have these too; and with such constructive suggestions as may be accordingly.

In this way our conjoint presentment of sociology in general, & of cities in the concrete may be worked towards a more orderly one, & alike in Bombay & New York, as in Edinburgh or London, or in Jeruslaem, etc.: each of course with its local perspective, yet thereby all the more truly in harmony.

As we can do this, we shall gain in understanding & in co-operation first among ourselves, but also increasingly with others. You in America particularly are rich in many & varied centres of sociological teaching & research, of surveys & interpretations, of endeavours toward action perhaps most of all. Well, cannot we increasingly provide [...]

Dear Professor Geddes,

100 West 94 Street
New York City

5 December 1920

[Ts NLS]

I was rather annoyed by the perfunctory quality of my last letter to you, and by the fact that it had taken me such a long time to write it; and on looking around for the causes I discovered that I had been concealing, or rather averting attention from, a series of personal problems and conflicts that first came to my attention clearly during my journey to London. Perhaps if I had been fully aware of these problems a year ago, and if I had been able, as it were, to lay them on the board at the time, you might have been able to help me to a better understanding of them, and we might have been able to adjust our several aims and purposes more intelligently. I did indeed tell you something along these lines in a letter that I sent to Palestine shortly before leaving for London: the Zionist commission did not hold it however and it was returned to me in London, and when I read it again I found that it did not express precisely what was on my mind. The other night I was noting some of my problems down on cards – a trick I learned from some notes of yours I found in London! – and it occurs to me that these notes summed up all the things that have conditioned my actions and have kept me from responding as freely as I would like to, to the proposals that you and Branford have been making. Any effective collaboration between us, or even the beginnings of it, is dependent upon my ability to break through these conflicts and to steer a single course of action. Perhaps as quick a way as any of finding out what I am driving at is to read a transcription at haphazard of these cards:–

1. Status. What am I? A journalist? A novelist? A literary critic? An art critic? A scholar? A sociologist? An artist? By native capacities, an emotional, by

equipment, an intellectual. MUST I TAKE A DEFINITIVE LINE?

2. Responsibilities. Mother Nurse[1]

How far do they influence career? Movements? Marriage?

3. Marriage. Marriage vs. career. Can they be reconciled? Should one be sacrificed to the other? Which one?

4. Place. Is Manhattan the right place for me to work and live in? Is the Manhattan region? Should I try to get out into the country? Break away entirely?

5. Finances. Present assets: $2750. How much can I afford to reduce this if pressed? For Travel? Present Liabilities:

   1. Household, per year: $840
   2. Personal expenses: 400
   3. Insurance, per year: 160
   Yearly minimum: 1400

Item two rises if I leave New York. Can I?

6. As Sociologist. What are the weak points in my intellectual armor? The preliminary sciences? History?

7. Travel, 1921. Join Geddes? (London? Dublin? India?) Join Branford in London? Paris? Rome? Copenhagen? Bergen?

8. Work. Lines of investigation for the next five years. Where should they lead to?

Is my own work more important than any I may be able to do for Geddes? Have I found mine yet? Will P.G. help me?

9. Plans. The next five years. How far are plans contingent upon bachelordom or marriage. How far has latter been settled in negative?

10. Opportunities. Given my opportunities, what should I be able to do?

Here in brief are the questions I have been tussling with and trying to settle. You will easily see how they have entered into all my decisions during the past ten months, and how much my ability to answer them will govern my future actions. I have set them before you partly in order that you may better understand my problems and partly in order to enlist, if possible, your experience and counsel in dealing with such of them as can be dealt with from the outside. Now that I have put them down in black and white, instead of repressing them every time they came to the surface while I was thinking of our possible relations, I feel that something of a load has been lifted from my mind and that I shall be able to talk about other matters with a great deal more freedom and directness.

I trust this letter will reach you near enough to Christmas to convey not too incongruously my hearty greetings for the season and the new year.

Always faithfully yours,

Lewis Mumford

1. Nellie Ahearn was Mumford's beloved "nurse."

Lewis Mumford AND *Patrick Geddes*

THE CORRESPONDENCE

1921

Dear Mumford,

University of
Bombay

10 January 1921[1]

[Ms UP]

Yours of 5 December just received. It is no easy matter to answer such questions, the more since we have never met! I'll try – but with due warning not to take my attempts too seriously!

(1) *What are you?* that you have essentially to find out, as time goes on: for only that can reveal: even you cannot foresee![2] To me the rule has been to go on quietly, working up to as full standard, *daily*, as health allows, & with ample change of occupation, and thus keeping interests & activities alive in a wider circle than most are contented with (or say rather a wider *arc*). For if life be six-sided, & thus at once {Ethical-Econonomic (conduct) / Psychological-Organic (Behaviour) / Eu-technic (Activity)} the emotional, intellectual, & imaginative life are all needed, & must not be weakened by disuse through undue concentration in any single field. All are needed for true efficiency in any line that turns up with work worth doing!

(2) *Mother?* Of course a first duty. (I do not understand "nurse.")

(3) *Marriage?* Why, marriage is career! and with the right person (as mine was) the best possible! "Sacrifice one to the other!" nonsense! – think of H. Spencer settling down to edit his youthful essays into his many volumes, instead of marrying George Eliot, & being stirred up to write better ones![3]

(4) *Life in Metropolis?* Recall the formula (from Babylon to Thebes, from Rome to Petrograd, Vienna, etc., etc.): Place/Polis (the city) > Work/Metropolis (its function) > Folk/Megalopolis (its cockneys) then Folk/Parasitopolis (its developed life) > Work/Pathopolis (its functional condition) > Place/Necropolis (its culmination accordingly). You may say even Berlin has hardly come to that, still less London & Paris, & least of all New York City. But the handwriting is on the wall. All have overgrown their usefulness.

(5) *Finances.* $1400 = say £400 or so: not impossible I should think to earn anywhere? say half (or a little less) as correspondent & writer for American press, & half (or a little more) with me? – or Branford & me?

(6) *Weak points? as Sociologist.* These again you have to find out for yourself; but I imagine geographical basis wants widening & concreteness, with travel & observational sciences. And the psychological & ethical grow with corresponding reflection and activity. (Teaching helps the one, & social science the other, as in civics, etc.)

(6a) *History* – is best learned in intelligent travel, with such relevant reading as it suggests. My historical imagination & clearness (such as they are) are not so much from reading ordinarily, as from places – say *Aachen Dom* & the *Walls of Constantinople*, Paris, the *panoramas*, from Stirling & Edinburgh to Rome, Athens, Jerusalem, Benares, & so on. (*Masques* like mine should be outlined for people to play for themselves, e.g. after reading Wells' *History* in many ways excellent, though in others very deficient.)[4]

(7) *Travel 1921.* Unfortunately my plans still contingent on others'. Thus I *may* be detained in India, I *may* be wanted in Jerusalem: I can't yet say! Send cable address by return. (But not likely in London till 1922.) I think it is also well worth your while (if you can't come to me, or if I can't ask you) to go on with Branford again. You can learn much from him still (& help him from breaking down besides, a task worth any man's doing). (I think either of these alternatives preferable to a fresh start in other cities alone; though of

course I admit that is how I have had to educate myself. Still I lost time often, as well as learned much.)

(8) *Work*. If and since Branford & I are out for nothing less than outlining the *Encyclopedia Sociologica* (which we can never accomplish!) & the corresponding though far still fainter outline of Making of the Future you will see more clearly in collaboration & going over our papers (middens though they be in their incompletenesss & untidiness!) than you can do in the same time alone. We need younger partners in the concern, & to carry it on as we drop out altogether.[5] You seem to me, & I think to him also, one of the likeliest men (if not even the only one since others got to their own work, & my son was killed!).

(9) *Plans* (again, naturally, back to marriage). If your projected mate can help you, things will be easier, despite needed larger income. If not, try Branford or me for a year, or two at most. After that, we might continue collaborating, though you were home in America again. Your Universities have all more or less sociological chairs. Would you not stand a better chance of one, with more experience, such as above? (No doubt our school, or view, is not in power; but the others are not so well-oriented to coming needs, we think; but that is for you to choose.)

(10) *Opportunities*. Again, from my point of view (1) & others above.

Lame answers, you may feel: but if & when I can settle my own programme, I can be more definite regarding possible collaboration, of which there is only too varied choice: e.g. Books on Universities & on Cities, graphic treatments of notes of all kinds, from occupations in particular and Economics in general, & from "Cloister" Squares to "City" ditto (as per Life-scheme of 36, which in outline I think you more or less know?).

Your paper in *Freeman* (on retardation, etc.) was very good, & capable of expansion, with fuller explanations into more than one review article.[6] If you can write such papers without too much toil & time, you can certainly help us greatly, & we can soon be making volumes together. True, they don't pay! We have still deficits on them, but it is worth trying, we think, to hold on till things mend a little & demand grows, as it may. Your generation of active spirits will not have so long a struggle as had we, though the hostile worlds of {LIF/RSA}, now all breaking down, & from internal poverty of ideas, as well as mutual antagonisms.[7]

Send me other of your *Freeman*, etc., papers from time to time. Remembrances to Fuller, to Mr. Kallen also, from whom I had a nice letter (not yet answered, amid so many changes & uncertainties, & now I fear mislaid: give me his address).

In summary, all professions are crowded round entrance & from bottom, but thin out from middle to top! You have thus to struggle along, as others have done before, but fit yourself boldly for the larger world, coming later!

For instance (though in the Jewish medley & mêlée of ideas there is no predicting) it is at least a chance to be preparing reports like that on University, raising more boldly than heretofore all questions of Higher Education, & towards the *post*-Germanic University from the present *pre*-Germanic (old scholarly & examination compromise of Oxford) & *sub*-

Germanic types (Cambridge (England) or Chicago).

Again, you hear something of Mr. Gandhi's immensely spreading Indian "non-cooperation" – a wholesale "*strike*," which is affecting colleges & Universities here appreciably, & perhaps increasingly.[8] Two of my best students of last year are now teaching with him; & one came to ask if I would open my Jerusalem schemes to his principal of projected new university. Of course!

But at same time I asked him – "You fellows are not simply leaving for Mr. Gandhi's sake, but because these cram-exam schools & universities are sickening you?" "*Certainly so*," said he.

If so, however, even these Indian governments are not so entirely fossil in educational matters as they seem; & Lord Reading as new Viceroy (who escaped the University & public school machine) *may* be able to modify their attitude, as I trust he will also do the political governing apparatus (as of Rowlett Act, etc.) & the attitude of his Services in some small measure also, hard though that be. In short I do not despair of bringing some educational influence to bear on improvements, & beyond Sadler's too timid University Commission lately.[9]

But this means University Militant in action. How far can we co-operate internationally in such matters? America is surely more open than this old world? Let me know of any definite progress you can! Send me at least the particulars & critique of the New York Sociological School (Dewey, Veblen, etc.) do you teach in it? Also any other sociological teaching schemes you can lay hands on without too much trouble. (Reply as to Stanley Hall. I hope the rumour of his death is not correct?)

I send this letter *via* Branford (for we are such close partners, you will not, I trust, mind my doing this?). Reply soon, & I'll also write by return.

Yours very cordially,

P. Geddes

1. Mumford's annotation dated 1963: "How thrilled I was by this!"   2. Mumford's annotation of 15 April 1969: "As I used to put it then – a sociologist or an artist?"   3. Herbert Spencer (1820–1903), the philosopher-scientist, was a particularly close friend of the novelist George Eliot (Mary Anne Evans, 1819–1880) between 1851 and 1853. He contemplated marrying her but, apparently with considerable egotism, decided to remain a bachelor.   4. Aachen Dom is the great cathedral in Aachen, Germany. H.G. Wells, *The Outline of History* (1920).   5. Geddes and Branford edited "The Making of the Future" series of pamphlets and books, published between 1917 and 1926. The series showed how Geddesian methods and ideas could address the problems of reconstruction following World War I.   6. "The Adolescence of Reform," *Freeman* (1 December 1920): 272–273.   7. Geddesian formulations: "LIF": liberal, imperial, financial; "RSA": radical, socialistic, anarchistic/communist.   8. "Mahatma" Mohandas Karamchand Gandhi (1869–1948), leader of the Indian National Congress party who led popular non-violent protest against British rule; Geddes corresponded with him.   9. Rufus Daniel Isaacs, Lord Reading (1860–1935), served as Viceroy of India, 1921–1926. The Rowlett Act (1919) allowed crown judges in India to try political cases without juries and allowed provincial governments to imprison suspects without trial. Sir Michael Sadler (1861–1943), distinguished educator and friend of Geddes; in 1917 he served as president of the Royal Commission to reform the University of Calcutta; Geddes lobbied the Commission to organize the University for India at Indore along the lines of his sociological theories.

100 West 94 Street
New York City

3 January 1921

[Ts NLS; Mumford's
annotation on UP copy:
"Important see full
letter."]

Dear Professor Geddes,

Your last two letters, one dated the thirteenth of November, came during the last week, and I am answering them both together. (They were my most welcome Christmas presents!)

First to answer some of your questions. Stanley Hall's decease, like Mark Twain's death, seems to have been exaggerated. So far as I can gather he has not even been in ill-health, but upon this latter point I am not positive. Last year he published two books: *The Recreations of a Psychologist*, and *Morale*, a book consisting of a series of lectures which he delivered, I believe, during the latter part of the war. Last year I was in contact with one of Hall's students, a young sociologist named Harry Barnes, but I have lost track of him, and so I am unable, for the present, to give you any more personal information about Hall.[1] There is something of a gap in the library catalog between Hall's latest work and the earlier ones on adolescence and the interpretation of Jesus Christ. Chamberlain died in 1914, and unless he has some fugitive papers to his credit, had published nothing since *The Human Side of the Indian* (1906) and *The Child* (which seems to have been published only in an English edition) in 1907. Hodge, apparently, is still living. In 1918 he published *Civic Biology: A Textbook of Problems, Local and National, that can be Solved only by Civic Co-operation*. I have not had time to examine any of these books personally and so I am unable to give you any appraisal of them. Hall's personal influence at Clark University, like that of Adams in the formative years of Johns Hopkins, has probably been of far greater importance than any particular contribution, with the possible exception of *Adolescence*, that he had made to thought.[2] If only our smaller towns would awaken to a more robust life there is no reason why the smaller universities in America should remain the sickly, half-baked institutions that so many of them are. For the most part our colleges and universities still are in their regions, but not of them: they are like potted plants that have been stuck in the soil and their roots have never had a chance to tap their whole environment.

I shall ask the Sunwise Turn for a copy of the list which they sent you, and shall then add to your present order a few recent books which count, more or less, in the literature of sociology. R.H. Lowie has summed up in *Primitive Society* the results of American anthropological researches on primitive institutions, and because it is one of the few American books that covers the field that Lewis Morgan opened up almost two generations ago I shall send it to you, even though it falls only partly within the domain of sociology proper.[3] Dewey, as you possibly know, has been in China for more than a year and published last summer a book on social philosophy. But there have been no striking departures in American sociology during the past decade. Small has been slowly developing a conception of sociology which more and more approaches your own, in that he makes the control over the social process the final state in sociological research, the preliminary ones being survey, evaluation, and projection; but as you will doubtless suspect, Small's formulation is still in the stage of abstraction and it is doubtful whether he or his school are aware of the necessity for beginning in Chicago to harness up sociology to specific interests and tasks. Ross published his magnum opus last summer – to my shame I have not even glanced at it – and unless this is already in your list

I shall send it on.[4] When I was in Pittsburgh the Carnegie Institute had a very interesting set of sociological courses, largely regional, sketched out in their catalog, and I am rather eager to go back to the heart of our black country and discover whether anything is likely to come out of those courses. Again and again I have the feeling that it is our ignorance of what other people are doing, rather than their failure to do it, which handicaps the development of ideas, plans, and projects. How far is the Institute Internationale de Bibliographie a genuine clearing house for the literature of the world?[5] How far has it developed to a point that other institutions might pattern their organization after it? Beyond doubt these clearing houses, both general and special, are what is needed if we are to cope with the growing body of literature in every department of thought and action, and except for the so-called Economics department of the New York Public Library I know of no institution where not merely the formal publications, but news accounts and, so forth, may be consulted, so that one may actually have a sense contact with remote and inaccessible groups. Branford planned to have a guide to the intellectual resources of London at Le Play House: we need however, not a guide merely to London but to the world. For lack of such a guide we have to rely too completely upon the intercourse which is developed by the super-concentration of resources in a few megalopolises whose disadvantages for living vastly outweigh the aid which they give to intellectual organization. Here, then, is one task that the Outlook Tower and Le Play House might undertake to the temporary subordination of every other scheme. Without this mechanism the business of getting in contact with other groups and influencing them seems to me almost insuperable, for it cannot well be left to chance encounters with books and pamphlets. One of the things that struck me at Le Play House was the need for a definite plan of campaign. Mann did indeed spend considerable time and energy in devising an elaborate program of action; but his program rested on the assumption that the Sociological Society was to become a temporal power, or at least to direct relations with the temporal power, whereas I believe that it is first necessary to build up a coherent intellectual organization whose efficacy will in time be recognized and taken advantage of by the various temporal groups.[6] Mann's departure automatically swept away the plans which he had laid, and which Branford had never specifically approved, and while we were able to devise a makeshift program for the coming winter, a program that has attracted more attention than one could reasonably hope for, there nevertheless remains the major problem of making Le Play House count for something more than a clubhouse in which intellectually amiable people may occasionally meet. The weakness of Le Play House at present seems to me twofold, perhaps three fold.

First, as I have said, there is no general conception of what has to be accomplished in sociological research and civics propaganda. There have been sorties and even costly attacks at one point or another of the line, but there has been no attempt to set some definite objective, or series of objectives, to which these actions might contribute. The result has been that many things have been started and little completed. Much energy has been spent in doing things that did not have to be done, and leaving undone things that should have been done.

Second, as no definite plan of campaign has been developed – or at any rate, *adhered* to, it has been impossible for the staff (Mrs. Fraser Davies, Miss

Loch, Mann, Farquharson, myself, et al.) to carry on effectively such work as they might have performed within the limitations of space, time, money, and personal intelligence that have prevailed.[7] Last summer Branford told me that he wished the staff at Le Play house to take over the work, devise a scheme of action, and proceed with his sanction to carry it out. So far good. But as Branford himself would doubtless be quick to see, any working society involves the co-operation of C.P.I. and E., and since this primary division of labour was not established it happened that everybody was in the position of attending to everybody else's business, and as a result there was confusion and uncertainty.[8] The lack of clearness as to what has to be done is coupled with the fact that Branford, in the isolation of New Milton or Richmond, develops plans of his own which sometimes tend to upset the frail scaffold of purposes that someone has raised up.[9] Now this is a specific criticism which I should not be so frank to make if it were not for the fact that it points, I think, to a more general failure to make the sources that already exist count at full value. In a *general* plan of campaign almost every grade of worker can find a place for his particular talents and something to occupy them with. With a series of *specific* plans it may be very hard for anyone to find a place for himself or to be able to fall in step with his leader. Furthermore, in so far as the conceptions of sociology for which we stand are looked upon as a closed system, which must be taken over en bloc by neophytes, it is doubtful whether they will make rapid headway. The mistake has been, I believe, to lay emphasis upon the system itself, instead of laying it upon particular fields of investigation and particular problems of social reconstruction, in which the methods that you and Branford have developed are the soundest and most fruitful in application. From this two things have followed: 1. the people that have been attracted to the "system" and have faithfully taken it over have been too frequently people of second-rate capacities who were content to repose intellectually in a bed of down already prepared for them; and 2. people of first-rate intellectual capacity have tended to resent the assumption that the thought with which they were presented had been already poured and moulded for them, so that nothing remained to be done except to reinforce, as it were, the walls of the structure that had been raised. The natural tendency of intellectually honest people in this situation is to break out of the house altogether – or never to enter it. If there is any timidity as to what will happen to the structure when rude outsiders begin to occupy it and adapt various chambers of it to their particular purposes the house is bound to remain empty. I believe firmly enough in the outlines you have laid down to be willing to have them take their chances without protection: Branford, however, is at times resentful of work that does not happen to fit into the system, and he is rather inclined to keep his idea-complex intact than to let it fight for itself in the rough-and-tumble of experience. I perhaps exaggerate Branford's distrust of heterodoxy, but I still have the feeling that he has become more interested in the preservation of the system than in its exploitation, and this tends to exclude people who approach Le Play House from other backgrounds and who have different situations to face and different problems to solve. This is necessarily a rough and ungenerous criticism, and in all conscience I should have addressed it to Branford rather than to you: Branford's extreme sensitiveness and his miserable health combine to keep one from saying anything directly to him however, and the only way in which I could convey to him my criticisms or divergences last summer was by my silence. I feel much freer in talking to you because you have called for my criticisms, as Branford never did, and because I count, I hope, upon your continued good health!

When it comes to constructive suggestions only one or two points at present occur to me which may be of use.

First: the establishment of a regular bi-monthly correspondence between Bombay, Edinburgh, London, New York (and Brussels and Paris if possible) so as to keep each group properly informed as to the progress of work, the publication of new literature, and so forth.

Second: the building up of an adequate Index Museum in London, with a section to take care of current developments (associations and congresses).

Third: the use of the *Sociological Review* to lay down a comprehensive intellectual program and to stimulate sympathetic workers to explore in detail the various fields that have been thus opened up. The *Review* is now a hodge-podge, scarcely worth publishing. Rather than continue it in its present form Farquharson and I advised Branford to scrap the magazine altogether and publish his third alternatives as pamphlets. F. broached this to him and reported to me that the subject could not be discussed.

I shall postpone any further suggestions until I get the sequel to your last letter, so that I may work upon your lead.

Before closing I must say a word or two about personal affairs. In fishing around for means with which to join you should that prove possible next spring or summer, I hit upon the idea of trying for a fellowship at the New School for Social Research. (It offers two thousand dollars a year with residence at the School, and I think I can get special permission to spend part of that time away from New York City.) I should have liked to have had your suggestion of some thesis to work on which might directly serve one of your interests; but I shall probably have to submit my suggestions to the committee before I can hear from you. My plan at present is to make an ecological study of three communities: a primitive seaport village, a small crowded quarter of New York or Boston, and one of the new small industrial towns, like Framingham, Mass., in which the principal industry has just escaped from the metropolis, for the purpose of establishing the conditions upon which decentralization may effectively take place. If your reply to this letter will reach me before April first I should be indebted for any recommendation to the Committee that you may be able to make on my behalf.

One thing more. Whitaker, of the *Journal of the American Institute of Architects*, has asked me to write a short history of the modern town planning movement, for their yearbook, and to write an article on the work that you have been doing in Palestine and India. The only thing that I have at hand is your Indore report: can you give me any other data or refer me to any other? Sketches and photographs woud be very valuable, if they can be reproduced. Also, have any of your Indian reports, or any part of them, been acted upon? This is a splendid opportunity not merely to make your work better known in America – many people, of course, are already acquainted with it – but to give American city planning a push in the direction of civics and city design. I am sending you herewith one of the *J.A.I.A.*'s supplements which shows both the qualities and defects of American thought on this subject. The most significant part of the paper is that one of the authors is a sociologist. Also I am sending you a copy of the *Freeman* with an interesting review of some German town-planning – or at least town-dreaming.

Gratefully yours,

Lewis Mumford

P.S. I do not know which of us more keenly regrets our failure to get together in Bombay this winter. Were it not for the fact that I still feel that, on the premises, I could not wisely or honorably have acted otherwise the sting of disappointment would bite even deeper. My love-dream was speedily shattered on my return to New York. I left Artemis and returned to – Aphrodite-with-the-mirror.[10] The mirror may represent only a phase, but as far as our relations go it is a decisive one. I wanted a mate, not a mannikin.

1. Harry Elmer Barnes (1889–1968), sociologist and historian at Clark University and the New School for Social Research; contributed articles to the *Sociological Review*; see Mumford's letter of 9 December 1930.   2.   Herbert Baxter Adams (1850–1901), European-trained historian on the original Johns Hopkins faculty; organized a seminar on the German model.   3.   Lewis Henry Morgan (1818–1881), pioneer American anthropologist and author of *Ancient Society* (1877).   4.   Edward   Alsworth   Ross   (1866–1951),   *The   Principles   of   Sociology* (1920).   5.   Geddes' Belgian friend Paul Otlet (1865–1944) in 1899 established the "International Office of Bibliography" at Brussels to compile a central international bibliography. 6.   William W. Mann was a member of the Sociological Society–Le Play House group. In 1920 he wrote from Berlin to Mumford, who was then in London, about the possibility of working on the Reparations Committee. He later served as Editorial Secretary of the International Federation for Housing and Planning at The Hague.   7.   Members of the Sociological Society–Le Play House group.   8.   "C.P.I. and E.": chiefs, people, intellectuals, emotionals (a Comtean formulation).   9.   Victor Branford kept a cottage at Broadley Farm, New Milton; he also resided for a time at Richmond, Surrey.   10.   Artemis is the Greek goddess of the wilds and the hunt; Aphrodite is the Greek goddess of love and beauty.

Department of
Sociology and Civics
University of
Bombay
Bombay

11 February 1921

[Ms UP]

Dear Mr. Mumford,

Your scheme of social surveys – *Seaport-New York quarter – & New Industrial Towns* – seems to me very promising, & likely to be fruitful. I sincerely hope that you may obtain the fellowship you mention as possibly available; and if my support can carry any weight with the electoral body of the New School, you are welcome to pass on this letter.

But why not also co-operate with us of the Regional Association?[1] I am going home next month for the summer, & partly for the purpose of helping to get City Surveys – now at length spreading throughout Britain, & as preparation for the general preparation of town-planning schemes, as well as for academic & educational use – into better order. If we could compare & organise our methods, that might be of mutual suggestiveness, & usefulness too. Is not that so?

But my main errand is the still more ambitious one, of trying to develop a more or less regular co-operation in sociological studies – Bombay & Indian centres (like those of Gilbert Slater in Madras, Stanley Jevons in Allahabad, etc., etc.) with the London & Edinburgh Schools of Sociology, & these with Paris, Brussels, etc., on continent.[2] Of course if possible also with your very large American interest and activity, as per New School in New York City, & older ones too – Chicago, etc., etc. The *Sociological Review* (& perhaps others) will give us the needful space. In this way we can have a succession of symposia on the main questions & moot points, which should help toward clearer understanding, sometimes even substantial agreement.

Now, if you get your fellowship, & with leave to come over for at least part of your tenure of it, it would be of great help in this work, and you would act as a link, and be of enhanced value on both sides thereafter.

Indeed, when we are at it – why not come on with me to India in October? This would give you an excellent *Wanderjahr*; and you will return to your own home problems with freshened eyes and keen zest.

Yours faithfully

P. Geddes

1.   The Regional Survey Association was an informally structured, sporadically active organization founded by Geddes and associated with Le Play House and the Sociological Society during the 1920s.   2.   Gilbert Slater (1864–1938), principal of Ruskin College, Oxford, (1909–1915) and professor of economics at the University of Madras (1915–1921); co-author with Geddes of *Ideas at War* in "The Making of the Future" series. Herbert Stanley Jevons (1875–1955), professor of economics at the University of Allahabad; author of *Economics in India* (1915).

Dear Mumford,

Bombay

11 February 1921

[Ms UP]

I don't know that I have much as yet to add to my last, in this fairly busy life. But I am trying as best I can to think out the complex collaboration I would like to initiate, with your help, if as I hope, you can get over by say end April. I'll try to send you notes of this; but in the meantime be applying your own wits, alike to the problems best suited for discussion, & to the likely men willing & able to take part. This will be a fund of matter for the *Sociological Review* – & which will I hope help to strengthen & organize it, beyond this too desultory character you see. Also the Sociological Society itself. (Your suggestions are practically mine too, but the trouble is to develop them.)

As to my town-planning recommendations, leave that till we meet. I can talk over the Reports, of which I'll bring what I can, & meantime rather be doing what you can do better in New York City.

Your scheme of social surveys (Seaport – New York City quarter – & new industrial towns) is well-chosen; but I take it that coming to London will upset it – for a time only! I write, as you suggest, & I hope the enclosed letter may prevail with the gods who decide!

If not, I abide by my economic suggestions of former letters. (I am sorry indeed not to increase them, but I worked last summer in Palestine for "Zion," & have also only expenses & no earnings before me this summer again.)

I have used a cold & business-like tone (or as near business-like as I can!) in the letter; but am writing Branford to send you one also in case you have not asked him; as I'm sure he'll welcome your return.

Yours faithfully,

P. Geddes

I sail 19 March – get to Marseilles in a fortnight & spend a fortnight or so in France before getting to London: so if you can come before end April, you'll be in time for our real start in London. P.G.

Among the literature you are gathering for us, pray include representative City Surveys. Some of these might be given to Sociological Society Library? Some too to this (University) Library? Try them! (If not, of course we'll pay.) P.G.

I have written to Mr. Nolen the town-planner & an old friend (Harvard Square, Cambridge man) & asked him to think of any town planning material for my Bombay Exhibition on return here.[1] If you can come, write him to send it by you, as it will be less smashed than by post. Indeed if you can gather any other material of such use, graphic as may be, pray spend $100 or so, & I'll remit or reimburse you when we meet. Beg too what you can – for Cities & Town Planning Exhibition!

1.   John Nolen (1869–1937), American city planner; he served as one of the judges with Geddes of the housing projects competition in Dublin, 1916; author of *New Towns for Old* (1927) and *City Planning* (1929).

---

100 West 94 Street
New York City

15 February 1921

[Ts copy, unsigned, UP]

Dear Professor Geddes,

Just a note in answer to your welcome cable, in the hope of reaching you before you leave Bombay. I shall make every effort to join you in England at the earliest possible date this spring. You are already aware of the difficulties that I shall have to overcome, and if I am able to make the trip it is doubtful whether I can remain away much later that the beginning of September. However a great deal might be done in three or four months, and with this assurance I shall go ahead and make my plans. I shall probably not know anything definitely about the New School Fellowship until the first of May. I am looking forward to your letter and trust that it will somewhat clarify my problem. Pray excuse the haste of this note and believe me,

Gratefully yours,

P.S. Have traced my debility during the past year to septic condition of some dental work in my mouth, and am now improving rapidly under treatment.

Dear Mumford,

I have been working away towards long-dreamed co-operation with Victor Branford & yourself, in *Sociological Review*, etc., etc., but have nothing clearly enough written out to enclose. Besides, you are yourself alert to all that you can carry off from your large environment! Be ready, as clearly as may be to tell us (& why not in *Review*), what are the schools & thinkers of sociological productivity; bring such books as we are not likely to have seen, etc.

It would be of great interest if you can be jotting down for *Review* such a *census sociologicus*. We need also, for European readers & in French translation as well as English, a book, outlining the essential ideas, of Veblen & other progressive minds still far too little known among us; for when in England you must have felt how little American books reach us as yet. Similarly, & on your return home, you may interpret British & continental ideas to your own public, & thus be carrying pollen on your proboscis in both directions, towards who knows what hybrid seed!

Still, with all respect to Veblen (whose books I've just got for department library & been re-reading), I don't feel he is making so much progress as I expected, beyond those in *Leisure Class*, nor even in the full elucidation of these; some of which I doubt not you will be able to help me with.[1] But is he not arrested by preponderance of his *physics* over his biology, like other men?

See enclosed diagram which you'll readily decipher, (especially if you fold & draw it out for yourself!) & with your own illustrations, & developments as you write!

I hope you are to get your fellowship, & that my letter (enclosed with another) was not too specialised to pass on?

With great anticipations as well as all good wishes.

Yours Cordially,

Pat. Geddes

P.S. Just on off-chance, it occurs to me to ask if there be any use in us of Sociological Society asking help (towards publication & diffusion for instance, of proposed Sociological collaboration scheme, when clearer, & of its gradual development) from any of your great recent foundations towards research? You know these better than we can.

Again what societies might be interested in collaboration with Sociological Society? e.g. American Academy? or which, if any?

Department of
Sociology and Civics
University of
Bombay
Bombay

(arriving if all's well
c/o Sociological
Society by *at least*
April 25)
25 February 1921

[Ms UP]

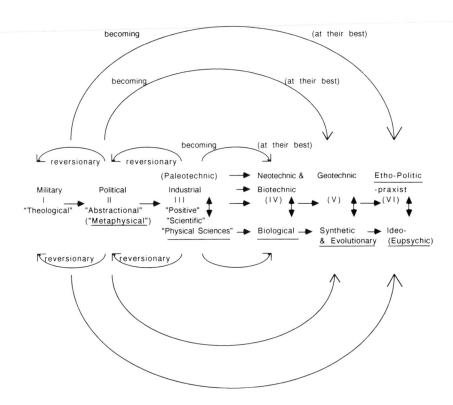

Comte's "Law of Three States" & its Needed Development in Current Progress

1.   Thorstein Veblen, *The Theory of the Leisure Class* (1899).

---

100 West 94 Street
New York City

22 March 1921

[Ts NLS]

Dear Professor Geddes,

This letter divides logically into two parts: an 'if' and a 'then.'

The first part has to do with the possibility of my getting to Great Britain at all this spring. The solution which I had hoped for has unfortunately evaporated: there has been an unfortunate split in the faculty of the New School for Social Research, and under the pretext of being without funds no fellowships will be offered for the current year. My illness this winter, and my consequent inability to earn current income has done away with all my available cash resources, and unless some unforseen commission comes my way, or unless I can secure a contract for some articles or a book against which I can draw for current expenses, I shall not be able to pay my passage over and back. I have small doubt about being able to make my way once I am on the other side: but at the present moment I do not see how I am going to get over to the other side. I have been very busy during the last fortnight, seeing one person or another who might assist me, but so far I have not succeeded in clearing the way. There is no use discussing the "if" any further. I shall endeavor during the next month to remove it, and by the end of April I hope to be able to tell you whether I have been successful or not. In any case I shall not be able to start until the end of May.

Now as to the "then." I have been looking back over your letters and Branford's and my own, and I shall now bring the various proposals in them together, and indicate in what particular fields, and in which specific manner, I am likely to be of most service.

There is first of all the Opus. This, it seems to me, is a matter for joint collaboration between you and Branford and Thomson, and I doubt whether anyone else should be invited to participate in the initial work of getting it drafted. There is, however, quite a gap between setting the material down and getting it ready for publication, and with the second part of the task, if given a free hand, I might be of considerable service. How much there would remain for me to do would depend upon how thoroughly you had been able to finish your task, and to what extent you were willing to have the more or less esoteric vocabulary in which you have for the sake of precision and swiftness expressed your thoughts be translated into popular, pedestrian English. If Branford's help is not available there is a question as to whether it would be advisable to undertake the opus at this moment: might it not be better, in that event, simply to assemble the materials, as far as possible, and decide upon scope, material, and treatment.

There is, in the second place, the two remaining books in the Making of the Future Series, "The University and the City," and "The Land and the People." The materials for these, as Branford noted in his letter to you of September 4, 1920, are pretty well in hand. I urgently plead, however, that none of the articles or reports be incorporated in the proposed books as they stand. Each book, it seems to me, is a unique attempt to address a certain group of people at a certain time in order to accomplish certain results, and in order to be effective the material must be thoroughly revamped in order to accommodate itself to these three conditions. The practice of using existing papers and endeavoring to piece them together is a poor economy: it is much more effective to melt all the coin down and mint them again with a fresh stamp. Here is a job which I am prepared to tackle, and with my fairly long training in journalism I believe I could, with your criticism and advice, turn out one, and possibly two, books during the summer which would have a fairly wide circulation. Your task would be that of formulation – already for the most part accomplished – while mine would be presentation. An excellent division of labor.

Third, in order to add to my current income, there would be the opportunity to do a certain number of articles on your town-planning work. This, in addition to what I may be able to do for the *Freeman* in little sketches and essays, would keep the pot at least simmering.

Besides these specific plans for publication there is the opportunity to gather men like Farquharson, Fleure, Thomson together in order to discus the possibility of laying down certain lines of research for the younger sociologists to follow, the publication of the results to take place in the *Sociological Review*. Farquharson will be particularly valuable in this department, and since he is the obvious candidate for the editorship of the *Review*, once Branford resigns the helm, there is no doubt that the continuity of the plans decided upon will be secured. This last task, together with the preparation of the Opus, seems to me the most pressing one. As for concrete suggestions in this department, I have as yet none to offer; but I hope before long to submit a slight memorandum, which will perhaps come to your attention as soon as this letter.

Gratefully yours,

Lewis Mumford

---

100 West 94 Street
New York City

9 May 1921

[Ts NLS]

Dear Professor Geddes,

This letter has come a little late, for the reason that it has little of good report to tell, and it is not easy to confess failure. It has proved financially impossible for me to break away from New York this spring – I have given the tedious detail bit by bit to Miss Loch and Branford and will not trouble to go through them again – and I am therefore pressed to do what I can by way of collaboration from this side of the ocean. I had hoped to have something tangible on paper by now, but the long essay on "The City" which I have had to contribute to a book which a group of New York critics and scholars are getting ready for publication next fall (a criticism of American civilization) has kept me pretty steadily engaged and all my surplus time has gone into remunerative trifles.[1] Now that a certain portion of my time is free I shall be able, I think, to send off to you by June first a "Census Sociologicus," such as you suggested in one of your recent letters. It will be in a form fit for publication in the *Review*. I have been in correspondence with Mr. John Nolen and have obtained from him the addresses of two organizations which may be of some use in getting the plans and pictures which you need from them. As for getting help in laying down and building up a scheme of sociological collaboration, from any of our big American foundations, I have my doubts: as far as I know most of their moneys go to the perpetuation and security of that which is already established. There might be some hope of getting a yearly endowment of two or three thousand pounds from some individual – indeed my old colleague, Helen Marot, is endeavoring to get such a foundation for a group of biologists, psychologists, and anthropologists to work together on a common basis – but such subsidies depend upon one's individual connections more than on the validity of one's plans and my own circle of acquaintances includes no one of the necessary financial girth. Miss Marot, by the way, is in a sort of half-way house toward the completer synthesis for which you and Branford are working: she is a penetrating psychologist, as much by intuition as by research, and is an incisive analyst of the human side of reconstruction, but professes complete ignorance and helplessness with regard to human factors.[2] Helen Marot plus Thorstein Veblen would make a mathematically complete sociologist: but the problem of putting the two together is not one of mere arithmetic, and although they must derive considerable inspiration from each other it is doubtful whether they will ever be able to do any active work in collaboration. Dewey is still in China – learning, I hope, from the Chinese, and there is nothing new on the horizon except the foundation of a school of geography at Clark University.

In my analysis of the American city one of the chiefest puzzles for interpretation was the prevalence of masonic orders, dating back from the middle of the eighteenth century, and now proliferating in great numbers and varieties. Structurally, of course, the masonic orders are probably degenerate craft-unions: but what is their immediate function in the modern city? In the pioneer days in America they were the only institutions – outside the partial, because sectarian churches – that promised fellowship and security. Why, however, should masonry flourish in France? For the same or a different reason?

I have just received a letter from Miss Loch which tells me of your projected itinerary and at the same time of the final date of Branford's operation.[3] The effect that the latter event may have on your plans – and upon the general work of the Sociological Society, for that matter – increases in my mind the uncertainty of what can and what ought to be done by way of collaboration. There have been times when I have thought that Branford spent a little too much time, perhaps, in laying the pieces on the board and neglected the opportunity of opening the game, and in particular, of encouraging younger men, and specialists in particular fields, of making their own moves. It seems to me much more important that particular researches and lines of investigation should be *infused* with the sound sociological method, and illumined by the general outlook, that you have developed, than that should begin literally with an acceptance of the entire schemata. A great many people, who have neither the experience nor the background nor the bent for taking over the system as it stands, are nevertheless sympathetic enough to do valuable work along the right lines if they were once put on the right track. Instead of searching for a general preliminary agreement among sociologists, as to scope, method, aim and so forth, it seems to me more expedient to center attention upon getting work done in particular fields – following the broad line that you have laid down – and then trust to obtaining a general agreement after the efficacy of the Edinburgh school had been demonstrated. Is this not what Demolins and De Tourville did in France; and did their work not demonstrate at least the limitations of their method, as well as its good qualities, so that sociologists now have a definite body of data, instead of an outline, to draw upon for suggestion and criticism.[4] The weakness of the Edinburgh school so far has been the weakness of the Aristotelian school after Aristotle: the work of the founder has been so comprehensive and magnificent and inspiring that it has in appearance left nothing for the scholars to do except to go over and annotate and dilute the master's work. I have no doubt that with the group which is now accreting around Le Play house this tendency will be largely overcome, and a genuine basis for co-operative work will be established. And no doubt the lectures you are now giving will help to this end.

I have not sent over any books so far because there has only been a handful worthy of attention, and I have been waiting for a few more before making up a parcel. Dewey's *Reconstruction in Philosophy* is the most noteworthy.

Misfortunes, someone says in Hamlet, come not singly but in battalions. When they cease to trample over me – as I trust they shortly will – I shall resume my plans for taking a more active part in your work. For the present I remain, with keen regret,

Yours faithfully,

Lewis Mumford

1. "The City" in *Civilization in the United States: An Inquiry by Thirty Americans*, ed. Harold S. Stearns (1922), 3–20.   2. Helen Marot (1865–1940), influenced by Geddes, friend and correspondent of Mumford; author of *The Creative Impulse in Industry* (1918).   3. Dorothy Cecila Loch worked as Victor Branford's secretary at Le Play House. Mumford had met her in London in 1920, and they became close friends. He dubbed her "Delilah" because, as he writes in *Sketches from Life*, "in gently pressing me to conform to English ways, she was behaving as a post-biblical Delilah, shearing the rough locks of her young American Samson" (257). She later married the American sociologist Luther S. Cressman and moved to the United States. She and Mumford maintained a frequent and lively correspondence which continued through 1969.   4. Henri de Tourville (1842–1903), author of *The Growth of Modern Nations* (1907).

The Sociological
Society
Cities Committee
65 Belgrave Road
S.W. 1

24 May 1921

[Ms UP]

Dear Mumford,

Well, well! – we must each make the best of it, & hope for some co-operation later; though each life tends to get caught in the surrounding stream! I thank you sincerely for the promised *Census Sociologicus* – which will be a sort of outline also of the essential types of the University Militant. (By the way, do you ever hear what has become of Charles Ferguson, who had many good ideas?)

Now as to sociological collaboration, you contrast – rather too strongly I think – the extremes of completely synthetic presentments (if these were possible) and personal & special work of individuals, no doubt stirred by some general sympathy. Our scheme, if we could float it, would be that of scattering as widely over hopeful quarters as might be, a variety of main theses, so far as we can distinguish these, and with statement of them from their initiators onwards. Thus, to use the (historic) illustrations of my former letter, (if I remember it rightly) Comte's Three States, Classification of Sciences, Differentia of Sociology, etc., or again Le Play's Occupational & Family Types, or again Tarde's & Durkheim's main theses – and so on.[1] Each should thus elicit its own discussion & contribution from those who think it worth while, and we, as editors, might then sum up as best we could, & hence leaving open ends for further workers. The general harmony would thus become clearer with time, as you say.

Your criticism of Branford is so far true, & it applies to myself also. We need some colleague who has more gift of stirring others to work. Our various groups seem working, but I do not yet know what may come of them. And in the press of interruptions, in which London life seems to me worse than most – since from *vaguer*-minded people on the whole than I find elsewhere – it seems hardly possible to get more than a very few real workers of any distinct & definite bent, & of original powers & working habits. The School of Economics, etc., etc., still keep the keener students, whatever they make of them. This place too is out of their way. Branford came here by persuasion of his old friend Norman Wyld, on account of the growing aggregation of Engineers' Societies in this quarter; but these, though no doubt in a way possible allies, are not yet aware of our existence, much less of our meaning anything to them. (Veblen has particularly pointed out this relation, so far as I know; but we have as yet no such writer here; & he is not yet read by them. Nor is Wyld himself an adequate link; though friendly.)

Branford had his operation a fortnight ago, & came through it well. He is progressing to the great satisfaction, I understand, of the surgeon as well as nurse, & Mrs. Branford brings in good reports daily, though of course there are slight ups & downs. I have seen him twice, & shall soon again: he is able to read a little, & be read to a great deal. He will doubtless come out of this much better than for years past.

You always surprise me by your generous appreciation of our "Edinburgh School"! Pray indicate in a paragraph or two why you feel it of value – what you think it has done or is doing in theory, and wherein it differs from others. *But pray also add what you feel it needs* – (1) for further clearness & completeness, & (2) towards wider acceptance? For in the extraordinarily unreceptive attitude of the London–Oxford–Cambridge attitude, apparently as solidly befogged as ever, in the main, we sometimes lose hope of making ourselves understood at all: nor does this environment stimulate us to further attempts. Perhaps I shall feel more cheerful when I am back to native air for a little next month, & then come back to London for July to fight again. Then back by Edinburgh to British Association there, & work before & after with J.A. Thomson.

Tell me too what you feel the urgent modern (i.e. contemporary & incipient) problems to which sociologists should be addressing themselves, (a) in the interpretation of current events, & (b) in the presentment of forecasts & ideals? Of course I have some ideas; but think of me also as having been very much out of Europe these seven years since 1914, on a desert island, etc. (if not asleep in the Catskills with Rip van Winkle!) & also much occcupied with plannings, etc. – which, however suggestive, are yet so local, & personal almost, as not to be intelligible in this vast city, nor of the slightest interest to it – & not adequately illuminating it for myself either! So do not fear to tell me what you might suppose I should know!

I have not heard before of Miss Marot. Pray sent postcard to Sunwise Turn bookshop to send me to Bombay any book of hers you think characteristic, & I'll read it on return. Thanks again for any such suggestions.

I am sorry I have no knowledge of masonry or its influences. I have only thought of them in this country as mildly sociable clubs, & in France as groups of more or less political wire-pulling. Yet I have sometimes dreamed of joining; with the ambition in view of their sentimental interest in masonry, towards seeing whether they could be interested in town-planning, & especially through that of Jerusalem? Perhaps you can tell me if there be any chance of this? (Their interest in symbolism too *might* indicate an openness to graphic methods, such as we work with?)

Always yours most cordially,

P. Geddes

P.S. Herewith a letter from Miss Defries, who wrote a life of me or rather an appreciation of teachings, with preface by Israel Zangwill, & has had it accepted by Mr. Kennerley; but is now afraid he has forgotten her.[2] (I fear myself it may be too long? even if of any interest to sufficient body of readers?) Anyway, can you see him for her, and write to her to *Barnes* address on head of her letter? (She is ill, & will be encouraged by an early reply – or at least may worry less, even if answer be unfavourable.) P.G.

1. Important figures in the history of social thought whose ideas influenced Geddes: Auguste Comte (1798–1857), Frédéric Le Play (1806–1882), Gabriel Tarde (1843–1904), Emile Durkheim (1858–1917).   2. Amelia Defries wrote *The Interpreter: Geddes, the Man and His Gospel* (1927), a compilation of his conversations and letters. Israel Zangwill (1864–1926), author and Zionist.

The Sociological
Society
Cities Committee
65 Belgrave Road,
S.W. 1

26 May 1921[1]

[Ms UP]

Since above written, I have been trying to stir up Sociological Committee at Council meeting in various ways. (1) I asked them to empower me to invite for (paying) *membership* 9 or 10 distinguished Indians; & then (2) used this as an argument for doing the like all round our respective acquaintance, & in all countries more or less – thus too getting a larger circle for *Sociological Review*. I therefore suggested asking you as to Americans, & so now do so. R.S.V.P.

Next, *Sociological Trust* came up – the Branfords' gift of this house till expiring of lease 17 years hence (though I trust we'll get out of it long before then) which I proposed to use as basis for an enlarged appeal, with dinner (indispensable to John Bull), if possible in a City Company's Hall, & with appeal to them to endeavour to stir up public opinion towards a society so usefully investigating the essentials of society, & linking all classes into its surveys, discussions, etc., etc. This rather tempted them: we'll see.

Next the *Sociological Review Committee* (of Branford & White, etc. – losing more money on it!) I advised a bigger start – and with change of name, more pages, etc., as *Sociology & Civics* or some such name – perhaps even *Society & City* or even *Civics & Education* (the Education being of course more sociological).[2] What say you to this? Suggest other titles? A cover with the word Sociological on it 3 times is enough to frighten most people away! Why not also put on cover a group of collaborative societies, if this could be arranged? Can it? What do you say – e.g. of your New School or its divided parts, one or other? What others?

Next, I have arranged with Town Planning Institute to talk in July, on wider bearings of Town Planning, etc., & get the discussion beyond the usual technical limits of roads, house-groups, etc., & to Civics proper. (They want this, & suggested it.) I saw to Pepler our Honorary Secretary who was talking hopefully of his recent attempt to interest University & City people at & around Oxford towards Regional Survey, etc.[3] Suppose we organise this appeal throughout all the Universities as far as possible – can't we make some of them take it up? Here of course Sociological Society, Regional & Civic Education people, etc., will come in, & I trust others. Why not Geographic, Economic, & Anthropological Societies, etc.

How far also would this go down in American University Cities, more open towards leading than ours, I should think?

Get your head warm over this, & give us your schemes – and dreams!

Yours *P.G.*

1. This letter was apparently enclosed with the previous letter of 24 May.   2. James Martin White (1857–1928), wealthy patron of Geddes; invested in several of his projects; endowed the Chair of Botany at University College, Dundee, especially for Geddes, which he held during summer terms, 1889–1919.   3. George L. Pepler (1882–1959), architect, city planner, and advocate of Geddesian ideas; a founder of the Town Planning Institute (1913).

Dear Professor Geddes,

100 West 94 Street
New York City

26 June 1921

[Ts NLS]

This is part one of a letter which I expect to finish this week as soon as I get up in the country, in New Hampshire, where I am going to teach literature for a couple of months in a little experimental school.

Miss Defries' note I have already replied to, with the best advice I was able to give; namely that she had better prepare herself to give up the idea of getting Kennerley to publish her manuscript. Kennerley is no longer a serious publisher; and he makes these 'contracts' in a fit of absence of mind, in order to persuade himself that he is still in the game: once in a while he lives up to his promises, by accident as it were. I have offered to take charge of the manuscript and present it to other publishers. To Miss Loch I have already, in answer to the double request from her and you, given my advice about inviting American sociologists into the society, although, because of my lack of contacts at the universities, I was unable to furnish any names which the Sociological Society did not possess. In my memorandum on the *Sociological Review* I heartily seconded your suggestion to change the name of the publication; and alter – or rather, *openly* alter – its policies. It will be well if the appeal for funds brings forth any response; for a certain minimum is needed in support of the staff of the society and the review in order to ensure that the work shall be carried on regularly and consecutively.

Now for a couple of queries about yourself. The editor of the *Menorah Journal* – a catholic Jewish review – has asked me to see whether you could not be persuaded to write him an article, long or short, upon town-planning in Palestine, or your conception of the University of Jerusalem. Failing that, he suggests that I might get your permission to set forth your general line of thought about universities, and in particular that in Palestine; but of course he had much rather have the article from your own hands. May I have an early answer about this? Perhaps, in spite of the unfortunate breach that has been opened between the American Zionist group and the World Organization, something might be done to further the cause in Jerusalem through such an article in such a medium. You need have no fear of offending either party; for the *Menorah Journal* is non-partisan. Second: as to your papers before the Town Planning Institute. I have not had the opportunity to talk to Mr. Whitaker; but I am sure he would be eager to have your papers in whole or part for publication in the *Journal of the American Institute of Architects*. It would be a very vital contribution to the American city planning movement; for it is the sort of stimulus for which we starve here even more than they do in England.

Finally, one more matter of business? Is there any possibility of your coming over to America in the spring of 1922, and have you any inclination to do so. Walter Fuller and I were discussing this matter the other night; and we were both sure that there was an audience here for you, and that, in city after city, we could get together small groups of people (social surveyors, architects, engineers, sociologists, etc.) who would be very eager to sit at your feet. It is not anything on the Chatauqua scale that we plan; rather a series of similar public conferences, from California to New York.[1] There is still a large body of people who are already acquainted with you; and a still larger body who are thinking along similar lines and are ready for further stimulus and cooperation. Should you consider seriously this proposal, Fuller and I will gather together an invitation committee and secure a financial guarantee, upon obtaining which we would send you a definite invitation. This might all be arranged, one way or the other, by the beginning of October. Fuller said that, although he would beg money for no other cause, not even for that of the paper with which he is associated, he would do it for you: so deeply does he feel his debt; and I of course echo his sentiments. We are assuming that you would be able to reach California early in April and spend the next four or six weeks in lecturing. Perhaps this is remote from any of your plans or desires; in which case, pardon our urgency! As a further attraction I pledge myself to meet you in California and personally conduct the whole trip!

This is all I have time for at present; more later.

Gratefully yours,

Lewis Mumford

1. Originating at the Methodist Episcopal camp at Lake Chautauqua, New York, in 1874, Chautauqua refers to a popular adult education movement; the widely imitated Chautauqua Institute offered home study courses, lectures, and summer school programs.

The Sociological
Society
65 Belgrave Road,
S.W. 1

14 July 1921

[Ms UP]

Dear Mumford,

Thanks for yours of 26th June or rather "first one" of the full letter promised.

Thanks too re Miss Defries' book of me! I wonder how she'll get the manuscript out of Mr. Kennerley? Perhaps she'll have to ask your help? but we may leave that for the present till she writes you.

Thanks again re Sociological Society & *Review.* What name do you advise for New Series. "Sociology" – "Sociology & Civics" – "Society" – "Society & City" – etc., have all been discussed; but there are objections to each.

Yes, I'd like to sit down & write hard regarding Jerusalem – but never get time! I'll try to do what I can & send it you for *Menorah* – but dare not promise a date.

Paper, Town Planning Institute, will be printed & sent you – I am sure neither the Institute nor I can have any objection to reprinting as much as may be of interest on your side.

Thanks once more for your & Fuller's kindly thoughts of bringing me over. But (a) I am due here & in Aberdeen with Thomson (& in south with Branford if strong enough for work) until to India in October–November, (b) all winter Bombay 1921/22, (c) *all next summer, probably, planning towns in India,* (d) then *perhaps* back in Bombay again for winter 1922/3. This however is not certain. I am free to leave Bombay after this coming winter, as I only accepted positively for the three of the five years of my invitation; and if more useful work turned up – with some pay towards compensating the loss of £1000 or thereby (for these 4 winter months work) I should certainly be much tempted to consider it – for the arousal of Indian Universities (out of the present exam-machines) will be a longer & broader job than my small-endeavours can materially contribute to!

But remember too that I am not much of a lecturer – fluent enough to small sympathetic audiences such as I get here, but without the "popular gifts" wanted for pecuniary return by the larger audiences of lecture agencies.

Again the Town Plannning Exhibition is rather a big cumbrous affair, needing large space, & much exposition to make it interesting, & then only to small numbers at a time as they go round with guide. It needs subsidy from governments, cities, etc., as it earns 0 at doors! And then of course it has always had at home & in India alike, & from £300 per visit in former case to more than double that in India – at pre-war rates too – & yet yielding practically 0 to capital after clearing off! – for the expenses have been serious, & transit, etc., now dearer a long way!

My son Arthur is to come to me in Bombay this winter – since you cannot; & we may probably have a final Indian Exhibition.[1]

1. Arthur Geddes (1895–1968) assisted his father in India but was not content in that role; he eventually became a lecturer in geography at the University of Edinburgh.

Dear Professor Geddes,

Peterboro
New Hampshire

31 July 1921

[Ts NLS]

Here I am in New Hampshire and here is my letter, a month later than I promised! It was not till I got up here that I realized how sadly I was in need of a vacation, after the painful, neurotic winter that I had come through. (I gave you a glimpse of my predicament once or twice in my letters perhaps, but didn't convey the fact then, didn't in fact more than half realize then, that it was the most distressing period I had ever passed through in my life.) I am in smooth waters again, and the earth once more is good; but the waters are quiet, too, and one of the things that has kept me from answering your last letter is that I have been so constantly enmeshed in personal entanglements and difficulties that I have not had the energy to dwell as much as I ought to be dwelling upon the great world that lies outside my private concerns, and upon whose decent ordinance so much of the success of my private world, and all our private worlds, ultimately depends. I miss the stimulus of the discussions I used to enjoy with Branford and Farquharson in England, too, and my mind has so long lain fallow, at least with respect to sociological things, that the ground should bring forth good harvest at next planting, if only it is sown again with the right sort of seed. My month up here in the isolation of a six hundred acre estate in the hills of New Hampshire, not far from the famous Mt. Monadnock, has if anything increased my diffidence in dealing with sociological subjects: while the Peterboro library is the oldest free town library in America, and a reasonably good one, I am many miles away from it, and the only books in my possession are a copy of Walt Whitman, some of the more apocryphal parts of the Bible, your *Masque of Learning*, and your University of London lecture syllabuses. Not exactly the right place in which to write a scholarly Census Sociologicus! I am much more in a mood to write something about the school in which I am teaching: McMurray, of Teacher's College, ranks it as one of the twelve most interesting educational experiments in the United States.[1] It is a summer school, run by a rich young Boston woman for the children of the village; its distinguishing features are (1) a simple balanced curriculum which recalls the Greek ideal, namely, science, gymnastics, music, and literature, and (2) a teaching staff selected, especially in literature and music, with a view to the teacher's capacity to do creative work and without any regard to the question as to whether he has ever taught before. There are four teachers and thirty-eight pupils; and for my own part, I have been learning steadily. Also, it is a great recreation.[2]

But to go back to your letter, which I have been reading and pondering a good part of the afternoon. You ask what I think the main contribution of the Edinburgh school, namely, in good measure, yourself. What seems to me the great thing about the school after having spent many weary, fruitless hours on what is called "sociology" is (1) the fact that it is comprehensive, whereas the other sociologies are partial, dealing like Cary and Le Play with industry, or with Tarde and Cooley with social psychology, or with Comte and Giddings with historical interpretation; in other words, the "Edinburgh" sociology makes a point for point contact with life itself in every aspect. To say the obvious, it is synthetic. (2) Its basis is in concrete observation and factual analysis, as over against the "systems" of sociology, which use facts in order to bolster up rationalizations, projections, wish-fullfillments and myths; cf. Ward, Giddings, Gobineau, Houston Chamberlain, and minor writers who seek decently to uphold the status quo.[3] Concreteness and synthesis are accordingly the main achievements which make the Edinburgh school tower head and shoulders over the other schools in relevance, applicability, and immediacy. What are the weak places? First, the main deficiency, I think, is that the material of your syllabuses, lectures, reports, and book chapters has not been gathered up into a comprehensive and systematic outline, on the lines of a Home University book, which would provide a definite basis for study and discussion and criticism. This is the most important present task, as I conceive it, and it should take precedence, because of the need for it, over any more compendious attempt to sum up the grand total of existence, physical, organic and social. At least get the analysis of society articulated, and the method itself established. From this starting point the anthropologists, social surveyors, city planners, educators, psychologists and the rest of them may set forth, each following the direction as far as the method will carry him. Poignantly interested people, like Farquharson, myself, and others have done this already; and more people would if there were a convenient book to guide them. The second weakness, which I noted occasionally in the Making of the Future series, is the tendency to stress points like the primitive filiation of individuals, which is more an anticipation, a pregnant aperçu, than a scientific demonstration, and which does not throw the light upon social problems that a more detailed contrast between rustic and urban types would have done. I realize the temptation of the Making of the Future series; the temptation to have your will upon a book, and let him who might, understand it! But, with due allowance, the criticism, I believe, holds good. (Please forgive the brusqueness of a young man who intellectually speaking is scarcely fit to wash out a beaker in your laboratory.) There is another point, which I am not by any means sure is a weakness, but which needs at all events to be reckoned with. This is the psychological analysis that has played a part in all the diagrams from those in the

Sociological Papers onward. What is necessary here is either to point out the connection between your psychological schemata and the psychology that is now taught under the aegis of Dewey, Freud, Jung, Watson, Myers, Woodworth, and the rest; or to elaborate, in greater detail than hitherto, your criticism of this psychology and to point out the steps that are necessary to bring psychology into useful relations with sociology.[4] I confess that in this section of your diagram I have always had a sense of missing something; of not having been present at some previous explanation in which this all was made clear. Now in the bookcase diagram psychology is grouped with ethics and esthetics: on what grounds, however, other than historic accident should one separate psychology from biology?[5] Ethics and esthetics themselves are but the penumbra of sciences; and their place in the bookcase has still largely to be filled. Psychology on the other hand thrives apace, and as you know it has been a great driving force of education in America particularly. Has it been properly reckoned with; have you perhaps now discovered a new science, and are confusing us with the conventional name? I do not pretend that this is a gristy criticism; but it calls attention to a hiatus which I at any rate feel.

Now as to the contemporary and incipient problems of the sociologist. It seems to me that you and Branford have made many admirable contributions to those problems that arise out of place and people, and in spite of everything, investigations in these departments and practical work in housing, city planning, education, and so forth are moving at a pretty good speed, and along the lines which would be indicated by a genuine sociological analysis. The department which you have been inclined to slur over, and which sociologists have not dealt with effectively, so far as I am aware, is that of work. Industry and the labor movement need sociological methods, sociological outlooks, and sociological ends more clamantly perhaps than any other field. Yet just because it is so difficult to reconcile conflicting positions, just because the interests involved are so manifold and delicate this is the last field to be touched by any sort of sociology except that of disreputable apologetics. There would be a keener interest in your whole work by the younger men in the labor movement and in the school of economics, if they thought it linked up with their immediate interests, and was not merely some thing that "came after" when the hard, sweaty task of redressing grievances was done. This is a big order of course, and it is a hard one to fill in a scientific, sociological fashion; but impartiality does not consist, as our American phrase has it, of sitting on the fence. I know that this occupies your personal attention, as in your introduction to Mukerjee's *Foundation of Indian Economics* and elsewhere; but it does not occupy the place in your reconstructive program that finance and housing do. Vide Branford's "what to do" program.[6] The overshadowing problem of modern times, after all the specific ones have been accounted for, is the adjustment needed as a consequence of meeting of discrepant cultures, like the North European and the Hindu, and different races, like the white, yellow, and black. Has not a staggering commixture begun? To what extent is the negro problem that exists in so many regions of America an accidental one, due to the associations of slavery, with its terrific burdens of evil, and to what extent is it an essential problem in the mingling of all races? I have just been reading two of Tagore's latest books, *The Wreck* and *Glimpses of Bengal*, and I wondered how near we were to a genuine humanism in a world where regional difference and a long historic tradition have made the great communities in so many respects worlds apart.[7] The injunction, cultivate your region, would of course, if heeded, reduce the movement of population somewhat and thus lessen the abrasive effects of inter-emigration; but failing this what remains? Why shouldn't Hindus and Chinamen come to proceed apace? Are there any scientific answers to these questions. Are sociologists ready to formulate any? As a presentment of the difficulty I wish you would read the article on "Racial Minorities in America," in the book on *Civilization in the U.S.* which I will send you as soon as it is published.

As for the future, the presentment of forecasts and ideals, I find myself diffident; for the reason that, as before noted, I have been personally too muddled to think generously. More of this later perhaps.

I was sorry to be able to send to Bombay a barren handful of city reports. The one society which was rather favourable in its reply assured me that it had already sent you all the material at its disposal. In the enclosed letter you will find a list of men who have done some particularly interesting work; and a note from you might bring forth some of their work much more quickly than anything I could do. Veblen has been in very poor health and I have seen nothing of him; Ferguson has quite dropped out of sight, and is not writing the way he did two years ago; in fact, I have not come across a single article of his. Miss Mayer tells what a joy it is to be in your company and work with you; and I duly envy her. I remain, as always,

Gratefully yours,

Lewis Mumford

1. Frank M. McMurray (1862–1936), professor of elementary education at Teachers College, Columbia University; proponent of Herbartian pedagogy; later, following Dewey, adapted an experimental outlook.   2. Mumford was teaching literature at a summer school on the estate of Arthur and Joanne Johnson near Peterboro, New Hampshire.   3. Joseph Arthur de Gobineau (1816–1882), author of *Essai sur l'inégalité des races humaines* (1853–1855), in English *The Inequality of Human Races* (1915), a study of the importance of race in history which argues for the superiority of the white "Aryan" race. Houston: the reference is obscure.   4. John B. Watson (1878–1958), psychologist and founder of behaviorism; author of *Behavior: An Introduction to Comparative Psychology* (1914). Perhaps John Linton Myers (1869–1954), Oxford historian; author of *Who Were the Greeks?* (1930). Robert S. Woodworth (1869–1962), psychologist and proponent of functionalism; author of *Psychology* (first edition 1921), an influential textbook.   5. "The bookcase diagram": Mumford is referring to Geddes' "Classification of the Sciences" chart, which is reprinted in Boardman, 467.   6. Radhakamal Mukerjee (1889–1968), professor of sociology and economics at Lucknow University who had studied with Geddes in India; prolific author of such works as *Regional Sociology* (1926); Geddes wrote the Introduction to Mukerjee's *The Foundations of Indian Economics* (1916). "Vide": see.   7. Rabindranath Tagore (1861–1941), Indian writer and poet; awarded Nobel prize in 1913; founded the International University at Santiniketan, Bolpur.

Dear Mumford,

Outlook Tower
Edinburgh

28 September 1921

[Ms UP]

Just a line of greeting & remembrance before leaving again for India (University of Bombay always finds me) & to hope that you will continue correspondence as occasion allows. I expect to be away for two next winters, & not return here till 1923: but one never knows! It may be longer or shorter. If you ever see your way to arrange a meeting pray try again to do so: & let us see what can be done.

Branford has been very ill – prostate operation too long delayed – then pneumonia & pleurisy – 3 months in bed – but now up & about again, & I hope with gathering strength – though now I fear but an old man's. We'll see when we meet (at his new home at Hastings) in a week or so. Thus I have missed the collaboration with him I hoped for in coming home – but have had a useful month with Arthur Thomson in Aberdeen planning *The Vital Revolution* – of much more someday, I trust, if we old fellows hold out. Otherwise time much broken up, as by lecturing & by British Association here most recently, & by attempt to keep my poor old Tower alive, after war depression & post-war too.

Pray give me any encouragement you can: we need it all: but I think there are signs of progress again, & in intellectual circles especially, though these too slowly.

I am performing another Town Planning Exhibition at Bombay next winter: any fresh indications will be gratefully received. My surviving son Arthur (after a long war & post war illness & slow convalescence) is coming out as my assistant for it: & I should like then to send him round the world with it, & especially (if there be any demand manufacturable) to U.S.A. & Canada too.

Always yours cordially,

P. Geddes

Please send enclosed to Veblen, whose address I don't know – but as acknowledgment for a volume only now found in Tower, but happily already read in India. What other lights are dawning? What are you doing yourself towards incandescence?

Dear Mumford,

Department of
Sociology and Civics
University of
Bombay

10 December 1921

[Ms UP]

Horrid to have allowed 2 letters to you to be snowed under when I thought them posted! Forgive my untidiness; I have two halls here each 63 yards long all littered & heaped & deep-piled with papers & plans & books! And such things happen too at home – whether at Tower or with Thomson or Branford. But it is not from want of goodwill. Don't be discouraged – next time I'll reply by return! Here till 10 March, after which I'll send another address, though this will (or *should*!) always find me.

Thanks for occasional copies of *Freeman* – with which I generally agree – though too hard on France – where I have seen the corpse of the victim – regions of battle. When that wound is healed, France will be herself again – but of its depth, John Bull & Uncle Sam have no adequate idea!

Yours ever,

P. Geddes

[1921]¹   *Note as to Possible Collaboration*

This as at present & hitherto considered, was of {Principles/2 Outlines of Biology} with J.A. Thomson & Ditto of Sociology with V.B.

But why not rather a sort of 2-3 volume novel? Say *The Sciences of Life: Organic & Social* or *Biology & Sociology: Their Scope & Main Ideas* or *Life and Society in Evolution* – or (? ? ?).

All more or less in collaboration & co-adjustment, & with sections perhaps written by other collaborators – like yourself for me? We are fairly ready for such group action – & though we should be assailed all round we could do some assailing too!

Such volumes would need to be fairly big & costly: but the presumable success of e.g. Thomson's Gifford Lectures (despite 2 expensive Volumes) may encourage us – & such bigger books command more attention than do little ones, like *Making of Future*.

Again, might we not take a leaf out of Wells' book; & like him publish in parts – with some illustrations beyond graphica & diagrams – pictures of life & of cities? Yet what publisher would not think us too technical? What say you, with your experience of England & your knowledge of U.S.A.? P.G.

1.   Undated manuscript page in UP f. 1811.

Lewis Mumford AND *Patrick Geddes*

THE CORRESPONDENCE

1922

Dear Professor Geddes,

143 West 4 Street
New York City

15 January 1922

[Ts NLS]

I had finished a letter to you a couple of days ago, when your several letters came to hand; and I must therefore rewrite it, asking fewer questions and telling more news. In the meanwhile I had received a note from Miss Josephine MacLeod, and in the course of a very interesting hour learned of her plans to go to Jerusalem and thence to follow your trail to India.[1] I wonder whether Lasker has sent you the little notice of your work in Jerusalem which he printed the other month in the *Survey*? At any rate, I am sending it on to you now, along with a few other things. The *Menorah Journal* is now printing a little expository essay by me on your philosophy of universities, in relation to Jerusalem – in lieu of any more specific treatment which I could not give.[2] Thinking about the problem of the Jewish community, not merely in Palestine but in the rest of the world, I am wondering if the time has not come to turn one's back flatly upon nationalistic projects and nationalist ideologies, and to salvage what is valid in nationalism by way of regionalism and humanism. I know that we are headed this way; but we are slow to make the break; and nothing short of a dramatic movement, it seems to me, will break up the nationalist complex. Whenever I attempt to analyze what a nation is I discover that the idea is as slippery and tricky as a globule of mercury: nationality does not correspond to any observable social or anthropological fact: it is rather a procrustean mold – useful for reasons of state – into which, during the last three centuries, we have attempted to force the social and political life of particular regions. Particularly with regard to the Jews, I see no way of maintaining their spiritual integrity while they are either swayed by or subject to the idea of nationality. If I can demonstrate this in an article in the *Menorah*, and convert the Jews to the regionalist-humanist outlook, I feel that the cities movement would begin to gather momentum at a faster rate. In America there is little that is hopeful to report. Mr. Henry Ford, however, has offered to take over the government's nitrate plant at Muscle Shoals, Alabama, finance a project to develop the water power, and build a city, or rather a series of cities, seventy-five miles in length. Here is a first-rate neotechnic project which gains by its juxtaposition to the Birmingham coal-fields, where coal-industrialism is now dominant. If Mr. Ford's bid is accepted by the government there is still the danger that Ford's city plans will be one-sided; that they will touch only mechanical and material developments; and that the Muscle Shoals conurbation will in a short time be as badly run down and as inadequate as, let us say, Pullman, Illinois.[3] So I am now taking it upon myself to write an open letter to Mr. Ford, which I hope to publish in the *Freeman*, and in which I shall try – with due deference to your ideas – to show what is needed before this new urban district becomes a City indeed. I feel that it is not enough to urge Mr. Ford to look at Letchworth: Letchworth itself appears to lack some thing which the medieval cities apparently had – it lacks a fully developed spiritual power – and while its physical environment is attractive enough and sound enough, it remains culturally too much on the same level as the metropolis from which it has escaped. How Mr. Ford is to give a true spiritual power the opportunity to function is another question. Branford, I think, would not hesitate to answer it; he still has hopes for a renovated church. Here and there, no doubt, revitalized priests and renovated churches may come into existence and add something to the spiritual stock of the community; but it seems to me that the original impulse of Christianity has almost dwindled out, and there is at present more energy in the words of Tolstoi, Thoreau, and Nietzsche than there is in the New Testament. It is plain enough, at any rate, that Mr. Ford himself can not found a militant church or a militant university: that would be a contradiction in terms; and yet if his city is to be humanly successful he must somehow prepare the soil for their growth. What are we to suggest? How are we to alter Mr. Ford's plans as to what must be done? My only answer to this, so far, is to show him that the city is not merely a vehicle for commerce and industry, but a place where the social heritage is preserved and re-shaped. Even if nothing can immediately be done, without a renascence in our cultural life, it will be somthing to convince my fellow-countrymen that they have not done everything when they have provided metal roads, bathtubs, and washing-machines for the entire population of Muscle Shoals.

Civilization in the United States has just come out; and I am sending you a copy.[4] It is a pretty fair gauge of the post-war mind in America, the mind of a disillusioned generation whose most generous hopes have been mocked and flouted. During the last two years we have come to see that the war has not brought us two steps nearer Eutopia. Those war-years were a misery and a waste, and so far from being able to garner any of the thistles that Mars has sown, it will be necessary, in order to make an advance, to return to the projects and the tasks that occupied us when the war broke out. Yet the force of those projects has been dissipated during the war; and therefore, in order to gather enough energy to react creatively upon our environment, we will have to return to ourselves, and stoke up our own furnaces. In short, on the most hopeful view of the situation, this is an introvert period; and while considerable advances may come in art, science and philosophy – and there are plenty of indications of these in America – the immediate effect upon our social life is likely to be a more confirmed and aggressive reaction which will reach levels below those that we had attained in 1914. So it proves that the change of the old reconstruction *Dial* to the new monthly of art and letters was symbolic: the extrovert became introvert: no longer strong enough to mold our environment we returned to the business of remolding ourselves. Perhaps this is an inevitable rhythm; and the war collapse has but hastened it. Perhaps in the end our despair will save us!

I see that Thomson is about to publish an *Outline of Science*; and I wonder what part you have had to play in it, and whether the sociological sciences will be adequately represented. In the mood that I have described above, I am planning to write a critical history of Utopias; not a formal treatise but one that would appeal to a wide audience; a history not of actual places but of the regions in which men at various stages have wished to project themselves. The renascent interest in history itself would give such a work an impetus; and since the general moral of it would be "from Outopia to Eutopia" it might do its bit to prepare for the period that is still beyond the horizon. A new universal history, called *The Story of Mankind*, by Willem Hendrik Van Loon (written for children) has just come out; and is already a best-seller. There is an intimate flavour about Van Loon's work that I think you would like; and his chapters on the Medieval City – he was born in the Low Countries and got his first view of the world by climbing to the tower of Old St. Lawrence in

Rotterdam – are particularly good. He has more of the juice of history and fewer of the dry bones than Wells.

During the past six months my fortunes have bettered not a little – I told you, did I not, in my last letter, that I was married? – and my plans for the future are all conditioned by the fact that New York offers me a living at present, and other regions do not. Criticism seems to be my forte: at any rate I have written no plays since I was twenty-two, and have not been able to finish the novel which I had gone more than half way through at twenty-three. The prospects of teaching are so slight that sociology as a special subject has somewhat fallen into the background; although I am occupied with socio-logical problems. My friend Brooks, on the *Freeman*, is perpetually plaguing me to write a book; and I have no doubt that my career would go a little more smoothly if I had a book to my credit.[5] But on this subject I am inclined to follow Tolstoi's injunction, and not to attempt anything until the creative urge can no longer be denied. In lieu of a definite book I should, it is true, have some definite line through which I could canalize my energies; and for the last two years I have lacked such a line and have allowed much of my energy simply to spill over and become dispersed. This disintegration was however, I believe, as much the outcome of sexual conflicts and maladjustments as of anything else; and I am now, in this respect, on a fair way to recovery. The problem of complete continence is not an easy one for a normally passionate young man after he has reached twenty; and I can not pretend that mere physical cleanliness came within miles of solving it for me. Sexual reactions are disciplined through use rather than denial, it seems to me, and the first real approach towards discipline came with my marriage. There is a sense in which a person who has never had sexual experience is as unfit for marriage as a person who has never handled money is unfit for a big inheritance; and "purity" does not steer one through the first difficulties any more than honesty does through the second. We overlook this by our emphasis on negations, do we not?

Branford wrote me an encouraging letter about himself recently – the first in a year – and to judge by his handwriting alone he is in much better shape. Please tell me of your goings and doings; and if I can be of service to you in any capacity, command me. I have sent no sociological books to your department for the reason that the literature of the past year is scarcely worth the postage; and unless you wish to have your library formally complete it isn't worth the trouble.

I trust you continue in good health; and that some of your burdens have been lightened by Arthur's help.

Yours in discipleship,

Lewis Mumford

1.   Josephine MacLeod, American friend and follower of Geddes whom he had met in 1900.   2.   "The Hebrew University: The Vision of the Architect," *Menorah Journal*, 8 (February 1922): 33–36.   3.   Henry Ford (1863–1947) founded the Ford Motor Company in Detroit. Muscle Shoals is a town in northwest Alabama on the Tennessee River; in 1916 the United States government built a large nitrate production plant there which was never put into operation.   4.   *Civilization in the United States: An Inquiry by Thirty Americans*, ed. Harold S. Stearns (1922); Mumford contributed an essay, "The City."   5.   Van Wyck Brooks

(1886–1963), prolific American literary critic whose friendship was very important to Mumford; writings of Brooks which particularly influenced Mumford include *America's Coming-of-Age* (1915) and *The Ordeal of Mark Twain* (1920).

Department of
Sociology and Civics
University of
Bombay

17 February 1922

[Ms UP]

Dear Mumford,

I have been remiss in writing to you as to all others. My stay at home last summer was much broken up though of all the more varied interest. First some visits to French Universities, Montpellier for old friends, Toulouse & Bordeaux for regional activities; at the latter I met Duguit, the abolisher of "rights" from law – a hopeful sign from a senior jurist & Dean of his faculty.[1] Then some time in Paris, and a few days through the war-zone devastations from Château Thierry to Lille, with my old friend & former Assistant Marcel Hardy, now one of the board in Berlin for study of possible reparations.[2] Land nearly cultivated again; but farms, factories, etc., far from rebuilding, as mines from working order anew. A woeful experience! – too much forgotten by other nations.

Then home, save for a run to Brussels to my old friend Otlet, whose International University & Museums have now vast space. In London, lectures at Sociological Society & frequent visits to V.B. in private hospital (where they let him get pneumonia & pleurisy while specialising on their particular point of repairs!). North to my grandchildren & daughter in south of Scotland, lectured at Edinburgh, back to London for another spell. Then to Aberdeen with J.A. Thomson, but we got no book written, though mutually stimulating talks. Then British Association at Edinburgh, back to London, & so via Paris here.

Very busy time here, arranging Cities Exhibition with my long invalided son Arthur, now out here with me, & slowly getting on towards ordinary working day: so this has been rather over-straining to him, since so to tireless old me – who have got run down sufficiently to be in my first touch of malaria – happily taken in time with ample quinine! ¼ mile of screens is a long job to arrange, & even more are here.

The idea will interest you, beyond plannings. For if this city be the gate of India for the west, & of the west for India, it is the best of cities for a double panorama, of the cities of each. With the exhibits run tables of books, some model houses and villages & of course all with explanatory lecture-talks peripatetic. So arises the general idea of a new type of educational Exhibition – essentials of atlas, gazetteer, History, Encyclopedia, Museum, Pictures, Library & Lectures together, and of University Extension in a more than usually popular form.

With this comes the Teaching & Research Department arranged with gallery on one side & general library on the other. It is divided into a dozen rows of tables (separated by screens in east half).

This [following] is a very poor diagram but if you can visualize it from this, you will see a graphic presentment of the way in which all specialisms are so far oriented towards the City from which they came, yet how the theoricians turn their backs on this from the Library. Nor can any on the east side see beyond their own Department: only on the west side, as the separating screens are got beyond, can they see the others, and the City & Library at each end & beyond.

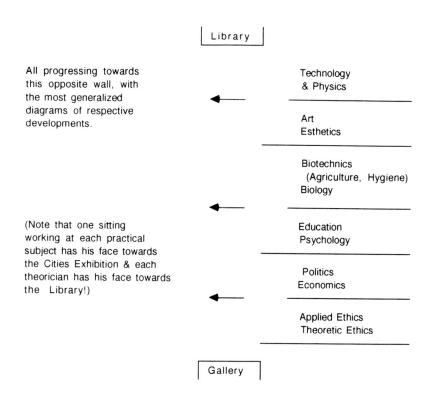

All progressing towards
this opposite wall, with
the most generalized
diagrams of respective
developments.

Library

Technology
& Physics

Art
Esthetics

Biotechnics
(Agriculture, Hygiene)
Biology

Education
Psychology

(Note that one sitting
working at each practical
subject has his face towards
the Cities Exhibition & each
theorician has his face towards
the Library!)

Politics
Economics

Applied Ethics
Theoretic Ethics

Gallery

This is a way of protesting against the various treatments of sociology from one or other viewpoint, physical, biological, psychological, political or ethical alone, yet of recognising the contribution of each & expressing it upon its own appropriate line of table & screens.

Through this arrangement next run boxes with my own untidy manuscripts of accumulations of books & papers in all stages of incompleteness. But now I am more ready to see where I am, & also to associate in various collaborations as with the few best students here. So if this be not a real Institute of Research, in principle, how shall we arrange one better? I have done my best for the present: so now devil's advocate it! I wish we could form a group working together on such lines. If there were only some way of bringing you here next winter! Certainly for the present. I mean this coming year to be my last in India.

Is there anything in the idea that if this plan of Exhibition & Social Studies Department together can be made intelligible, it might attract attention in U.S.A.? Enough to have you sent out for a spell at it and a report on it?

In fact should we not aim at developing on one side towards a great city & University Institute, & on the other hand condensing & adapting it to the small town, & even village Institute, should we not be producing a real contribution to the Making of the Future?

I was glad to hear from my friend Josephine MacLeod that she had met you, & had talk. I shall no doubt hear more when she comes here, as she intends doing after visit to Palestine.

I must stop now for post, but have still to write regarding points in your last letters. Another time however.

I hope you go back to summer school as pleasant & stimulating outing? Best wishes for the coming summer.

Yours cordially,

P. Geddes

1. Leon Duguit (1859–1928), French jurist, sought to discover rules that should control those who govern; author of *Etudes de droit public* (1901–1902) and *Traité de droit constitutionnel* (1921–1925).   2. Marcel Hardy (b. 1876), Geddes' assistant at Dundee; later served as War Reparations Commissioner in Germany; author of *The Geography of Plants* (1920).

Department of
Sociology and Civics
University of
Bombay

25 March 1922

[Ms UP]

Dear Mumford,

It is long since I wrote, for all correspondence has gone into arrears, during the toil of recent months – getting up my Cities Exhibition on a larger scale than heretofore – now more than ¼ mile of screens, crowded close & high. It is thus more than the simple Town Planning Object-lesson with which it began its wanderings a dozen years ago – of course from the first origins long before in the Outlook Tower it was of definitely sociological & general civic purposes: but these are now clearer. Moreover a new type of educational organization has been evolving, & is now fairly intelligible & usable – a University & civic gallery & library, lecture-room, etc., in one, & addressed to all – student or citizen, from University heads & city fathers, experts & students, to schools, etc., & thus applied to all levels of education at once – from university research & teaching to university extension, & to schools down to youngest age that can look at pictures.

Of course it is very incomplete – each panel but a door which should open into a room of further detail; and thus practically there is no end to its possible extension, were rooms & funds & time available. Nor yet to its possible condensations – as for smaller town, even village & at length into many illustrated books – encyclopedia to primer.

I am thus beginning to think of its popularisation beyond the given city where it happens to be (always far too short a time for much real usefulness) & to its adaptions as basis for forming permanent collections on various scales, "to suit purchasers" – i.e. city to town, village, schoolhouse & to libraries, great & small.

To make something of this scheme would be more than the desultory lectures or writing, of an ordinary educational propaganda. But how to do this? How it means a realistic sociology & civics, not beginning with philosophy or psychology, with economics or with history, with politics or religion, with hygiene or education, with art or industry – but with the survey of the concrete lives of men as they have grown & are changing today – & with visions of the possibilities, again in no merely verbal utopias, but with definite plans, and perspectives for these possibilities in typical places small & great – that is what such an Exhibition indicates. And in such ways all the various aspects of social (& individual) life appear, but now in these mutual relations, and inductively displayed, with minimum of personal equation, instead of maximum, as in sociological literature too much hitherto.

Now if all this be more or less clear to the varied visitors of such a collection as they give it some study, as some are doing, what next follows? If we can go on clarifying the results of such studies, & see libraries in general or in particular, with the departments of universities (in widest sense) as all so many drops of honey & bits of comb from the social hives (given hives, & of these too few in past or present) there begins also to emerge not only the perpetual dream of synthesis but the hope of increasing synergy – and even something clearer than heretofore of the needed policy – of the Making of the Future. Policy emerges from politics, & is materialized, vitally, as Ethopolicy – (pro)Synthesis tends to (pro)Synergy, & imagination clarifies towards the more clearly designed lines of achievement. More simply stated – there thus emerges a conception of group-formation – of constructively-minded people, seeing possibilities of co-operation when wherever 2 or 3 are together; & from simplest betterments towards more fully regional & civic developments. Conversely also, from large schemes to their needed details.

In all sorts of ways then, these things have been growing clearer. Between city galleries on one hand & library on the other, this needed University Department of Sociology & Civics is now getting arranged, and with its various sub-departments as long parallel double pairs of tables, Logical & Mathematical, Physical & Esthetic, Biological & Psychological, Economic & Ethical; & in each with theory & practice confronted: the theoretical looking towards the Library, & the practice towards the Cities. So, in very definite ways, we are making this Department as a miniature, indeed a microcosm, of the University itself, with its various Departments viewed as cooperating towards the cultural & social wholes. We cannot, & should not wish to, interfere with the *Lehrfreiheit und Lernfreiheit* of the Universities, nor introduce any pressure: but we can increasingly classify their departmental work, & express its essential ideas, as progressing towards more & more unified thought, & action also correspondingly. So this work has been good practice towards the University plannings, at Jerusalem & elsewhere, which have so long been in progress. Of course, an enormous amount of definite work remains to be done, but it is something to be able to locate say Einstein, Bergson, Freud, & so on, as well as earlier thinkers & workers, on the relevant tables, & their illlustrative walls & intermediate rows of screens. The Encyclopedia Graphica is thus in progress as an *Exhibition of Ideas*: the University intelligible to itself.

It is extremely difficult, of course, to write all this as I have found all my life, even with its relatively clearer fields, so far defined & investigated or pro-synthesised. But I have more hope of writing now in the next year or two – & so am beginning to see how after a few thousand pictures are at length passably arranged, to tackle many more thousands of pages of notes & fragments in all sorts of disorderly accumulation in stacks of boxes, here & at Tower, etc.

I wish you had been able to come – but I see how difficult – I fear impossible? Still, it will interest you, I think, this report of progress, so far. My son is with me this winter, & gaining health – & helping in some directions, though with education arrested & broken up by war years & invalid ones, he cannot easily take hold of the larger issues: nor have I any students yet able to do so, & though one, if not both of my best men, now at Aberdeen & Cambridge, may be prepared towards this, I fear I shall have to pull all this down again, if not also even leave India, before they get back. However I work away. But now off shortly for the summer: first to planning 3 towns in north – & then perhaps to my old friend Bose at Darjeeling, & if so probably with a visit to Tagore, now busy with his new University endeavour, on the way. Then back here for rest winter, if all's well.

Department of
Sociology and Civics
University of
Bombay

25 March 1922

[Ms UP]

Dear Mumford,

Your letters raise many points of interest, & I do not miss them, even if I don't (or can't) answer! e.g. As to masonry – Albeit symbolic now adays, it is no doubt spiritually helpful to its members so far. But why not now interest them in *real masonry*, via Town Plannning & Civic Movements? I have never become a mason, nor indeed joined any definite bodies I could help; but I am tempted, the more since the upbuilding of Jerusalem might be more or less made a name to conjure with! Is that so, do you think?

No parcel of books ever arrived, nor any plans, etc. I presume none got dispatched? But if you have been at any expense, let me know. I have only indirect news of Branford this long while. But others tell me he is ever so much stronger & happier now that he is well again, & of course busy in his country home, in a well isolated & pleasant locality. I think the *Sociological Review* keeps up fairly in quality, is indeed improving – don't you – but the larger changes, etc., will be needed before it wins a circulation!

Sorry press of work has kept me from writing for *Menorah* and all other journals, etc., too. But if editor notes my paper (of October 21) in *Contemporary Review*, he may find some passages worth extracting, & I'll write something if possible next winter, if not sooner.[1] My paper to Town Planning Institute herewith in duplicate. It may be passed on to any quarter caring to use it for extract or review.

Very interesting your educational experiment of last year – I hope being repeated this? I have been struck by vivid interest of my town students here in a short visit of Village Survey to outlying country: just as of old at Summer Schools at Edinburgh & am thus encouraged to precede my next winter term with a fortnight in the open.

Your criticisms of limitations of our "Edinburgh School of Sociology" are very just, & need utilisation: & I quite recognize some of course, the lamentable deficiency of writing – "*Le mieux – c'est l'ennemi du bien*," & I have gone on then forty years always planning & re-clarifying, but seldom getting a fragment into print![2]

Did I send you a number of *Indian Journal of Economics* of nearly 2 years back – (if not, see it in Library) with Part I of my *Outlines of Sociology for Economics*, or some such title? Part II, still very incomplete, will soon be out.[3]

Must stop for mail, but have written you another letter for Branford to send on to you, regarding Exhibition & other developments. (*No: here it is!*)[4]

Congratulations on your marriage – but tell me more.

Yours cordially,

P. Geddes

Miss MacLeod arrived here yesterday via Jerusalem. She speaks warmly of her meeting with you.

1. "Palestine in Renewal," *Contemporary Review*, 120 (October 1921): 475–484.   2. "*Le mieux – c'est l'ennemi du bien*": The best is the enemy of the good.   3. "Essentials of Sociology in Relation to Economics," *Indian Journal of Economics*, 3 (July 1920): 1–56, (October 1922): 257–305.   4. Apparently referring to the previous letter, also dated 25 March 1922.

Dear Professor Geddes,

143 West 4 Street
New York City

29 March 1922

[Ts NLS]

Your letter of 17 February came a week ago, and I am answering it now before plunging into the thick of a book that I have contracted to write. During the last couple of months I have been plannning and re-planning the immediate future; but as yet nothing seems assured except that I shall devote the next quarter to writing; and beyond that there are the alternative possibilities of joining the *New Republic* as an editor, or going abroad for a few months with my wife, chiefly in England and France, and keeping on at the old and uncertain business of desultory journalism. Branford tells me that he has been trying to awaken interest in a Bryce Memorial Fund to support a travelling fellowship for sociological reserch in England and America, with a view to bringing me over sometime as a fellow; but apparently his appeal in the *Nation* has not met with a very fervent response. In the matter of the Exhibition and the Social Studies department I don't know whom to approach: most of my friends in the New School are leaving it in June, and I am quite out of touch with any other academic circle; indeed, working alone like Teufelsdröckh[1] in top story of an old tenement that looks up town over the roof tops, I am alas! out of touch with almost every circle. It is hard to strike a balance between the cloister and the marketplace, and I can't pretend that I have succeeded.

The book came about very unexpectedly. My friend Van Wyck Brooks suggested that a book upon Utopias, which he had once thought of writing, would be just the thing for me to take up; and immediately I saw that there was a possibility of combining an historical sketch with a present day criticism of all the social philosophies which have so miserably collapsed during the past decade. I took the idea to Mr. Horace Liveright, and he offered to accept the manuscript without reading and give me $300 in advance royalities – enough to enable me to abandon, for the next three months, my regular literary work.[2] If it had not all come about so suddenly I should have written to you for criticisms and suggestions; as it is, I can only send you a chapter outline of the book as it now stands, and trust that, if you are stimulated to criticism, your answer will not come too late for me to take advantage of it. The general thesis of the book will, of course, follow the lines you have laid down: from Ou-topia to Eu-topia: it will attempt to establish the validity of the Eutopian method (all-round survey, plus projection into the future), and incidentally it will aim to give a coup de grace to the various abstract and partial philosophies that have hitherto held the field. A big order, of course; and the book will be nothing more than a rough outline – a sort of reconnaissance of the terrain. Among other things, I have discovered a remarkable Utopia by J.V. Andreas, a friend of Comenius, which I think ranks much higher for sociological insight and constructive criticism than the work of either Bacon or Campanalla.[3] This

"Christianopolis" was exhumed from Latin by a young American Ph.D. in 1916, and it is about time that it was more widely noticed. Andreas seems to have been responsible, through his correspondence with Samuel Hartlib, for the founding of the Royal Society; and it is interesting that he warned his colleagues against the dissociation of literature from science, a warning which the hard-headed English physicists alas! failed to heed. In the treatment of Coketown[4] and the Country House I am going to suggest that each great historic period has a real, and to a certain extent, a realized Utopia, implicit in its habits and its institutions and its experiments; a Utopia which is, so to say, the pure form of its actual institutions, and which may therefore be abstracted from them and examined by itself. To write a history of these pragmatic Utopias would be to present the historical "world-within" and thus supplement the conventional historian's account of the world-without. Until psychoanalysis claimed the field we did not sufficiently realize the importance of the world-within; or at any rate, we did not see that it had a directive function. (I realize that *you did* see this; what I mean is that psychoanalysis gave us the tools to explore this field more fully.) So it comes about that a great many of our Utopias are infantile, in that they seek to entrench what Freud calls the pleasure-principle, and deny the reality-principle: so the Utopia of the Country House, with all its comforts and luxuries, is an ideal whose fulfillment means a distinct loss of vitality. This is just the briefest hint of some of the trails that are opening. If nothing more comes out of the book, it will be great fun.

There is little real news about America, except that we are now in the throes of another huge technological jump. The success of the radio-telephone in long distance communication has now reached a point where it is a distinct rival to radio-telegraphy: and as a result thousands of people are buying radio-receiving sets, from twenty dollars to two hundred in cost, for the purpose of getting the weather reports, sermons, lectures, health advice, stock market reports, and what not that are broadcasted by central stations in Newark, Pittsburgh, Chicago, etc. The whole countryside is now in direct communication with the city: even in the remotest districts it will soon be possible for the farmer to get storm warnings at much shorter notice than the present service. Will not this probably give a new turn to rural life?

I am sorry that malaria has touched you; and I trust that you are now well-recovered. I still look forward patiently and hopefully to seeing you; and trust that the physical distance will not always be an obstacle.

Gratefully yours,

Lewis Mumford

THE STORY OF UTOPIAS[5]

1. The World as it is; and as it might be.

Man walks with his feet on the ground and his head in the air; so he lives in two worlds, the world of fact and the world of ideas. The history of each age is not merely the story of men's acts but the story of their projects. History tells us about what happened in the communities where men actually lived; how they worked the land and loved and built buildings and made themselves at home in an imperfect world. Utopia tells of the communities in which men have wished to live; it tells of the places in which all that was incomplete or ugly or harsh or difficult in their daily lives was improved by one means or another; in short, it tells of the world as it might be. Each great period in European history has had its Utopia, along with a great many half-Utopias. To see what these utopias have expressed, and to see what they have failed to reckon with, is to have a better grasp upon the drift of things at present. So the story of Utopias begins.

2.   The Classical Utopias.
The next three chapters deal with the Greek Utopias of Plato, the Medieval Utopia of More, and the Renaissance Utopia of Bacon. There will be excursions into other sources, and an attempt will be made to build up a picture of a complete society, putting in the things that Plato and More took for granted.

3.   The Lesser Utopias before the 19th century.
Harrington, Hobbes, de Mandeville, Fénelon, Campanella, Andreas, de la Roche, Mercier, et al.

4.   Modern Utopias.
Industrial: Owen, Bellamy, Fourier, Ruskin, Hertzka
Human: Morris, W.H. Hudson
Scientific: Wells, et al.

5.   The Criticism of Utopias.
Utopias of Escape and Utopias of Reconstruction. The first group offer a temporary world – like the world of literature and art – in which one may live a more interesting and vivid life than is possible in the "real" world. The second group has had definite effects upon human institutions and communities. Examples of realization.

6.   Utopia and the Half-Worlds.
The reaction against Utopias in the nineteenth century did not halt the efforts to create better communities: they simply encouraged men to substitute "half-worlds" for Utopias. The orrery of the intellectual life may be represented as follows:
Half-Worlds: *Utopia*
              Partial Ideals in separate compartments
              The World
Half-worlds: *Degrading Ideals*
              Hell
Criticism of present-day half-worlds.

7.   The Foundations of Eutopia.
The world as it is: Science
The world as it might be: Art
The application of art and science in a particular region: Eutopia.
"Dreaming" and "knowing" as ends in themselves; and their possible utilization in founding Eutopias. When entirely cut loose from social life art becomes irrelevant to the point of becoming in the end little more than the phantasy of the lunatic, whilst science, similarly unharnessed, contributes to

the scarcely less private world of the specialist. In the humanist tradition art and science, while pursued independently for their own sakes, are also carried over into the life of the community. This juncture of art and science, which has only recently become possible, is the pledge of Eutopia in a world which has hitherto been at the mercy of ignorance and the undisciplined expression of primitive wishes. From No-Place to the Good-Place. Examples: Geddes, Branford, Kropotkin, AE George Russell, Ebenezer Howard. Anticipations.

BEYOND UTOPIA

I.   The World as it is; and what the Utopians have made of it
II.   Plato
III.   More, Bacon, & Campanella
IV.   Johann Valentinus Andreae
V.   Vairasse
VI.   An Excursion into the Romancers
VII.   Fourier, Owen, Buckingham, Ruskin, and the Associationists
VIII.   Cabot, Bellamy, and the Collectivists
IX.   Spence, Hertzka, and the Land-Animals
X.   Mr. H.G. Wells and the Criticism of Utopias
XI.   How the Modern World Broke Away in search of New Utopias, and where the Trail Led
XII.   Megalopolis: The Utopia of Nationalism
XIII.   Coketown: The Utopia of Industrialism
XIV.   The Country House: The Utopia of Imperialism
XV.   The Utopians versus the Partisans
XVI.   Partisania: the Utopia of the Single-Track Critic
XVII.   A New Inquiry into the World as it is; and a new Route to the World as it might be
XVIII. The Foundations of Eutopia

1. Professor Diogenes Teufelsdröckh ("born of God Devil's dung") is the hero of Thomas Carlyle's *Sartor Resartus* (1836).   2. Horace B. Liveright (1886–1933), head of the publishing firm of Boni & Liveright which published Mumford's first three books.   3. Johann Valentin Andreae (1585–1654), author of *Christianopolis* (1619); John Amos Comenius (1592–1670), Czech–Moravian educator.   4. "Coketown": from the Industrial city depicted in Charles Dickens' *Hard Times* (1854); both Mumford and Geddes use Coketown as a generic name for the "paleotechnic" city at its worst.   5. This outline, with the handwritten annotation "original sketch," was apparently enclosed with the letter. Most of these utopian writers and thinkers are discussed in Mumford's *The Story of Utopias* (1922): Plato, *The Republic* (*c.* 375 B.C.); Thomas More, *Utopia* (1516); Francis Bacon, *The New Atlantis* (1627): James Harrington, *The Commonwealth of Oceana* (1656); Thomas Hobbes, *The Leviathan* (1651); Bernard De Mandeville, *The Fable of the Bees* (1714); François Fénelon, *Les Aventures de Télémaque* (1699); Tomasso Campanella, *The City of the Sun* (1637); Charles-François Tiphaigne de la Roche, *Giphantia* (1760); Louis Sebastien Mercier, *Memoirs of the Year 2500* (1772); Robert Dale Owen (1801–1877), active in the New Harmony (Indiana) and Nashoba (Tennessee) communities; James Silk Buckingham, *National Evils and Practical Remedies* (1848); Etienne Cabot, *Voyage en Icarie* (1848); Edward Bellamy, *Looking Backward* (1888); Thomas Spence, *Description of Spensonia* (1975); Charles Fourier, *Le Nouveau Monde Industriel* (1829); John Ruskin founded the philanthropic "Guild of St. George" in 1871; Theodor Hertzka, *Freeland: A Social Anticipation* (1889); William Morris, *News from Nowhere* (1890); W.H. Hudson, *A Crystal Age* (1906); H.G. Wells, *A Modern Utopia* (1905); George William Russell ("AE," 1867–1935), Irish poet and essayist; Denis Vairasse d'Allais, *The History of the Sevarites* (1675).

(I suppose this includes too much reflection of last, but it is well to focus clearly. PG)

Dear Mumford,

Here I am after a busy winter at Town Planning Exhibition, plus beginning of arrangement of my Department, no longer simply as Department of its subjects, but as also miniature of University, mobilizing its essentials, from the scientific side of course primarily, but not exclusively.

H.H. The Maharaja's
Guest House
Patiala, Punjab

15 April 1922

(but address
Department of
Sociology &
Economics as
usual, marked
urgent & please
forward)

[Ms copy Oslo]

Thus:

With each line of practice facing the Cities Exhibition.
With each line of theory facing the Library.

- Logic in theory & practice
- Math in theory & practice (metrics, graphics, etc.)
- Physics & Technics
- Esthetics & Art
- Biology & Biotechnics (Agriculture, Hygeine, etc.)
- Psychology & Education
- Economics & Politics
- Ethics in theory & in Practice (including Religion)

Activity — Behaviour — Conduct — Life

The above as you see is a large order, needing next winter's work to make it clearer; & infinite toil (& collaboration) to make it approximately efficient.

Here too a *very interesting set of problems of town planning practice.* An old Indian city grown up very irregularly within its fortifications round palace-citadel, & extending at various times with growth, up to 50,000, but not yet adjusted to railway & other modern conditions; & strongly mingled of wealth & poverty, magnificence & squalor, chaos & confusion! Outside is a stately ellipse of irrigation canal-boulevards, & inside are labyrinths of lanes between a few main thoroughfares – among these a multitude of old tanks filled up & drained off by engineers, now available towards three or more miles of park system, & so on. But with money to carry out improvements, this again being swallowed between urgent needs of government & magnificence of H.H.'s policies & equipment on all sides from ancient elephants & modern cavalry to Rolls Royce staff as well as stud of motors (fifty & more!).

Then 120 miles westwards a big railway junction, six or seven lines (Bathinda) growing towards modern considerable importance to be developed. I suggest the old world fair (which became Leipzig, Novgorod, Paris, etc., in Europe) & nether side to culture-gatherings (against Leipzig, Paris, Montpellier, Oxford).

Next a hill-station up towards Simla; perhaps reducing its congestion – & finally an old garden of the Moghal Emperors, with their noble terracing, for such renewal as may be. You see a varied & interesting programme for the next month or two, & a good change for my son, after the winter at Exhibition in Bombay.

But time is passing, and this country is not yet enough in modern world of ideas. Consider this great & terrible Mahatma Gandhi, who has perplexed government, & even alarmed them into imprisoning him. What are his ideas? Hence – viz.: (a) *Mazzini*:[1] *Nationality* 1848–70 (say 1860 at climax). (b) *Ruskin*: *Criticism* of mechanical revolution & wistful return to spinning-wheel, etc. (also 1860 in climax). (c) *Tolstoy*: *Peasant non-resistance* (same date). i.e. all two generations dead!

Why then government so alarmed? Because its ideas all *three* generations back & more. (a) *Liberal Parliamentary Representation* (1832). (b) *Pax Britannica* (1775–1830–57). (c) *Railway Development* (1846–). (d) *Commerce* (East India Company).

But now are just all revolutionary movements of the first order of antiquity (& senility) even Bolshevik Russia (Marx 1860 also) [...] governmental ideas of the second? Not simply of old European Empires, but of France & Britain too. Hence even President Wilson was of such startling novelty to them, with ideals of 1848, renewing earlier ones.

Where then is remedy & line of progress? & towards period of orderly construction? Obviously for me in taking later lines – e.g. Denmark after 1864 spoliation with Agricultural Development followed since 1890 or so by Plunkett in Ireland, by French syndicate agriculture, etc., & in Garden City developments (1910 ff.) with Civic & regional renewal increasingly organized together, as in these beginning here, as elsewhere. See again Ireland, between Carson (old Protestant fanaticism – 1618–48 & 1688) & de Valera (Republican fury – 1789 onwards).[2] And how much further on is even U.S.A.?

Do not all ideas need 2 generations to mature into wide public acceptance & practice? Recall Isaac Watt, twenty years after inventing before selling first engine – & slow development for long or Lister's surgery – "Scotch-fad" in London in 1870 & rejected by French army in war.[3] Then efficient for Japanese in 1904! but not for Russians, so that recent war was first in which all combatants were fairly equipped & expert.

Now if we have here as seems to me (cf. "Law of Generations") a fair approximation to rate of process by which ideas (of "cloister") come into action (in "city"), is it not up to us to be (a) clarifying our ideas; & (b) pushing on application in practice, so as to shorten the latent and germinative period? Rural (regional) development & civic renewal are thus, to my mind, the most intelligible, and the most advanced; moreover, they carry with them the various more specific developments of the Hexagon of Life = PWF: FWP. & thus the essentials alike of 'City/University = objective/subjective' in perpetual interactions. (*Vivendo discimus*.)[4]

Now I think we need to make our *sociological teaching*, our *civic & educational endeavours*, more simply intelligible in these ways, before we can reach the public we have been aiming at, e.g. in *Sociological Review*, in "Papers for the Present," in "Making of Future," etc. And further that with this sort of clearness we should accomplish something. But where are the men? And the women?

Our Sociological Society group is very small. Here (in Bombay, etc.) far smaller. My *students* are sadly spoiled before I get them – in *verbalistic empaperment*, *abstractional essay writing*, etc., worse than can be any European or American students & writers, trained though they mostly are in the same way. And (speaking quite straight) *how far do even you feel adequately possessed of (and by) the essential conceptions of the sciences – energic, evolutional, psychologic* – to apply them in the social fields? Keep clearing them up!

Of course we are all more or less in the same incompleteness. *But among the blind, the one-eyed are kings*! So come along, and let us do what we can!

In short then, I now expect more clearly the suggestions of my last letter – that – despite the temporary separations from home (which I have once & again had to make, as in first journeys to America & to India, & which were always rewarded by the happiest of reunions) – you should come there for next winter!

Miss Macleod (whose intuitions I have more than 20 years experience of trusting) writes me from Calcutta, that she has written to such relevant friends & acquaintances as she has in U.S. (as of Russell Sage connection?, etc.), urging that you be sent on Commission to report to such bodies as Town Planning Exhibition, & Sociological & Civic Department with a view to utilization of what may be found suggestive.

That is of course a much stronger position to be in than as Wander-student simply, as press correspondent in India, though both facilities remain, & better than alone. It may also remove economic difficulty. If not wholly, then I'll do all I can to meet balance, e.g. arrange for your pleasant quartering in college in Bombay, like son & self, & contribute what I can to expenses of journeys. When you tell me (a) if you can come, & (b) if & what you'll need beyond such help as you may find from bodies above indicated, I can then by that time (c) cable clearly what difference I can make up.

Of course you will understand all this as a business proposition & towards co-operative advantage. Also as towards fuller co-operation. (You need not fear my son will be in your way! You are older than he, & he is still younger than his years (26½) having lost his regular education through War years & prolonged convalescence still incomplete: but having of course some compensatory experience & entry towards sociology & civics in reconstruction work in France, labour in Highlands, etc., & music in Lowlands & South. His plan is to work for the Doctorat Etranger at Paris (which admits by a good thesis, despite irregular academic record & so may appeal to you also); & also qualify to tote round the Exhibition wherever it may be wanted for a year or two. But you are ready for writing, as he will not be for a while yet. I think too you will get on together, the more since *your qualities & outlooks not a little correspond to those of our lost Alasdair*.)

I should value a talk with Branford & Farquharson also, if it were possible, and as nearest approach to this I send this via Branford. I think to begin with we should arrange for a mutual editorship on two sides of water, so as to adapt our stuff to these different markets. (Has the Home University Series, so successful in Britain, much market with you? Is Professor Brewster, its American editor, effective in this? I have never heard one way or other.)

Your criticisms of our "Edinburgh School" sent in reply to my pressing questioning, are I think *quite sound*. The needed economic presentment, for instance, is however too large to talk over at end of a letter; but I think we are more or less alive to it; and if you can come, as I hope, we can go into this. I have plenty of notes towards it!

I think I can also arrange that you give such course of lectures as may be; not I fear as an appreciably paying proposition; but useful alike towards clearing up ideas, and towards academic record. For I have found the alternating life, of professor & practitioner, which I have always managed to keep up these 40 years & more, very much more suggestive than can be either alone, and I submit the principle for your consideration, as well as an element in University reform. The medical Faculty has always done this more or less, & the engineering, agricultural, architectural, & other schools also so far express it. Why not sociological & civics most fully of all?

Reply soon to yours.

Cordially,

P. Geddes

1. Giuseppe Mazzini (1805–1872), Italian patriot and republican leader.   2. Eamon de Valera (1882–1975), Irish nationalist; leader of Sinn Fein party.   3. It was James Watt (1736–1819) who invented the modern steam engine; Isaac Watts (1647–1748) wrote a number of popular hymns. Joseph Lister (1827–1912), British surgeon who introduced antiseptic surgery in 1867.   4. "*Vivendo discimus*": we learn by living (Geddes' motto). Geddes sketches a hexagon with the sides labeled as follows: "Sociology, Ethics, Psychology, Esthetics, Physics, Geography"; on the side he has listed in column: "Conduct, Behaviour, Activity."

H.H.'s Guest House
Patiala (but always
University Bombay)

5 May 1922

[Ms copy Oslo]

Dear Mumford,

Your budget of 29 March just to hand, & to escape delays, I reply at once. First thanks for interesting printed matter, & picture to mount & hang in Cities Exhibition – Cathedral of God & Mammon, Ltd. (or rather Mammon & godling!). Of course the world has been living in this, for a hundred years & more increasingly; but it is most cheering to see it as an actual design, & I hope to be erected! For it has been a prime function of Architecture history to build the tombs of the departing – witness Pyramids & Fortresses (of the previous war experience) or again the liberal crop of new churches during the culmination of the Victorian & kindred pre-war period – or most imposing of all, the New Delhi – a vast series of palaces (sarcophagi) of centralizing bureaucracy, still incomplete, though decentralization to Provinces has been established, & with further decentralization of course in view.

Thanks too for papers, extracts from your own writings, & others. Will you allow me this further request? Herewith a small cheque (which will I suppose realize anything between $15 & 20) to cover expense of a few more such little purchases & postages – just what you happen to come across – for in India we see 0 of American publications! I suppose I ought to subscribe to *The Survey* & other papers – which do you advise? Pardon my thus troubling you.

Utopias – excellent: there is certainly scope for such a book, & I hope it will be taken up on our side of water, & translated to into French & German – why not even Russian!

With my little recent reading, & much forgetting, I do not recall any important Utopia you omit from sketch-plan. Save perhaps *Nova Hierosolyma*, which was reprinted in England some twenty years ago, & attributed by its editor

(I forgot on what grounds) to John Milton! (I can't say if it is important, for though I bought it, I fear it got snowed under. I'll ask my daughter in Edinburgh if she finds it, to send it to you.) Not much in Harrington's *Oceana* (but Hobbes' *Leviathan* has peculiarly been realized since.)[1]

Ruskin's *St. George's Company* you will find described in *Fors Clavigera*. At any rate the Sheffield Museum is identified with it, & the Curator might tell you of it. (But the small adherence gave a sad sign of how little the cloister can really accomplish!)[2]

You have had of course a great crop of Utopian communities in the U.S.A., from Brook Farm, & from Mormons to infinite sects; & these you will no doubt do justice to. (I think one Hepworth Dixon (?) made two volumes on them long ago in England.) Was not Owen, with his New Lanark, & then New Harmony, the great pioneer? Also Fourier with his Phalaristeres – that of Swiss is still going on, at any rate I found it moderately prosperous & busy with iron work about twenty years ago.[3]

I fancy Ebenezer Howard was stirred by these, as well as by Ruskin.[4] (He was good enough to tell me once he had been encouraged by my beginning long ago in Old Edinburgh.)

There are no doubt a good few papers in old reviews, etc., on Utopias (cf. Poole's *India*, etc.) – but probably not very rewarding. (Thus S.W. Slaughter, I think, wrote a paper for Sociological Society.) Your treatment will be fresher – as the older criticisms will mostly be pre-psychological; and your line is probably here practically new. As associated with it, of course, you will note Sorel's [...] treatment of *myths*, as working conceptions & collective representations. Your political treatment also sound: each phase realizes its Utopia as far as it can.[5]

I don't agree with the ordering of your chapters – though doubtless you have your reasons for it. Thus the Country House is not really *modern Imperial*: but of *Renaissance* – & from plunder, etc., of Monasteries at Reformation. "Coketown," & its *Manchesterismus* are of the *Industrial Age*, as much as Owen's more idealistic endeavours, as at New Lanark. (Recall too, in fall of this, the "Gothic Renewal.") This normal decadence of the Country-House in face of this industrial age is well described in Wells' *Tono-Bungay*. And the present revival of it in England, etc. (& I fancy with you also), is but the snobbish romanticism of the profiteers & financiers, for whom with us it is the proper thing to call in Lutyens for the house, like Rolls Royce for the motor-yard.[6] (Hence surely these chapters (except last) should come in before Wells, etc.'s, criticism of Utopias?)

Again in the succession of political changes there are endeavours towards specific Utopias. A great example was the Commune of Paris (1871); & the reaction & disappointment from this was a great factor in the cynicism of of classes & the bitterness of masses thereafter – the decline of conscious Utopias accordingly.[7]

Megalopolis is latest, great development & is not merely of *Nationalism*, as we use this, e.g. Dublin, Prague, Warsaw, in limited scale, but more of *Imperialism*, as for London, Paris, Berlin, & Vienna, Petrograd, Constantinople, etc., before War (and has not New York City an ever-enlarging world-empire!). *Cosmopolitanism* with its large Jewish factor, has also played its part very considerably.

Branford in these *Whitherward* papers has put very clearly, I think, the economic succession of Ricardo & Marx, & thence passes to his Financial Utopia – of "Sabbatical Year for Finance" (& former papers): then again he goes on to Regional & Civic Eutopia, with its survey & service.[8] But after all, beyond these, two great Jewish economists, we should give credit to Hertz & his Zionists as the third: for it is in Zionist Palestine that we have the most definite & concrete endeavour of today – uniting the dreams of Israel in Egypt & in Babylon, & the many reconstructions of Jerusalem even long before Plato's day![9] Central indeed to all Utopianisms is this Jewish endeavour & ideal in one!

(Here too a word for "the Kingdom of Heaven," & its preacher as concrete – before side-tracked into post-mortem *"Heaven"* by Paul & by Augustine, in their despair of Rome.)

You speak, as I too so long have done, especially of knowing & doing, Science & Art. But *feeling* colours & dominates all these – and thus Utopias are *religious*, in the wide sense, of *"Religion"* as synthetic from the theoretical side, as *"Politics"* from practical: while Etho-Polity (or some such term) is needed for their union in life. (Hence of course Branford's insistence on *religious* point of view – though I feel it is deterrent to those sick of jazz-houses!)

In your final list *Kropotkin* (with whom pray name *Elisée Reclus*) is of course earlier & independent of Howard, & of town-planner generally. (Le Play too, though too concrete to speak of Utopias, was essentially of that mood, no doubt pretty paternalist, as with adherents of *"La Réforme Sociale"* (the journal & group – Comte, de Mun, etc., which preceded Demolins' fresh start of *La Science Sociale*).)[10]

(You will not of course forget the paternalist employers since Owen – Pullman, etc., with you. Cadbury, Lever, etc., on this side, & the like.)[11]

Before Horace Plunkett & A.E. (George Russell) was the bright endeavour of *"Ralahine,"* an estate with model landlord in Ireland; the book of that name (I forget author) may interest you.[12]

Note moreover the splendidly realized Eutopia of *Denmark* after the defeat of 1864, & loss of Schleswig & Holstein, by Bishop Gruntvig, & a squire whose name I forget, with their vow to revive the spirit of the people by their story & song, in Folk Schools – for adolescents male & female alike – & to make agricultural Denmark not only better than ever but even bigger (by reclamation of moors & marshes).[13] This is of course the great European example, though agriculture, banking, etc., have had (more or less) independent origin in Germany & Italy; & *"Sindicats Agricoles"* after the war of 70–71.

In my first student winter in Paris (78–79) my persistent "spiritual home" was largely formed in the atmosphere of *"Il faut refaire la patrie!,"* which was nerving our leaders like Lavisse, & explaining the energy of Pasteur, Baihaut, Lacaze, & many more.[14]

Finally, can we not define, even more clearly & fully than in the writing of the Regional & Civic movements hitherto, the Eutopian policy – this presentment of civism as the third alternative from "Empire" & "Socialism," from reaction & revolution?

Here for instance I am trying to envisage my present problem of an old city subsiding into decay, despite large expenditure of State Capital. For the (say) 50,000 people, there must be at least 5 or 6000 houses, great & small, & thus with accommodation which has cost several millions sterling, & would need more to build now. Yet over 1000 ruins, & much disrepair! So the problem is to repair, renew, revive, to re-appreciate the property, as a semi-derelict estate, & thus not merely or mainly from State Exchequer (as a mere over-strained Taxation Bank) but by arousal of citizens. Arousal for the commercial economy which is so weakening them to the regional & civic economy they had of old, & new again on the modern spiral.

But this is more or less the problem everywhere, particularly conspicuous in Ireland, where I would fain be, & I hope to go to next summer.

Still each town has its own Civic Situation (though these are largely akin, it is with their distinctive elements we may best work) & to descry this, as the interpretation of our survey, is the essential problem in any city – & the opening of its true policy. If we could work out clear examples of these – & not until we do (& in better "Reports" than mine as yet!) – we have the Civic Etho-Politics we are searching for, have we not? Your Appalachian project comes in well also as example of conurbational & rural policy.

Yours ever,

P. Geddes

(In the coming Utopia – world Utopia – *"there are many mansions."*)[15]

1. *Nova Solyma, the Ideal City; or, Jerusalem Regained*, anonymous work in Latin of mid-seventeenth century; attributed to Milton but probably written by Samuel Gott.   2. John Ruskin, *Fors Clavigera* (1871–1884), a series of letters to "the workmen and labourers of Great Britain"; in 1871 he founded the Guild of St. George, an organization whose members pledged to give a tenth of their fortunes to philanthropic causes.   3. William Hepworth Dixon, *New America*, 2 vols. (1867). Robert Owen (1771–1858), philanthropist and social reformer; established a model community for his mill workers at New Lanark, Scotland; his son Robert Dale Owen (1801–1877) was a leader of the less successful utopian community at New Harmony, Indiana. Charles Fourier (1771–1837), French socialist author who proposed "phalansteries," socialistic-communal units of 1800 persons.   4. Ebenezer Howard (1850–1928), originator of the garden city movement, founder of Letchworth and Welwyn garden cities in England; author of *Garden Cities of Tomorrow* (1902).   5. Stanley Lane-Poole (1854–1931), author of *Medieval India Under Mohammedan Rule* (1903). S.W. Slaughter, "Psychological Factors in Social Transmission," *Sociological Review*, 1 (April 1908): 148–157. Georges Sorel (1847–1922), French philosopher and author of *The Illusions of Progress* (1911) and *Reflections on Violence* (1912).   6. Edwin Landseer Lutyens (1869–1944), served as chief architect for the New Delhi planning commission, 1913–1930; designed buildings in a grand European style.   7. The Paris Commune of 1871 (March–May) was a municipal council which sought to decentralize Paris and to effect economic reform.   8. Victor Branford, *Whitherward? Hell or Utopia* (1921).   9. Joseph Herman Hertz (1872–1946), British Zionist.   10. Peter Kropotkin (1842–1921), Russian anarchist and social thinker; participant in the Edinburgh Summer Meetings; proposed "mutual aid" in place of Darwinian "struggle for existence" and "survival of the fittest"; author of *Mutual Aid* (1902) and *Fields, Factories, and Workshops* (1913). Albert de Mun (1813–1914), author of *Ma vocation sociale* (1908).   11. George Pullman (1831–1897) established the Pullman Palace Car Company and built Pullman, a factory town, south of Chicago. George Cadbury (1839–1922), director of the family chocolate manufacturing business; built Bournville near Birmingham as a model factory town. William Hesketh Lever, Baron Leverhulme (1851–1925), successful soap manufacturer who built Port Sunlight, Cheshire, as a model factory town.   12. Ralahine (1831–1833) was an experimental, cooperative agricultural association in County Clare, Ireland. The reference may be to Thomas Craig, *The Irish Land and Labour Question, Illustrated in the*

History of Ralahine and Co-operative Farming (1882).   13.   Nikolai Grundtvig (1783–1872), author of *Northern Mythology* (1801); established rural "Folk High Schools" in Denmark. 14.   *"Il faut refaire la patrie"*: We must remake the country. Ernest Lavisse (1842–1922), distinguished historian who led reform of teaching methods at the Sorbonne; Louis Pasteur (1822–1895), celebrated bacteriologist; perhaps Minister Baihaut (the name is difficult to decipher in the manuscript), engineer convicted in the Panama scandal (1892); Félix Henri de Lacaze-Duthiers (1812–1901), naturalist with whom Geddes had studied at the Roscoff marine biological station, 1878–1879.   15.   Adapted from John 14.2: "In my Father's house are many mansions."

Burford,
Oxfordshire

10 October 1922

[Ms NLS]

Dear Professor Geddes,

I have just come, by bus and foot, from the Oxford conference on the correlation of the social sciences, and I'm glad to breathe the clean air of the countryside after a weekend of metaphysical must at Oxford. Through attending most of the papers, through lunch in the study of one of the dons at Balliol (Urquhart) and through taking the rest of my meals at the Hall in New College, I got a clear insight, for the first time, into the great handicap on sociology, & scientific thought generally, through the dominance of the Oxford tradition. The fires in these Oxford studies are too comfortable, the ale is too musty, the books are too convenient to reach, and in this kind of atmosphere it is much more entertaining to go off on the trail of purely verbal hares than to undertake the very re-adjustment of physical posture that the scientific method, with its diagrams, tables, surveys, rambles, and observations, enforces. Sociology, as far as Oxford is concerned, is in the same position it was when the Society was founded; and alas! Oxford is England. Everyone said what he might have been expected to say; and no one had bent his wits to the problem of correlation. I am not sure that even Branford & Farquharson had clearly grasped the end that they might legitimately aim at, or had adequately envisaged its conditions. To Farquharson the problem of correlation seemed to mean a re-adaptation & revaluation of categories in leading ideas – in this fashion:

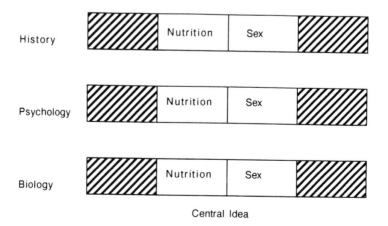

History · Nutrition · Sex

Psychology · Nutrition · Sex

Biology · Nutrition · Sex

Central Idea

Branford, on the other hand, wavered between two courses: (1) the suggestion that history & geography would come together, presumably with other sciences, through the detailed survey presented by successive city plans. (B. was sidetracked into this by a suggestion from H.J. Makinder.)[1] The other notion was that the correlation might be established through relation to the book-case diagram – How? Whence? What? Whither?

Both of these schemes of correlation – and no one else pretended to have any – seemed to me to ignore the context of each science in dealing with its grammatic structure. The problem of correlation is a problem, is it not? of making the "social" sciences, with their vast acccretion of factual data, genuinely sociological; and of including other sciences, in so far as they are germane into the synthesis. [. . .] Now, in order to make the social sciences come together [. . .] it is necessary to provide each science with the same unit of investigation, namely, a community; or society, for otherwise their results will be incommensurable, and the sciences themselves will not join up. I suggested, as well as I could in the five minutes at my disposal in the conference, that the most profitable subject for discussion was the unit of investigation – was it a valley-section or a city-region, or a city, or what other definite unit could be taken, by each and all of the sciences, as a field of investigation for social phenomena? For lack of such a unit the historian, for example, at one time deals with the inhabitants of a given area, at another with a certain ethnic stock, and at a third time with a miscellaneous group within the far-different boundaries of a state – a chaos which has resulted in a pretty full falsification of the history of various communities in western Europe, by the eloquent scholars and patriots who write with their eye on the founding, or myth-making, of the National State. Once we have agreed upon the nature of a community, the work of co-ordination becomes fairly simple; for what is relevant in each science or discipline is that which enables one to grasp the sociological complex; and the point of correlation is the society itself.

I pass from the immediate stimulus of the last few days to an account of the past six months. In order to finish *The Story of Utopias* before I left for England and in time for the autumn publishing season, I had to task myself a little beyond my ordinary energies; and by the time I set sail I was badly under the weather. An illness of my mother's during May & June had made matters a little worse by using up all my hours of recuperation. We landed in France early in August, spent a desolate fortnight in Paris, and another fortnight in the Austrian Tyrol; & since 8 September we have been in England. The chief obstacle to going to India is my precarious financial position, coupled with the fact that, unless something intervenes to prevent it, I am slated to become an editor of the *New Republic*, alternating in half-year periods with my friend, Robert Lovett, who used to be editor-in-chief of the *Dial*.[2] If I get this position it will mean both financial security and some leisure – also babies; and also the possibility of joining you with a clear conscience about my other responsibilities; and I could not, without a very generous endowment, make my way at present to India & stay there even four months. I realize all that I have to gain from you; and all that I am losing by these successive postponements – indeed I feel it more keenly after my talks with Gladys Mayer than perhaps ever before: but it is hard, when one has an individual orbit to bring it in relation to the separate system in which you move.[3]

One word more; and this perhaps the most important. Branford tells me that Unwin will try on his present visit to America to persuade the Russell Sage

Foundation to invite you over to make a reconnaissance survey of New York. About this I wrote to C.H. Whitaker, the editor of the *Journal of the American Institute of Architects*; and this is what he replied:

"As for Geddes – I am taking the matter up with Clarence Stein, Chairman of the Committee on Planning. He has considerable influence. At the last meeting of the Committee in Chicago we all felt that if the Sage Foundation was to lay out a plan for New York with 50,000,000 people, our job was to show the folly of such a dream. And I cannot imagine Geddes participating in so colossal a scheme of centralization. *What would he want to come over here and make a survey? We might get him over independently of the S.F.* and now that they have called in Unwin, they may not feel like adding another. In fact, it would seem that the choice of Unwin was a fair indication of their point of view – 'practical' – so to speak, & having nothing to do with such philosophy as that of Geddes."

To which I ask: (1) Will you come over? (2) If not invited by the Sage Foundation, would you come at the invitation of the American Institute of Architects & (3) What would you charge?

I return to New York on 4 November. So please send all letters, until I can give a new address, to 100 West 94 St., New York, N.Y.

And believe me in the meanwhile, dear Master, Faithfully & gratefully,

Lewis Mumford

P.S. Apart from all the other reasons for urging you to come to New York, there is this motive: it would enable me to put at your service the whole body of detailed knowledge of the city & region which at present lies buried in my portfolios – especially its history. What better beginning for a collaboration?

1. Halford John Makinder (1861–1947) helped to establish geography as an academic discipline and founded the Geographical Institute at Oxford; saw the significance of geography in political and military terms; proposed the "heartland" theory of power in Europe.   2.   Robert Morss Lovett (1870–1956), editor of *The Dial* and later the *New Republic*; professor of English at the University of Chicago.   3.   Gladys Mayer, disciple of Geddes and member of the Sociological Society–Le Play House group.

Regional
Association
65 Belgrave Road,
S.W.1

16 October 1922

[Ms NLS]

Dear Professor Geddes,

This note is to clear up various odds and ends forgotten in the letter I posted you last week from Burford; and first, to thank you for the letter which you wrote me last spring regarding *Utopias*; which came in time, and of which, I hope, I made good use. I used quotation marks or their equivalent whenever I was consciously quoting your thought, and for the rest I made amends in a final paragraph of acknowledgement – in short, I did everything but implicate you as a co-author. I've asked Liveright to send you a presentation copy; but when it will reach you I've no notion, for all my zeal and energy in getting the book ready has been annulled by the tardiness of the man who wrote the introduction – Hendrick Van Loon, the historian; and for all I know the book is not yet published. (Did you, by the by, ever get the copy of *Civilization in the*

U.S., which was also sent you?) Also let me acknowledge the cheque for five pounds upon which I shall draw for any material I may send you in future. (It was a good notion: my available current cash is so small that whenever there's a choice between buying & not buying, I don't!)

Our walk in the Cotswolds – Cheltenham > rail > Honeybourne > Chipping Campden > Moreton-in-Marsh > London > Oxford > Witney > Barford > Bibury > Cirencester > Chalford > Stroud > Gloucester > rail > Cheltenham was perfect: clear skies, green fields, mellowing foliage: warm autumn days, and a deep rhythm of companionship; it was all just this side of Paradise.

We go down to the Branford's for a few days this week; and then back to London for the final three weeks. Of which – more later.

Faithfully & gratefully,

Lewis Mumford

Dear Professor Geddes,

Aboard S.S.
President Monroe

17 November 1922

[Ms NLS]

On the way home I'm trying to collect my impressions of England; and since many of them concern the Sociological Group I'm sharing them with you: so that you may have the benefit of whatever my detachment may give – a detachment due to the fact that much water has swirled under the bridges on both sides of the Atlantic, and that I found myself – unexpectedly – something of an outsider. One of my surprises was to find that Le Play House had become an independent entity: the Sociological Society has ceased to be the central institution: it has become one, rather, of a body of organizations housed under a single roof. This is both a loss and a gain. It is a gain in that it makes it possible to use the house as a propagandist organization and to permit the Society itself to function as a purely scientific body; thus doing away, for example, with the clumsy apparatus of the "Cities Committee." Thus arises the possibility of a Le Play House "School," as a means of putting forward unreservedly and unguardedly the whole body of civic doctrine for which you and Branford stand. The loss is due to conflict between the societies, and more particularly, perhaps, between the personalities within the societies. (What follows is in strict confidence, of course!) The mixture of personalities at Le Play House now is an explosive one. Miss Loch & Farquharson promised, in 1920, to work well together in harness: Miss Tatton, however, is an intrusive element, and during the past year, I gathered from her, she has been driving F. to put his own (economic) interests, and hers, first; and thus to subordinate the work he has been doing for the society itself.[1] The conflict manifests itself in ridiculously petty and inconsequential details, and it's a serious question as to how long Miss Loch can carry on in the face of it. She is devoted to the Society; indeed, she is much too fully immersed in it, and it absorbs her more completely than a nest of children: it is absurd of her to go on however against an undercurrent of opposition, and it is hard to figure what would happen to the society if she should give up! Le Play House would go on, of course, but the Society, as Society, would be pretty near the grave. Branford, I think, is not seriously interested in the society, except as a body which supplies him automatically with an audience – albeit at considerable expense! He is really much more the philosopher of sociology than the sociologist; and, like most philosophers, he requires a complete agreement in doctrine as preliminary to any actual work; whereas, its seems to me, the sociologist would require only so much agreement as would be necessary to carry on effectively any particular line of research. Thus his fine humility in personal contacts is compensated by a certain arrogance in thought which keeps him from being seriously concerned with sociology *as such*, that is, as comprehending the work of everyone who is engaged in sociological research throughout the world. Branford, I believe, is the victim of a deep personal conflict: his own tendency is towards Philosophy, rather than science, and his deep loyalty to you – who represented the aims & methods of science, has kept him, in a sense, from following his own deeper trend. In fact, it is only through your long absence in India that he has gained the strength to set up shop independently – instead of going on as a sort of Geddes *minor* – and he is still not a little encumbered by the duty he feels to express *your* thought at the same time that he is setting forth his own. Thus in his new book, *Science and Sanctity*, there is a new and fertile contribution to philosophy in his chapters on visions: and if he were content to set this forth by itself, it would stand out for what it is: these brilliant chapters, however, are lost in a complete re-statement of the Geddes–Branford philosophy; and the result is, alas! a vast conglomeration, with neither beginning, middle, nor end. If B. could overcome this internal conflict he would (1) either drop the Society or attempt to make it a real meeting-house for sociologists, (2) either give up supporting the *Sociological Review*, or would use the review as a general organ of sociology. As it is, his life-work is now a series of ineffectual compromises, and their total effect is one of frustration. Dr. Bruck's book on the sociology of medicine is, I feel, a very bad book indeed; and I advised B. that it would be impossible to find a publisher for it in America.[2] B. was inclined to agree; but because Bruck is, in his own lights, a faithful disciple, he is determined nevertheless to put the book in the Making of the Future series – to whose profit? I wonder how far your own judgment confirms mine on these points? Miss Mayer, I believe, – and she is a very penetrating psychologist – feels the same way I do about V.B.'s inability to face out a situation. I say all this, too, in all loyalty and love for Branford: I have gotten much from him: if you have been my intellectual parent he has been my muse; and it is because this net of cross-purposes and conflicts makes it so difficult for me to co-operate with him that I have had to unravel it for my own satisfaction. Again I feel that the Soso would either stand on its own feet, or would commit an honest suicide, if B. could define effectually his own relation to the society.

At our meeting in Hastings Branford suggested that, as preparation for your coming over to America (if you do come!) I might write a little book about your town planning work; and I cordially agreed; to which end he handed me five of your smaller Indian reports. So far good. Can you turn over to me any photographs relevant to these Indian plans; especially any that may show proposed or accomplished alterations. Also, can you tell me if it's possible to get hold of the report you handed in to the Zionists; and if so, where; and may I quote from it? If not a book, perhaps I can publish a few articles in Whitaker's *Journal of the American Institute of Architects*.

So much for the present. More from N.Y. when I'm settled. I trust you are in good health.

Gratefully,

Lewis Mumford

1.   Margaret Tatton served as Director of the Le Play Society and organized various educational tours.   2.   Dr. Bruck's book was apparently never published.

Dear Mumford,

Department of
Sociology & Civics
Esplanade Road
Bombay

18 November 1922

[Ms UP]

I am sorry we again must miss meeting, as I quite understand you and your wife could hardly face a journey here (though why not some day if things develop!). I hope you both enjoyed visit to Europe? A brief letter from V.B. came in yesterday afternoon, telling me to send you all my old reports, over India – but no time for hunt, as this is mail morning, & interrupted by usual lecture. I'll forward next week, with suggestions, and map of India, marking cities studied and reported on – now many. Also a summary by years (though you of course need not follow this, but arrange as you see fit).

One idea you might consider is to take first the Town Planning & City Development & (2) the University & College schemes (with criticisms here & there, destructive & also constructive). This might in fact make a second part? – as your American College & University developments go on so actively, & in different sections of the public so far, though doubtless more in contact than in England.

Last week I was at Tagore's School & incipient "International University" – to plan for developments. What he really has are good beginnings of an *Agricultural College* including development of village beside (2) *Art School & College*, headed by brilliant man – Naudalal Bose, well seconded. (3) *Music School*, of which I cannot judge, but no doubt good (Indian music of course). (4) *Modern Literature* – Bengali, English, etc. (5) *Classics* (Sanskrit, Pali). I am pleading for *Drama*, & laying out beside Theatre Hall for School & University uses (1) a smaller open air theatre outside this, & (2) a bigger open air theatre at Agricultural College & Village, a mile away. I have a genius of an actor & dramatic student in my class here – better than all the 1500 Edinburgh & London Masquers (broken up by war, but now reviving in both places) and after a preparation of dramatic presentments of historical & social character through this term, I hope to persuade the poets to take him over – & give him direction of these theatres & dramatic branch. (Such masques ought also to be done at one of your film-producing centres. It is a great pity the Masques of 1912–13 were not filmed. But matters were not then so developed – & the idea never materialised.)

Then also, returning to University planning, I have to go at Christmas vacation to Hyderabad (Deccan) where the Nizam's Osmania University has a hilly site of 1200 acres, thus promising for architectural effect & grouping (though how far modifiable educationally remains of course to be seen).

I have never been more disappointed with University developments than lately – (1) at Benares (which I lost through an officially minded enemy, the

Principal of Allahabad) & now with 20 miles of roads – i.e. impossible distances & useless outlay & vast buildings an abominable failure between European & Indian, without good qualities of either but 25 lakhs debt & heavy yearly deficit! All conventional university ideas too – despite Indian aim! Then in contrast (2) Lutyens plan for Lucknow University – super-extravagant monuments to himself, in coldest refinement of Georgian architectural indifference to Indian & European alike, and of fabulous expense – with, so far as I could make out, no useful accommodation for for any department – & expansion unprovided, indeed made as impossible as cold façading everywhere can do.

Now have you not something of these errors also in your huge developments of University buildings in U.S.A.? (Whereas at Jerusalem I hide possibilities of economic extension everywhere, & thus the effects not spoiled by growth, but complemented, while the additions will be inexpensive.)

Above is an outline for *argument of a separate treatment* (1) *Cities* (2) *University Developments*.

Palestine & Jersalem might even make a third? & *(possible) Culture Institute* a fourth? – i.e. simple & effective developments of Tower – to big Museums, Exhibitions, etc., with simple Botanical gardens as from Dundee to Patiala, which has mainly occupied me this summer, & of which I'll soon send report. (Its *Preface* may serve to help your *Introduction*. Also simple *Zoos*. Did you go to Edinburgh? I hope so, for more reasons than that.)[1]

Here I see is our long-dreamed collaboration, beginning in very effective ways & with promise of work, indeed also for my very effective colleagues like F.C. Mears, & I trust also my son Arthur (some of whose simple but effective drawings of Patiala actual & cleansed of rubbish might enliven your publication? I am seeing about possibility of a reproduction in colour for my report.)

The *Times of India* – our leading Bombay paper – makes a feature of coloured illustrations for its Christmas & Summer special numbers & has ordinary skill in these. It might be cheaper therefore to have such illustrations made here? & they might possibly also bring out your whole publication here as a special number, encouraged by the interest in such work in U.S.A. & thus helpful to our work & practice here, which despite recent engagements (all from native states, etc.) is now much limited by the opposition created in official circles everywhere by my frank criticism of their too generally Haussmannic[2] spirit & methods –

Must close today – Yours cordially *P. Geddes*

P.S. But here I see a difficulty. How can you, especially now as married, face such unremunerative writing? and with a journey just behind you which must have cost far more than you earned in the time?

I can understand that you may wish to preserve complete independence, even of me, & not to feel paid by me for writing as you propose. But the idea is yours, not mine; if you are short, why not accept a loan of $500 or $1000, which I can quite well spare – & as some job may not improbably come from your publication, you may then fairly keep it as your first commission as a colleague of this firm – or if you prefer, realise that it will be in my will as a legacy to a spiritual kinsman.[3]

Some such arrangement will set you free a little more, to get on more rapidly with such writing, and also give me a chance of remunerative work in U.S.A. before I am too old. (68 nearly two months ago, though not yet feeling it, save for a few days at Simla in rains, & that followed by convalescence & bardic yells! (which V.B. may have shown you samples of).) (So pray let me have your assent, & I'll send first $500 by return.)

Yours,

P.G.

1. Geddes had planned the Edinburgh zoo. 2. Baron Georges Eugène Haussmann (1809–1891), city planner of Paris who dramatically widened old streets and cut new ones: in contrast to Haussmann's radical, sometimes ruthless methods, Geddes advocated the planning principle of "conservative surgery." 3. Mumford's annotation: "not accepted."

Dear Mumford,

Pity that letter of yours to me was lost. But you & Branford tempt me to go to U.S.A., arriving mid or end April, & free there till (say end) September!

You must do what you can for me in way of lectures in various connections – architectural, etc., XIX Century Clubs, etc., as at Boston – & I suppose at some Chautauqua organizations in vacation? And what of courses in Universities?

I used to get maximum $100 per lecture & minimum $50 per lecture in 1899 & 1900: so, if possible keep up figure! Yet at same time I have to make it clear to you as you to others that I lack the "popular gifts" of political, religious, temperance, & entertainment lecturers, & have only appealed with any success to moderate audiences, interested in the class of subjects I can deal with, & thus not great beyond the range of my voice, which can't reach over great auditoria!

I have written by this mail to my old friends: (1) Richard Moulton late of Chicago, now in England. (2) S.K. Ratcliffe (much lecturing in U.S.A.). (3) George Hooker, late of City Club Chicago at Hull House. (4) John Nolen, Town Planner Cambridge, Mass. (5) Dr. Devine formerly of Charities Bureau, now I think Russell Sage Foundation. (6) Charles Zueblin, active lecturer. (7) Prof. Mavor, Toronto. (8) Jane Addams, Hull House. (9) President Vincent, Chautauqua.[1] So, if you are really to run me you might drop a line to any you know of these, & draw them out. Tell me result. (10) The Zionist Organization may send me on to their connections in U.S.A., as they want their movement supported & popularized.

But what of advice to Cities on Town Planning? Or any brief courses in University & Technical College Departments or any University Planning? Clinton Meyers Woodruff of National Municipal Review (703 N. American Building?), Philadelphia, might help with that sort of thing, if you sent him a note of my proposed visit.[2]

Yours,

P.G.

Department of
Sociology and Civics
University of
Bombay
Bombay

25 November 1922

[Ms UP]

1.   Acquaintances of Geddes, several of whom he had met during his 1899–1900 visit to the United States. Richard G. Moulton (1949–1924), British proponent of the university extension movement; helped establish the American Society for the Extension of University Teaching; professor of literature at the University of Chicago, 1892–1919. Samuel K. Ratcliffe (1868–1958), British journalist and publicist; his 2 June 1931 letter to Mumford (Penn f. 4032) describes Geddes' tastefulness and hopelessly unrealistic plans in old age. Perhaps George Ellsworth Hooker (b. 1861), author of *Through Routes for Chicago's Steam Railroads* (1914). Edward T. Devine (1867–1948), American social worker who helped shape nationwide welfare policy; served as Secretary of the New York Charity Organization society; founded New York School of Philanthropy (later the Columbia University School of Social Work). James Mavor (1854–1925), professor of political economy at Glasgow University and later at the University of Toronto; close friend and correspondent of Geddes; discusses Geddes' Edinburgh urban renewal work in *My Windows on the Street of the World* (1923). George E. Vincent (1864–1941), sociologist and educator; President of the Chautauqua Institution, 1898–1907, where he developed programs to study contemporary social problems; became President of the University of Minnesota in 1911.   2.   Actually Clinton Rogers Woodruff, author of *A New Municipal Program* (1919) in the National Municipal League Series.

Department of
Sociology and Civics
University of
Bombay
Bombay

25 November 1922

[Ms UP]

Herewith (Alas, delayed till next week! without fail!) (Registered) Large Book Packet of Reports. (Addressed as this letter & posted at same time.) Pray enquire for it, if not delivered same or following day.

SUGGESTED ORDER OF TREATMENT

1. *Madras Cities & Southern India*, with (1a) *Temples of Southern India*. (The first is already marked for abridgement for republication; but never have I had time for this. You will of course abridge far further but keep this old copy for me in case I need it. Note as to Madras, that H.V. Lanchester President R.I.B.A. (a very good planner and friend) has continued there (as also at Lucknow, since I left. His work there is for New Town, as mine was for old).)

2. *Calcutta*, Bara Bazaar. (Principal Business quarter, proposed for costly desolation through widening & rebuilding, by Calcutta Improvement Trust. This alternative leaves old streets as far as may be, cuts new ones through inferior central property (saving two millions sterling, or more).)

3. *Dacca*, old Metropolis of E. Bengal, as separated by Curzon.[1] Government Buildings never used almost; now since reunion turned into a new University! (Hartog is principal & doing all he can.)[2]

United Provinces

4. *Lucknow*. Two reports – old city mostly. Cawnpore I don't send. There "Trusts" scheme demolished by a Report too drastic for publication & Lanchester brought in to do it sensibly.

5. *Balrampur*. Small capital of a non-ruling Maharaja, but immense land owner. (Agra – Taj Mahal, Gardens) (Report not printed).

Central Provinces

6. & 7. *Nagpur & Jubbulpore.*

8. *Central India*: Native State & City of Indore (Maharaja Holkar). *Indore* is of course the most elaborate of series. See *Papers for Present* – George Sandeman's

(Editor of *Harmsworth's Encyclopedia*) Review – as "Gardens of Paradise."

9. *Punjab, Lahore.* Curious report for detection of more howlers than elsewhere (Golden Temple of Amritsar (Sikh's Cathedral) is on hand for improvements but report not written yet).

10. *Kapurthala.* Capital of that native state (Maharaja of French Culture).

11. *Patiala.* Capital of that native state (Maharaja of very English culture, general in late war, etc.) with (11a) *Bathuida* big railway junction on East of State & Pujanr Old Palace & Garden at Old Town near Khalka junction towards Simla.

12. *Bombay Residency.* A little done for neighboring towns, Thana, etc., but more northward. *Surat, Broach, Madras, Ahmedabad*; (but I have not copies of these; shall send them later if possible, all quite short).

13. Also *Baroda*, capital of native state. "Gaikwar" is very progressive prince (a small volume done in conjunction with H.V. Lanchester).

14. Colombo: *Ceylon* a large report being sent you from Edinburgh.
Accompanying map indicates all cities studied in past eight years. The majority (but not all) have been subject of Reports, partial or general, mostly the latter.

Viz in Madras residency	13
Mysore State	2
Bengal	5
United Provinces	6
Punjab	6
Central India (Indore)	1
Provinces	2
Bombay, etc.	8
Hyderabad	1
	44
Ceylon	2

46 – Reports, or rather more safe to say about 50, thus making fuller approach to study of India than theretofore, with other cities not reported on. Also classes for Engineers, etc., at cities not visited, say 50 or more again.[3]
Some minor papers & reprints, some of which you have already. Oldest are letters to Nivedita 1900 – but I think not obsolete yet (Blue cover – *On Universities*).[4] (Some you can distribute to possible lecture wishers.)
I shall of course be looking up plans & photos to send to illustrate articles; but I don't think you need use very many of these, as after all the essence you will boil down out of these reports is not so much descriptive as argumentative – homiletic even!
Jerusalem University & City Reports will be sent next week: but pray note that this had best be left for a separate book I hope to prepare: also I have not yet asked permission from Zionist Organization to publish it with plans, etc. – (I daresay they'll be quite pleased.) Of course a general mention is all

right, list merely in course of your introduction – which may also mention Town Planning work for Edinburgh since 1886, Dumfermline, 1904, Chelsea 1906, Dublin 1911–14, etc. (International Town Planning Competition organized, & adjudicated with John Nolen of Cambridge, Mass.) (I am asking him to write you regarding any lectures, etc., he can suggest.)

You may also mention planning for Cyprus, rural conditions – forests, farms, fruit, silk, etc., in 1895, for though not then very far carried out, they have not altogether been without influence. And Rue des Nations scheme of Paris Exhibition 1900 – if you know particulars, though again only a succès d'estime.

Town Planning Exhibitions as follows: London 1910 / Edinburgh 1911 / Dublin 1911 / Belfast 1911 / Ghent 1913 / Dublin 1914 / Madras 1915 / Bombay 1915 / Calcutta 1915 / Paris 1916 / Lucknow 1916 / Magpur 1916 / Jerusalem 1920 / Bombay 1922–3.

1. George Nathaniel Curzon (1859–1925) served as Viceroy of India, 1899–1905.   2. Philip Hartog (1864–1947) served as Vice-Chancellor of the University of Dacca (1920–1925) and Chairman of the Committee on Indian Education (1928–1929).   3. See Meller, 338–339, for a bibliography of Geddes' India reports.   4. Sister Nivedita, the Hindu name of Margaret Noble, disciple of the Swami Vivekananda and admirer of Geddes, whom she met in Paris in 1900 (Boardman 180).

University of
Bombay

8 December 1922

[Ms UP]

Dear Mumford,

Much interrupted this mail: but I write mainly to repeat point in last week's letter – viz. that before getting your present address (c/o Boni & Liveright) I had posted a big packet of Reports, registered, & I think a letter also, to your old home address, which I hope you received. If not, cable *Geddes. University Bombay. Lost, Mumford* – & I'll do what I can to replace them.

Work continues to come in. Thus besides Tagore's new University & the Nizam's new University at Hyderabad, I have next to tackle a big zoo for Lucknow. I'll leave my son however to carry on as best he can, & still keep to American wander-student project. But with all this varied planning – now for about 16 different Universities & with a good few zoos since Edinburgh & Botanical Gardens too, do you not think I may find some openings in American Cities, & in lines less competing with your city planners, but rather co-operating with these? For experience shows all these things need not be so costly as either European & American custom makes them – & thus with deterrent effect upon smaller Cities & Universities. My old Dumfermline book (1904) which I suppose you have (if not get it from V.B.) indicates such things on too costly a scale (Carnegie's) (& with the hurried drawings of a very old fashioned architectural assistant) but since then I have learned a lot, & may claim economy as my essential specialism.[1] So the theme – that small cities can do more than they think – may be worth expressing. Must close for mail.

Yours cordially,

P. Geddes

P.S. I believe too that there is a great deal to be said for uniting the studious & the literary life with the practical – as circumstances have no doubt helped me to do: so why not think of our relation as something of a partnership, in which you no doubt are at first more of "the selling partner" – but may perhaps also take on increasingly technical interest? P.G.

1. Mumford had been well acquainted with *City Development: A Study of Parks, Gardens and Culture Institutes. A Report to the Carnegie Dunfermline Trust* (1904) since 1915.

Dear Mumford,   20 December 1922

                                                 [Ms UP]

Your lost letter of 10 October turned up yesterday after being through long delays!

I am writing in train to catch this mail (on way to plan Lucknow Zoo, & then go to plan Nizam's new University at Hyderabad) so excuse shabby writing (awful carriage!).

Very interesting account of Oxford Congress & Branford's & Farquharson's contribution as also your own, which seems to have *gone most to the point*, & put it well.

(If Farquharson stresses *Nutrition & Utopian* in Sociology (with omission if not exclusion of essential sociological factors, he is in what Comte calls "Biological Materialism" for Sociology, as this above all, is moving *History (past, present, & possible)*, & as Comte again made clear has its *differentia* from Biology in "historic Filiation," i.e. the accumulation & transmission, generation by generation, & day by day, of socially acquired characters). (You'll find Comte well worth re-reading, despite irritating style, etc.)

You don't mention if you, or they, found any kindred spirit at Oxford, so I fear not! It will be interesting to carry the war into Cambridge where the difficulty is not so much of pre-Germanic & sub-Germanic philosophy but of the corresponding sub-Germanic & dis-specialized ideas of *sciences*.

By the way, did you ever reflect that the current specialisms are not only the thought & skill of the mechanics for physical science, & of the rustics (so much more rarely as yet) for biological science, but also have a large element in them of the outdated & trifling, but infinitely elaborate, *Dilettantism* of their betters in the Renaissance country house – of course with their working strenuousness thrown in, and their puritanic asceticism too – much as their form of wealth worship turned to saving, & thus to capitalism. Thus the first fine workmanship of these days was jewelling (Benvenuto Cellini, etc.) & the first ingenious mechanisms, after Nuremberg watches, were toys – automata, etc., as per Vaucanson, etc., or Marquis Worcester's steam-engine for his fountains & his "Centurie of Inventions" – mostly toys too.[1]

No! Oxford is (mainly) South England. Common sense & science survive more in North – e.g. Manchester & Liverpool, Leeds, etc. And you Americans & other visitors, like Londoners & even London Scots like Branford & Farquharson, miss or lose touch with Scotland. It is worth recalling that continental ideas come there first, & for many centuries back. Hence in recent times Adam Smith to Kelvin or Lister; & Positivism reached England via the old Aberdeen Comte Society (of Bain, etc.), & thus to Mill (whose father passed on his Aberdeen education to him).[2] (Of course however Oxford poisons us all, more or less, as does London too.)

On your next visit to our side, go to North England to Scotland to Wales, & above all to Ireland, which I hope may soon take the lead her peasants, singers, players, soldiers, etc., have so long been preparing.

Very interested to know you have got such a large job as *New Republic*! (What has happened to Walter Lippmann?)[3] Yes, I hope it will give you both scope for work & means for home & more travel too, as its correspondent in your off-time from editorial throne!

Utopias arrived just before I started & one of my assistants ran off with it! I had only time to glance before it disappeared but I look forward to reading it in return – I especially noted your excellent improvement on such books, of bringing in *realized Utopias*, like Thelema in Country House, Hell in Coketown, & its Vanity Fair & much more in Megalopolis – with correlation of these. (Note that with Country House goes city Palace as from Italy – English of Renaissance were out to be "Italinate gentlemen." (cf. Browning's contrast of City Palace & Country Villa, "Up in a Village; down in the City.") Hence Country House as "Castle Rackrent," with ruin of peasant, Irish beyond all, & evolution, of servile snobdom in growing Megalopolis, & back to country.)[4]

Thanks too for you generous dedication & your p.s. too. Yes, I hope we may come to collaboration.

Note that Dumfermline book (*City Development*) can be had of John Grant Bookseller & Publisher Edinburgh (as remainder at reduced price I think 12/-) & Indore Book (Town Planning, etc.) from Batsford (our principal Architectural Publisher) High Holborn London WC for 31/6. Neither is out of print (would they were!).

I have not heard from Unwin, or Russell Sage Foundation, so I suppose that is off. I don't know what fee he got, nor what American rates are. (Nolen once said to me low, apparently lower than in Britain.) Here in India our rates are better. As Professor I have £960 for four (4) months (including 10 days Christmas holiday) say £240 per month (& with seafare at beginning & end of 5 years engagement).

On planning, Lanchester & I charge regular rate of Rupees 5000 per month – which was £500 when Rupee at 2/- (or even when at 2/10!) best; but now at *1/4 per Rupee* = £333 1/3. So when we have a long job (as I lately at Patiala) or a succession, as just now, we are not badly off, as profitable classes go. Only that is uncertain of course. If I had not so many "dead horses" to pay for, & "white elephants" to keep, I'd be able to do more to help causes: & that is now ready to come – though late in life!

So by all means "bring me over"! – though I've decided to come anyway! It was not to do New York City (which is a vast job!) but to go on as *wander-student*, & on such lecturing, & such possible consultations, for Reports from smaller cities, Universities, & Colleges, for Botanical & Zoological gardens, for Outlook Tower, open air Theatres, & Masques, or what not, as might turn up!

Still, if you can arrange anything, I give you full powers as my agent & partner, & I accept what you can arrange, terms included. Only if there be anything to do in N.Y.C. – I think it should be end of stay, say August & September – rather than May, June, while Universities are going on & I could visit (& lecture) there, & occupy vacation in city study towards report. And, if with you, excellent!

Note however the big sociological difficulty about any Report on New York. If regionalism and Civics are coming on, Megalopolis' great day is ending! though New York as port, like London-on-Thames, will never be unemployed. Did I ever put this to you? As Schema:

Place: Polis, the city
Work: Metropolis
Folk: Megalopolis (Cockney Swell-head)
Folk: Parasitopolis (Bread & shows)
Work: Pathiolopolis (Culture of deterioration)
Place: Necropolis! (more or less)

Is not this the condensed history of the great Babylons of the past? of Rome above all – so that my old nineteenth Century plans have (the still unexcavated) Forum & Colosseum area marked *Campo Vaccino* – Cows Field! Note next however the history of Christian Rome at its best endeavours: Ethopolis (Idealopolis)/Sophiopolis (Psychopolis)/Eupolis (Temenopolis) though of course too much lapsing also into 4, 5, 6, above & 1, 2, 3 as well! See also Necropolitan stage now in full blast for Petrograd & Vienna – inevitable too for Berlin – ominous even for Paris & London. (And it is to Wells' sociological credit that he spoke out on like possibility even for New York!) (See his Washington letters or book.) So I should be like the fabled Balaam, who did not report as expected & *desired*!

Returning to business of more general tour, why not Russell Sage, or other body, give me some smaller area & city to report on (& especially if not wanted for New York City)?

I used to know Devine, 20 years ago, but have not kept up, so have not written him. He may however remember me. I wrote Jane Addams, George Hooker, etc.

And of course if work turns up, or seems likely to, of one sort or another, why should I not return the following summer? when free from India. Though work very interesting I can't do much with so few students & these spoiled in a sub-London cram-shop like an Indian University, & especially this, the most bureaucratized, etc., & least awakening (far behind Cambridge, etc.). Nor is working for British Bureaucrats or Indian Maharajas satisfactory, for different reasons, both so far fairly obvious. You have a freer & more open atmosphere to work in, by far. So while working powers last, I may be good for several seasons in States, & perhaps flee to California instead of Europe for winters.

What I'd like best would be a University to take me over, for say 3 years (half-years I mean) for *Sociology & Civics, Town Planning* – with freedom for practice as here (& always in Scotland) in the other semester (summer) leaving Assistant Professor for that. You do that, I suppose, with Professors of Architecture, Health, Surgery, etc.

But I'd need big space for this collection (Exhibition) which however would be found worth it, & always improving towards *Encyclopedia Civica*. This (& the lost) Exhibition have eaten up much of time & savings, just like

Outlook Tower, of which it is one of the developments, so I might sell it, after use, & some further improvements (though with regret at having to!). On other hand, it is just possible that Arthur may tote it round, as he is growing able to help me, as did Alasdair, & so can succeed to job. (I hope to leave him this summer coming whatever may be unfinished of my work for Tagore & Osmania Universities plus Lucknow Zoo – large orders to start with! – so if he can get through these, I shall not fear of him.)

On return to Bombay about 3 January I'll hunt up some more illustrative matter for your account of my reports – I think you'd better only mention my previous practice, e.g. in *Edinburgh* (Outlook Tower, Halls, Old Edinburgh, zoo, etc.), *Chelsea* (More's garden, Crosby Hall, etc.), *Dunfermline*, & Dublin Exhibition, International Town Planning scheme adjudicated, with Nolen, Catholic Cathedral (abandoned for time through Declaration of war – & alas followed by deaths of enthusiastic Bishop & Archbishop; & though my son in law Frank Mears is getting their successors on again, this new one may have to be on more traditional lines).[5] Paris Rue des Nations – 1900 & Exhibition 1916, & initiative from University Hall Edinburgh towards incipient Cité Universitaire, though they will have forgotten that. (Collège des Ecossais is an old dream of revival.)

Just as in U.S., no one almost will now remember Zueblin's enthusiastic account of Tower & Survey, or Dr. Johnson's (& Dr. Strong's) importation of Tower Surveys to former's book of Survey of New York, which I fancy had an impulse to your more recent & better ones. I don't go into priority controversies, but it may help relations with your movements of that kind if you express ours as from early 18 eighties, & as developing still as far as Regional Survey Association from Aberdeen to South, Exhibition, etc.; & resultant Regional & Civic Studies.

As to East, my first visit to Greece & Constantinople in 85–6, started my interest to activity – & my wife & I spent winter of 94–5 in Cyprus & with active endeavours, not very successful alas! But all that sort of thing was a preparation for India since 1914, also for Palestine in summer of '19 & '20. Keep however on your present yarn to India – big enough, is it not!

Enough today.

Most cordially yours,

P. Geddes

1. Jacques de Vaucanson (1709–1782), French inventor who made the first fully automatic loom. Edward Somerset, Second Marquis of Worcester (1601–1667), invented a steam water-pump.   2. Sir William Thomson, Baron Kelvin (1824–1907), Scottish physicist renowned for his research in thermodynamics and electricity. Alexander Bain (1818–1903), Scottish philosopher who advocated a psychology based on physiological principles.   3. Walter Lippmann (1889–1961), social philosopher and editorial commentator; served on the editorial board of *The New Republic*, which he helped to found, 1917–1922; author of *A Preface to Politics* (1913).   4. The reference is to Robert Browning's poem "Up at a Villa – Down in the City," in *Men and Women* (1855). *Castle Rackrent* (1801), novel by Maria Edgeworth, describes the dissolute living of eighteenth-century Irish landowners.   5. In 1914 Geddes submitted plans for a new Catholic Cathedral in Dublin; although the proposed site was purchased, the plans were never carried out.

Dear Professor Geddes,

7 Clinton Street
Brooklyn, New York

26 December 1922

[Ts NLS]

It is not easy to tell you how much pleasure and gratification your letter of 18 November gave me; and I shall do what I can to deserve the generous confidence you have shown in me. Coming on top of a rather desolate return to New York, in which complete financial prostration was topped off by a fortnight's illness, with a severe infection in my throat and mouth, your suggestion of aid in carrying out the work I had planned for the winter was something more than a tonic; and I do not scruple to accept it under one or the other of the heads that you propose. Before doing this however I wish to ensure the publication of the book; and so this morning I wrote a letter to my publisher, Liveright, outlining the book, setting forth its prospects, and proposing that he commit himself to publishing it. If I get a satisfactory answer from Liveright I shall cable you for the five hundred dollars, at any rate, immediately; if not, I shall wait until I have enlisted someone else's aid and interest. Needless to say, I can scarcely write about your work without writing about the course of modern thought and modern civilization; and so the book will necessarily be of a general nature, with all its particular points and applications derived from your work. This means that it will take at least seven or eight months for me to do the job properly, and so at earliest the book would appear in the winter of 1923. In my next letter, which will follow shortly, I shall send you an outline. Please let me strongly advise, however, that I have your permission to deal with your work in Palestine to the extent of at least five thousand words or so; for it is important to arouse an interest of a group like the Zionists in this general survey of your work; and a treatment as short as this would only pave the way for a more exhaustive account of your plans, when you succeed in putting this together.

Now as to your coming to America. I am going to have a conference with Whitaker, editor of the *Journal of the American Institute of Architects*, next weekend; and I will see what arrangements he can propose. I will also see Alvin Johnson, one of the editors of the *New Republic*, who is now managing lecture courses at the New School for Social Research, and see what I can do with him about arranging lectures or study groups. I strongly advise *against* coming over to America if you will have to rely upon popular lecture audiences for an income. This is a special field in itself, and it requires a degree of mediocrity that you do not possess, combined with vocal qualities which also, according to Branford's report, you do not possess; moreover, lecturing is an exhausting business, and if I arrange any at all for you it will be with groups of not more than fifteen, or at most twenty people. Put Chautauqua out of your mind, even if any one of the friends you have written to is able to arrange it. Unless you can get an honorarium as consultant from the Institute of Architects, from the Sage Foundation, or from some city planning commission it would be better to abandon the American trip altogether. If during the next month I can arrange for you to get as much as an initial $1500 dollars you can probably count upon other fees coming in as soon as you are on the scene; and in this case I shall advise your coming. If there is not sufficient support however I shall cable "No," and if you come in spite of that I shall piously wash my hands of the results! Again, I warn you not to be lured by lectures; unless they can be arranged for in special societies or in the university, they will only lower your status here and hinder your work.

So much for the present.

Yours in discipleship,

Lewis Mumford

Lewis Mumford AND *Patrick Geddes*

THE CORRESPONDENCE

1923

7 Clinton Street
Brooklyn, New York

7 January 1923

[Ts NLS]

Dear Professor Geddes,

Branford's letter announcing that you had decided finally to come over this spring, and your recent note of 2 December make me see that I had misinterpreted your letter of 25 November, in that I had taken it to mean that your coming was contingent upon your ability to cover expenses; and so my alarms and warnings and intimidations were quite beside the point. I still cling to the point, nevertheless, that lecturing at large, to audiences not yet educated up to your ideas, is a wasteful and devitalizing process, and in the contacts which I have just endeavored to make for you at the University of California, Wisconsin, Michigan, Dartmouth, and the Massachusetts Agricultural College I have emphasized the fact you would speak to *small groups of students.* Do not depart from this program, no matter how attractive the offer.

Whitaker and Stein, my friends in the American Institute of Architects, were both pessimistic about getting commissions; and they pointed out the fact that Adams, since leaving Canada, has had only a few odd jobs to do, in spite of the fact that he has spent a good deal of time and tact in building up connections in the U.S.[1] Last week, however, through Mr. Allen Eaton, a friend of Miss MacLeod's, I got in touch with Flavel Shurtleff, who is secretary of the Russell Sage Foundation Committee on the Plan of New York and its Environs; and he was much more encouraging.[2] They spent a considerable sum to get Unwin over here; and they wasted a good deal, I gather, on a young French town planner who was highly recommended; so they are a little chary about "going the whole hog" with you: at the same time, Shurtleff expressed their desire to get the best foreign advice, and he promised to bring the question of inviting you up before the Russell Sage Committee and the National City Planning Conference Committee, with the hope of obtaining a partial guarantee to obtain your services as consultant. The usual fee for a person of your authority is about a hundred dollars for a daily conference; but Shurtleff pointed that this would mean something like $9000 for three months; and he was not prepared to go to those lengths. On the other hand, he said, if it is a matter of covering Professor Geddes's expenses, I think that can be arranged. Furthermore, he gave me the names of fifteen city planners and heads of city planning departments at universities, from California to Massachusetts; and I devoted myself yesterday to writing to these men in order to see if they could use your services as either consultant or special lecturer on your way east. By the middle of February I shall have returns one way or the other from all these sources; and I shall cable you the figure guaranteed; in order that you may postpone your trip should your commissions in India seem more attractive.

As to the time of your visit. Come as early in April as possible; for according to Shurtleff all your work will have to be done in April, May, and June. The summer, unless you are able to pick up some odd commissions once you are here, is dead. I am taking it for granted that you will cross the Pacific and land in San Francisco: please give me the name of the steamer and the date of sailing and arrival, so that I may address letters and telegrams to you; also the name of your bank or your bank's correspondents, so that I may send duplicates there, to ensure delivery. Unless I can assure you an unexpectedly large number of paying commissions, I shall remain in New York, as the expenses of a trip westward and back are quite heavy; about four hundred dollars for a single month, hotel expenses included. A word about our American habits: should you obtain consultation fees plus expenses do not endeavor to economize on the latter: within reason, we expect a high expense account, commensurate with existence at the best hotels, Pullman travel, etc. In fact, a decorous amount of conspicuous waste is part of the ritual of being a leader in any field; and city planning is no exception. The business man is our Pattern; and the sign of the business man is that, as we say, he spends big. Always err on the side of expenditure and you will not err at all. The very purpose of all our efficiency engineering and industrial economy is the provision of a grand margin for decorative futility. This is particularly true on the west coast, I believe; as you go eastward, and especially east of the Alleghenies, you approach Europe, and come within hailing distance of European standards.

Shurtleff was interested in the Cities and Town Planning Exhibition; and so I promised him to ask you these questions: 1. What surface space is needed to exhibit it. 2. Is it cataloged? 3. Can it be sent to America, either with or after you? What approximately is its cubic content – so that cost of travelling and storage may be estimated?

The letter which you sent back registered was properly forwarded from the old address; and I am still waiting for the package confidently, knowing that packages are usually delayed one way or the other. If it does not reach me by the end of the month I shall cable.

As for the projected book, I have not yet been able to persuade Liveright to accept it in advance of writing. If nevertheless you think it worth while to go ahead in spite of the risk – a relatively small one now that my first book has been published and favorably received throughout the press – I shall gratefully accept the five hundred dollars you so generously offer, and proceed with the work, making it of as general interest as possible in order to ensure its publication. The five hundred dollars will carry me through, together with my other writing, till June: I then plan to give up our apartment in the city and take a little house, at ten dollars a month, in the foothills of the Berkshires, on my friend Mr. Spingarn's estate; my wife in the meanwhile remaining in the city and living with her mother, except for long weekends in the country.[3] The book should be finished by the end of the autumn and ready for early spring publication.

As for my personal condition, I exhausted myself physically by crowding eight months work on *Utopias* into a bare four; and I am not yet fully recovered from it; indeed, a rest in the country will be necessary if I am to go on with even the literary work which I have undertaken. Unlike Farquharson, and perhaps yourself, I have physical strength without stamina; so in spite of a surface robustness I am always ready to topple into a slough of debility. This is perhaps one of the reasons why I have confined myself so closely to literary work, and have resisted, in spite of the vast temptation, the desire to join you actively in Palestine and India: my present vocation gives me a power of control over my time and energy which any other work, with more active demands from the outside, would not permit; and it is therefore, a condition of my being able to do any effective work at all. I hope to work out of this condition; and during the last five years, with occasional relapses and mischances I have been working out of it; but for the present it is better that I should confine my energies to a province where I am quite effective rather than that I should disperse them over activities which would shortly bring me to grief. Every time I forget this resolution I find myself brought up in short order; and have reason to regret it. So count upon my active participation in your work up to the limit of my ability: but remember that this limit is for the present a narrow one, which I can only hope to widen by a long rest next summer. You are coming to America at a seasonable time, for there has not been such an active interest in housing and townplanning since the outbreak of the City Beautiful movement a couple of decades ago; and the present activities are, on the whole, quite healthful and sane, if also a little timid and superficial. You can count in New York on at least two dozen architects and city planners who are ready to sit at your feet and absorb all you can give them: they will not ask you to lecture: they had rather sit around a table and have a give-and-take discussion, they asking the questions and you answering. It was by this method that they got more out of Unwin than they did in his lectures. You can count, I believe, on a very active and hearty welcome at large, apart from my personal one.

I must close for the present; but I shall keep you informed of developments.

In the meanwhile, believe me always gratefully,

Your pupil,

Lewis

1. Probably Thomas Adams (1871–1940), city planner who served as general director of the *Regional Plan of New York and Its Environs*; author of *Modern City Planning* (1922) and *Outline of Town and City Planning* (1935). See Mumford's letter of 25 March 1923.   2.   Allen Eaton (b. 1878), compiled *A Bibliography of Social Surveys* (1930) for the Russell Sage Foundation. Flavel Shurtleff (b. 1879), also served as Secretary of the National Conference on City Planning; authored *Carrying Out the City Plan* (1914) for the Russell Sage Foundation.   3.   Joel E. Spingarn (1875–1939), professor of comparative literature at Columbia University, author of *A History of Literary Criticism in the Renaissance* (1899) and *The New Criticism* (1911), and later senior editor at Harcourt, Brace and Company. He owned a beautiful estate, Troutbeck, at Leedsville near Amenia, New York; on the estate were several cottages which he let to literary friends; the Mumfords frequently summered there during the 1920s, eventually bought a house adjoining the Spingarns' estate in 1929, and began living there year-round in 1936. See Mumford's letter of 7 March 1926.

Dear Mumford,

Department of
Sociology and Civics
University of
Bombay

Your analysis of the situation & latent tensions at Le Play House, etc., is acute & suggestive – but is there not another viewpoint? And more hopeful?

12 January 1923

[Ms UP]

Note first my present tasks, as regards Universities – (1) *Jerusalem* unfinished. (2) *Tagore's new Universities* planning beginning. (3) *Osmania University, Hyderabad* (Deccan) where Nizam is giving 1400 (!) acres, & about "a crore or so" for buildings – £666,000 will go a long way at our cheaper Indian rates of 6d. per cubic foot. Finest site in world, but Jerusalem, & in many ways superior even to that (though of course of less historic sublimity & cosmic range). (4) *Bombay University* also coming into my hands to plan great developments for, as it emerges from the former stage (London) of a mere examination machine crushing its colleges, to a higher teaching University, &

developing these. (This however is still *private* – or, if you mention it, merely as yet *another* large Indian University.)

Now beyond all these, as before them in time, the fifth & best of the lot in some ways, our own, & nascent, University Militant – e.g. Outlook Tower with Halls & Vacation Schools, Nature Study & Surveys, Exhibitions, etc., & kindred endeavours of Culture Insitutes here & there, successful or no – e.g. Dumfermline, Rue des Nations (Paris 1900), etc.[1] – Similarly Victor Branford's Le Play House with Sociological Society, etc., & Farquharson's Civic Society Tours, etc. With all manner of essentially kindred (or kindred-able!) endeavours over world – e.g. New York City School of Social Science & Economics – Urania Museum, Berlin, etc. – Deutsches Museum, etc., etc.–Educational Associations everywhere, Froebel, Montessori – etc., etc., not forgetting those of Artistic, Dramatic Musical Character, or of Social Betterment.[2] Now however the time is ripening for a coordination of these. And what more central ideas than those with which above (5) started? In short Outlook Tower, etc., but as "Civic College." Will that be serviceable name? Will it go down in States? Pray criticise, & find a better! But tackle the general idea – (1) that of an Institution smaller than College, etc., or Museums, etc. Yet central to them all and adaptable to even large village centres, let alone to non-collegiate towns. Yet (2) as the *Institute of concrete synthesis* for the University & College too. I am not without hopes of Tagore adopting this (& even of making this the great Tower of Hyderabad University!). The philosopher can keep shop next door, by all means, & fit his theory to the facts, & draw on them his cheques, really payable in these concrete facts, as not much hitherto!

Now return to initial point of letter. Apart from the difficulty of harmonizing active ladies, so often greater than for men in most experience (what say you with American experience as well as British, etc.?), where is the difficulty of harmonising V.B.'s, Alexander Farquharson's, yours, my, & other points of view & lines of work upon these kindred general lines? (The better our building & the more adequate, the more these would tend to work in harmony.) R.S.V.P. as to this, & state any difficulties anew.

Now returning to business. Herewith I send you my two Jerusalem Reports, posted as one packet, & registered. They are not yet printed, either by Zionists for University or for government of Palestine for City (which I think rather stupid, though of course they may turn out to have reasons).

I shall shortly send you my *Patalia* Report; but those of Lucknow Zoo, Tagore & Osmania Universities will not be ready of course, when I leave India.

But how far may these be made to lead to work of *planning* {as well as/rather than} mere *lectures*? (I may fairly pray you a commission, as selling partner or agent, on any you can find, though I know you would do it in any case.) The older I grow, though the more I see the verbalistic & newspapered world with its press, politics, etc., only responsive to words, & not understanding things, much less *germs, experiments, beginnings*, the more I care for these latter; as the more real elements of the making of the future – & though no doubt words then help people, as the catalogue to the pictures, even then they are dangers, as substitutes for too many!

Yours cordially,

P. Geddes

1. Various projects in which Geddes had taken the lead: his survey of Dunfermline had been published as *City Development: A Study of Parks, Gardens and Culture Institutes. A Report to the Carnegie Dunfermline Trust* (1904); he had established the "International Association for the Advancement of the Sciences, Art and Education" as a "Summer School" which was held in connection with the 1900 World Exposition in Paris. 2. Friedrich Froebel (1782–1856), German educator who emphasized the importance of "mother-play." Maria Montessori (1870–1952), Italian renowned for the method of children's education she developed.

Dear Professor Geddes,

7 Clinton Street
Brooklyn, N.Y.

13 January 1923

[Ts NLS]

The pamphlets and photographs and the note addressed care of Boni & Liveright have just reached me; and I am still hoping to get the registered packet of reports you sent me; but if they don't come, I shall add "lost" to the cablegram I am going to send you about your prospects a month hence. This will not cause any serious delay in working on the book, unless you find it difficult to replace on short notice the material you sent me – which after all has probably strayed rather than gotten lost. I note what you say about getting here in May and spending the summer; and I trust that my last letter will reach you at not too late a date to persuade you to change your plans, and if possible to come here more expeditiously, even if you do come via Europe. Nolen has made arrangements to see me in New York next week, and Frank Waugh of the Massachusetts Agricultural College, to whom L.H. Bailey had also written, has signified a certain interest; but the replies to my letters last week have not begun to come in, and so for the present I am hopeful.[1] Be sure to give me definite word as to your itinerary, in detail, as soon as you have made it up; in order that I may plan engagement dates for you. Also, what sort of accommodation would you care to have me arrange for in New York? I wish that I could offer you the hospitality of my own home; but alas, it consists of two small rooms, thrown into one, and even overnight visitors are a luxury that we can scarcely afford. Specifically, would you care for rooms in a hotel, or – if your engagements demand that you stay a good time in New York – a small apartment? Suites at hotels, even the more modest variety, are rather expensive; and if you require more than a single room with bath it might be well to consider living in more modest quarters with fuller accommodation. Or have you other friends in New York who would put you up? This detail is important, because New York is really a summer resort, and it is sometimes difficult to obtain rooms here except by engaging them in advance.

Please forgive the unavoidable brusqueness that seems to characterize these notes; it arises from a desire to be complete and concise. I am sending you herewith a few scraps that may interest you.

Hastily,

Lewis

1. Frank Waugh (1869–1943) wrote on landscaping and rural development in *The Landscape Beautiful* (1910) and *Country Planning* (1924).

---

7 Clinton Street
Brooklyn, New York

18 January 1923

[Ts NLS]

Dear Professor Geddes,

The posts are so slow that I'm writing you this letter forthwith, on the first receipt of favorable news. So far, I've had three favorable answers with regard to lectures, from Kansas City, Buffalo, and the School of Landscape Architecture at Harvard; at a minimum of fifty dollars per lecture plus complete expenses; and I have no doubt that more will follow.

Did I tell you that the New School for Social Research has been completely reorganized; that all of its courses are given in the evening; and that its present plan is to give a broad general survey in the Social science not for special students so much as for professional people who are conscious of these special interests, as affecting their own lives and work. The chairman of the committee on curriculum is Mr. Alvin Johnson, one of the editors of the *New Republic*; and yesterday he extended to you an invitation to lecture at the New School during the Summer quarter.[1] The quarter is divided into two sessions of six weeks duration each; one lecture being given a week; and you may either do two lectures a week for six weeks, or one lecture over twelve weeks. The subject is pretty much any range of topics you choose to treat under the head of city planning: I suggest that you make it a survey of city planning throughout the world, with reference to the special problems of Ireland, Palestine, India, France, and so forth. If you are so disposed, it may also be possible to invite you to give a course upon Sociology: please tell me if this would be agreeable and suggest an outline. The fee will be fifty dollars a lecture, six hundred altogether; but Mr. Johnson added that he would bring your coming to the notice of influential people (like the patron of the *New Republic*, I fancy, Mrs. Willard Straight) who are interested in architecture and city planning; and that possibly a greater sum would be forthcoming![2] I found that Mr. Johnson, who used to be an economist in Leland Stanford, had followed your work for a long time, knew a good many of your publications, and very heartily admired you. I should add by the way that he agreed to limit the class to fifty students; so that you should have no particular vocal difficulty in addressing them.

The eagerness with which all sorts of people have responded to the news which I have spread as to your coming has heartened me greatly; and I think I can promise you an audience that is already somewhat prepared to receive your message. So herewith I abandon all my canniness and caution and reserve and open wide my arms. Come!

Mr. Johnson had commissioned me to plan out for the fall term a course on architecture in its social relations; and if you can spare any time for suggestions, please send them along. The lectures are to be given by various specialists.

Yours devotedly,

Lewis

P.S. Please cable to "Mumford c/o Liveright New York" the probable date of your arrival in New York, so that I may plan engagements.

The summer quarter at the New School begins 20 June, so that I shall try to arrange that all city planning consultation take place before this time.

1. Alvin S. Johnson (1874–1971), economist and editor of the *New Republic* (1917–1923);

became director of the New School for Social Research in 1923.   2.   Mrs. Willard (Dorothy Whitney) Straight and her husband provided the financial backing for the *New Republic* when it was founded in 1914.

Dear Mumford,

Department of
Sociology and Civics
University of
Bombay
Bombay

19 January 1923

[Ms Up]

    Herewith, posted separately, registered, (c/o Boni & Liveright) 3 or 4 photos of this Exhibition (that of *this Department*, attached, proved impossible – too lumbered up with screens! – But I'll put it to you in conversation later). Did I tell you that besides Tagore's & Nizam's, this University is next coming up for planning (though this still private) & so being prepared to accredit me, like the rest, to search out elements of progress over your vast field. Yes, I think I mentioned this, but only in outline, this further point – that of mobilising Tower, Le Play House, Otlet's Brusssels endeavours, live elements of Museums, in Europe, & U.S., etc., & kindred endeavours of all kinds – towards the *Civic College*; & thus alike for cities wthout College & University, or with conservative ones, or for centres to live new ones. Think towards this Eutopia of the Spirit – & its applications!

    Two live Americans, who met you & Mrs. Mumford (& with charming accounts of you both!), the Misses Lewisohn of "The Play House," suggest that I stay at Henry Street Settlement, for a time at least.[1] Is that practicable? I have forgotten City Plan in detail too much to realise where that is, & where you may be? But Settlements were more congenial than hotels or boarding houses in former visits.

    R. Unwin's suggestion to his clients in New York City, that I help with Survey, etc., for a time, I have answered to him, & left him to arrange, if he can, without my writing direct, to Mr. Norton, Russell Sage Foundation, or anybody: it is better to leave it to him: though if you know any of them you may posssibly help, though of course with caution.[2] Don't say too much of Jerusalem & other Universities, lest that put them off: a more important argument, to their minds, may be that of many cities over India, as in former years at home, at Paris, etc., as also that I am reporting for Sociological Service League, an important body, on Bombay City Improvements by the present "Development Trust" (of engineers, not without ability & energy, but otherwise limited – in housing, in planning, in economy, & in esthetic results alike!). (Keep clear of Mr. Backus Williams, active on legal sides of planning & author of a recent volume on it (probably quite good); for it was he, with his account of the Exhibition (as "mostly of old cities each within a little round wall") who broke our contract with New York mayor for Exhibition some years ago!) And don't attempt to arrange anything with Brooklyn Museum – head of that & I fell out 20 years ago, & I think he is still there. I forget name. (These are the only two I did not get on with!) (Richard) Ely (is it not, I forget initials) but of "Town Hall" now – was an old friend; also Zueblin; also C.H. Henderson; also Dr. Felix Adler of Ethical Society – I hope all going strong? I remember meeting Dr. Frankfurter & I think giving a lecture under his auspices. Who else can I recall? – Percival Chubb – I hear now in St. Louis? But of new generation I hardly know anybody. Professor Seligmann, Professor Boas, etc., I met at Adler's and liked.[3]

Is Professor Wilson still at head of Philadelphia Commercial Museum? And is Mrs. Wilson (a live Professor of Biology) also living? They were real friends, & I'd like to meet them again when I go. What of Simon Patten in Philadelphia? What of Harvey Robinson (History, Columbia)? Is Professor John Dewey back from China? Is Veblen in New York City? Is Osborn always at Natural History Museum? I must get my notes of past visits from Edinburgh to look up old acquaintances. Is Senator Foster still living? What of Dr. Devine – formerly of Charities Bureau, etc. – now of Rockefeller or Russell Sage, as also Professor Vincent? I have not written to any of these, nor to President N. Murray Butler, whom I also met.[4]

Some of these above may now be conservatives, from our present viewpoints – that makes me cautious! But if you know or meet any of them in conversation, tell them I'm coming over, & asked for them of you; & judge for yourself (& me thereafter!).

Are Loeb & Carrel (the physiologists) both in New York City now – or elsewhere? Who replaces Graham Bell, etc., as sympathetic centre for others as in Washington when I was there? Is Superintendant Dutton still at Teachers' College? I liked him very much. My recollections bring back many friendly spirits, as also in Boston, etc., etc. But above all, I want to see Stanley Hall. I heard he has retired. Is he still at Worcester? or where?[5]

Is Mrs. Emmons Blaine of Chicago still living? (founder of Faculty of Education there, & sister of McCormicks, agricultural implement makers.) Don't take trouble over any of these personal questions: just answer what you know, & others will tell me what you can't.[6]

But be considering who are the men & women of large views, & large aims, in sciences, & in geography, in sociology & in surveys – in interpretative social thought & city development, etc. Indeed also in dramatisation. For to my *Masques of History* there is growing the dramatisation of social conditions in present – in generalised forms, as contrasted with the specific & personal treatments of Shaw, Galsworthy, Brieux, etc.[7] (though that old craft is still the more truly dramatic way!). Still there is room for devising though I have no time, etc., for producing – the more generalised types, and shadowing out in this way the current & changing order – actual & eutopian, cf. Inferno, Purgatorio, Paradiso – and even for Cities (as per my grotesque sketches of Edinburgh, put in form more grotesque still by F.C. Mears).

Glad to see from V.B.'s letter that you are now to be in a new home of your own – but as you don't send changed address, I send this still to Boni & Liveright also, with photos.

My daughter tells me of a Unitarian Church in Edinburgh which has *frankly* taken to Ethical position – of working & pointing to Kingdom of Heaven on earth – & dropped the whole supernatural & post-mortem City. Have you that much in progress in U. S.? I suppose so. I never asked you – nor does it matter much to my mind – what were your confessional origins – yet they do contribute something to the making – or marring – or both, of our modern types & characters, & alike at best & at worst – in tradition & in rebounds. (Thus John Knox was neither the stern old Calvinist nor the great & liberal educationist, according to his lights – but *both* in one – & so for others.)

Have you any impressions of the religious work(s) around you – or are you too out of touch with all? You'll tell me when we meet.

How goes your work with *New Republic*? I hope it has begun & is as you would hope?

Cordially yours,

P. Geddes

1. Alice and Irene Lewisohn, daughters of the financier and philanthropist Leonard Lewisohn, were connected with the Henry Street Settlement and founded the Neighborhood Playhouse in New York. 2. Raymond Unwin (1863–1940), British architect and city planner; designed Letchworth Garden City. Charles D. Norton, a banker who was instrumental in persuading the Russell Sage Foundation to fund the *Regional Plan of New York and Its Environs*, a ten-year, ten-volumed project launched in 1922 and costing over one million dollars (Mumford and other members of the RPAA were critical of the project). 3. American acquaintances, most of whom Geddes had met during his 1899–1890 visit to the United States: Frank Backus Williams (1864-1954) author of *The Law of City Planning and Zoning* (1922). Robert (not "Richard") E. Ely (1861–1948), director of the League for Political Education and organizer of the Town Hall of New York. Felix Adler (1870–1937), director of the Ethical Culture School in New York. Frankfurter: obscure, probably not the jurist Felix Frankfurter who would have been only eighteen in 1900. Percival Chubb (b. 1860), educator also affiliated with the Ethical Culure School and editor of textbooks for grade schools. Edwin R.A. Seligman (1861–1939), economist and author of *The Economic Interpretation of History* (1902). 4. Other American acquaintances of Geddes. William P. Wilson (1844–1927), botanist who established and directed the Philadelphia Commercial Museum. James Harvey Robinson (1863–1936), historian at the New School for Social Research. Henry Fairfield Osborn (1857–1935), paleontologist and President of the American Museum of Natural History. Perhaps Addison G. Foster who served as senator from Washington, 1899–1905. Nicholas Murray Butler (1862–1947), president of Columbia University, 1902–1945. 5. Jacques Loeb (1859–1924), physiologist at the University of Chicago and later at the Rockefeller Institute. Alexis Carrel (1873–1944), physiologist at the Rockefeller Institute; awarded Nobel Prize in medicine and physiology in 1912. Alexander Graham Bell (1847–1922) was affiliated with the Smithsonian Institution, 1898–1922. Samuel T. Dutton (1849–1919), superintendent of the Horace Mann Schools and professor at Columbia Teachers College. 6. Mrs. Emmons (Anita McCormick) Blaine (1866–1954), philanthropist who supported various educational institutions in Chicago as well as the League of Nations. 7. George Bernard Shaw (1856–1950), distinguished Irish playwright; the reference is probably to such plays as *Back to Methuselah* (1921) and *Saint Joan* (1924). John Galsworthy (1867–1933), British novelist and dramatist; his plays *Strife* (1909) and *Justice* (1910) deal with social problems. Eugène Brieux (1858–1932), French dramatist concerned with social reform; *Blanchette* (1892).

Dear Mumford,

I sometimes write my letters like this – but have not got much of a reply out of it – for the good old positivists do not readily move beyond their master & thus fall behind him! But do you, I beg, look over your Comte before I come: and you'll find, no matter how irritated by his style & mode of treatment, matters of great value, beyond subsequent would be sociological literature – and clearing away absurd criticisms of him, as for ignoring psychology – which he included in his *Biology*, using this in sense of *Bios* in *Bio*graphy – & thus avoiding the lapse into *Biology* as *Necrology* with the anatomists – Huxley even – & into Psychology as *Phantomology*, with more than poor dear old Lodge![1]

This law played with on [triad] principle, & yielding 9 squares, with first three states, & the like with second, gives a very vivid presentment on one side of the actual war & after war situation, & on the other side of the Eutopia we desire, & this in considerable advance of clearness.[2] Try it!

P.G.

Department of
Sociology and Civics
University of
Bombay

22 January 1923

[Ms UP]

c A	c B	C
b A	B	b C
A	a B	a C

1. Thomas Henry Huxley (1825–1895), prominent biologist and Darwinist; was Geddes' teacher and mentor at the Royal School of Mines in London, 1875–1878. Sir Oliver Lodge (1851–1940), British physicist who, after 1910, turned his attention to psychic research.   2. This chart is explained in Boardman, 468.

Department of
Sociology and Civics
University of
Bombay

25 January 1923

[Ts UP]

Dear Mumford,

In advent of my employment towards New York City Survey how far shall I find people open, able and willing to assist – and on more orderly lines – e.g. of Place–Work Folk – 36 diagram? and more?

How far has the Outlook Tower any credit in U.S.A. for initiating the Survey movement – (as well as National study movement) from 1887 onwards? I fancy none. Dr. Johnson came over (I think in 1896) to learn methods there, and thereafter published his first Survey of New York – but this he did not send me (as I guess he omitted any acknowledgment). This I do not mention as a mere barren "reclamation of priority," but only because it might so far make people more willing to consider the latter developments of survey methods I have since been working up and with. The Town Planning Exhibition at London Royal Academy in 1910 (which I had been trying to get my London friends – Unwin and others – to start, for a good few years before) gave a room to my Edinburgh Survey outline, and this became nucleus of all subsequent Town Planning Exhibitions. Hence regional survey and city survey movement in Britain, etc. But I do not suppose this had effect in U.S., since we publish so little. Zueblin however was strong on this, as per his old paper – "World's First Sociological Laboratory" in *Sociological Journal* Chicago, about 190(2?).

However, it seems to me that in the main your American Surveys, although more highly elaborated than ours, are still in our earlier phases as far as principles, methods and scope are concerned. Not but that "Spiritual Power" issues are often considered, and penetratively too; but that these are still handled, it seems to me, sporadically, without any correspondingly adequate & exact technique – whether mine, as of 36, etc., or any other? Is this so or not?

Not, I repeat, for priority's sake – but because I dread a little the rejection of my later methods and view-points, by those accustomed to earlier elements & methods of survey, and to carry them far (or too far) in detail. Indeed so far as not to see the tree for the leaves, or forest for the trees; i.e. not to see what sort of city it is they are disentangling from the medley of cities (city types) which make up "the city" as we see it. (Thus my first startling impression of New York, etc., was of its "*Militia Barracks*" – as to my eye Castles of Capital, prudentially as well as romantically designed – against Labour beginning to threaten.)

More generally of course, the American city is *more frankly Mechanopolis* and *Mammomopolis* than even our own. Its "West End" is more unashamedly – and so more joyously – *Scholopolis* (*Schole = leisure*) or Thelema of leisure class than ours, and so on. Less obviously, & less really, *Strategolopolis* than our old war cities, less of *Tyrannopolis* developed than Berlin; less too of *Eupolis* than Paris, Oxford, Edinburgh, etc. Less of *Biopolis* than Bournville[1] or Garden City, etc.; less of *Geopolis* and *Regionopolis* (Diocesopolis) than some, save in lower forms e.g. coalfields, towns, etc. (where we run each other hard): and less too of *Ethopolis* than the old sacred cities, despite their deteriorations. In short, too much huge larval cities (Larvalopolis) for future.

Is not a nomenclature of this kind (systematised from classification, still too undeveloped) necessary, to clarify the expression of what we are doing in (and as) cities, & where we are making for?

(Note the respective Swastika & Svastika are fundamental to the preceding. "Widdershins" of evil & "Deasil" of good.)

So, tragically suggestive to the dreams of those who see New York at 47,000,000 and the whole East coast conurbation at 500 millions, and so seek to "pull down barns, and build greater" – appears also this historic formula – of City Development:

Polis, the City (Place) > Metropolis (Work) > Megalopolis (Folk):
Parasitopolis (Folk) > Patholopolis (Work) > Necropolis (Place)

As per Babylon, Rome, etc., in past; and Petrograd, Vienna in present – and with handwriting on wall, etc. (as Wells indeed lately pointed out for New York City).

Yet beyond this again, there may be recovery – as of Christian Rome for its best days – (and also dawning in Vienna, etc.).

So the spiritualising of cities – in time – becomes their civic policy.

Yet beyond this too, in practical life their decentralisation – for as Railways and deep draught ships, etc., centralise, so electricity, motor, etc., decentralise. Cf. Industrial future of water power regions, Iceland and Norway, etc., Alps, Apennines, etc., and so for Rockies, etc., etc. In such conditions the spiritual process should be more natural from the first. (Are any working at these problems?) I mean from the speculative side, as well as the practical, so much in evidence.

Since above written yours of 26 December to hand.

You don't express nature of book in your mind – whether of general sociological character, or that more depending on Town Planning & Civics. I think the latter first. This you have materials for in my reports (& no objection to 5000 words or so regarding Jerusalem). For sociology generally wait till we can talk over; not long now.

I'll see about remitting what I can to account, but mail leaves this morning! (Unluckily 2 expected accounts not yet paid. So I won't now be much in funds till end March!)

Yes. I am quite prepared to leave out popular lecturing – not in my style or powers.

Must close for post.

Yours cordially,

P. Geddes

1. Bournville was the model garden town George Cadbury established near Birmingham.

---

7 Clinton Street
Brooklyn
New York

5 February 1923

[Ts NLS]

Dear Professor Geddes,

On the assumption that you'll leave India in the middle of March, from now on I shall address you in care of Branford: this last note however is to ask that you send me as soon as possible a list of lectures you'd plan to give, as inquiries have come from Michigan and from Dartmouth; and more perhaps will turn up. To these I have said, in a general way, that your lectures would come under the heads of Civic Survey, Regional Survey – there is an incipient regionalist movement in America, and you might give it a great impetus – University Planning, Zoo and Botanical Garden Planning, while in Sociology I mentioned the subjects of a few of your University of London syllabuses. In doing this I have not pledged you, it goes without saying, to anything in particular; and so I should like to have particulars at the earliest moment. It occurs to me that the departments of sociology and economics might benefit a little by your criticism of economics, as published in your Indian journal; and that this might be the subject of one lecture or a series of three. I have still to hear from the Russell Sage Foundation; but Nolen tells me that once you are over here, you'll probably get an invitation. In a recent review Professor Franklin Giddings, who is as you know the dean of our sociologists, confessed that the people who had called themselves sociologists, himself among them, had not deserved the title; for they were really "sociologians" or "socio-logizers"; and in view of this declaration I shall try to get an introduction to Giddings and see if there are any possibilities at Columbia! At all events, you'll find plenty here to keep you busy. It is also very likely that you'll get an invitation from the National City Planning Conference to address them (in May) at their annual meeting, on Regional Planning; but this still hangs in the air.

My affairs are not in such shape that I can work on the book yet; but I expect to set aside a certain portion of my time from the first of March onwards; and when the material is sufficiently digested the first draft should be a matter of only a few weeks.

Forgive this haste and believe me,

Always gratefully –

Lewis

As I sail from Bombay on 1 April, I may get letters arriving 30th or 31st March
– but thus you have no time to reply to this here. My address, till 22 April will
be Le Play House – or till 26/7 April 14 Ramsay Garden, Edinburgh (c/o Mrs.
Geddes Mears), I sail from Liverpool 25th.

Department of
Sociology and Civics
University of
Bombay

17 February 1923

[Ms UP]

Dear Mumford,

Yours of 7 January only briefly replied to, to catch last mail – so now
more fully.

Money matters re Russell Sage may be left to Unwin; of course the daily
consultation fee of $100 will be substantially reduced – to a little more than
half – on monthly engagement.

Who is Mr. Shurtleff? Did Victor Branford or I ever tell you that all was
arranged by him before war with Mayor of New York – Mr. McEnerey I think
his name – but Mr. Williams – his lawyer I think, & interested especially in
building regulations & laws, dissuaded him, by description as per record in my
Evolution of Cities.[1] (Alas, my beautiful old pictures of antique cities are not
so numerous now, though still a feature!)

Herewith on separate page some account of Exhibition in answer to
questions – I hope the whole lot of Reports has reached you? I assume you have
Indore – which can be got from Messrs. Batsford Architectural Bookseller
High Holborn London W.C. & *Dunfermline* (John Grant, Geo. IV Bridge
Edinburgh).

Now as to my visit. While I shall be embarrassed & disappointed if I
don't earn expenses out of trip – & these are fairly considerable, I'm quite clear
that I'm not a popular lecturer, having neither the voice nor the reputation
necessary, much less both, nor the "popular gifts" either.

Think of me partly of course as City Planner – & now still more as
University Planner – with experience of a score of college & university
plannings, great & small – as with Tagore, Nizam, Bombay, (& now
Ahmadabad) on hand just now, besides Jerusalem. Also keen on Botanical &
Zoological gardens. Three of latter on hand this year on lines of Edinburgh
Zoo – everywhere wanted. So in these lines no one else has so much experience.
And as accredited by these universities to inquire for them, I am also not
without hopes of finding contacts towards consultations of these kinds, for a
later summer, if not this one. (Then too, my colleague, Frank Mears, of
Jerusalem plans, etc., see dome, etc., I definitely think of, & work with, as our
brightest British architect – & very superior to Lutyens, etc. (The latter is more
than conventionally clever of course, but there is no real life behind his façades,
nor to me much in them.) So that while I make general plans & scheme out the
general effect, Mears can at once improve on the plans often, and above all
work up the effect – into what such buildings should be – unique creations, &
not simply good exercises in current styles, as most architects do, even good
ones.)

But now beyond all this, take me as the wander-student, synthetist &
sociologist who has been fortunate enough to interest you – and who is now
at once thinking of (a) increasing his experience by further travel as of old &
also (b) of putting up his theses and defending them – with purpose (c) of
getting them at last into papers & books (hitherto so much delayed by

speculative absorption in further developments).

It is the last (c & b) which most interest you – (c) I take it especially. So here, I should very definitely wish to make you such business proposition as need be, towards all the time & assistance you can give me during my visit, & towards help in writing up what I can *talk* so easily, though I write so little, & with result often of not being very readable when all's done.

This business proposition awaits indications of how trip may promise, but if Russell Sage job comes in as Unwin *may* cable, you may be considering your side of it, or indeed if you see my money difficulties passably met at all, in other ways. Write of it what you can to London address. I shall consider myself very fortunate in such collaboration – and so shall be very disappointed if it does not come off. Hence I am alarmed by your proposal to return into hills, to write on my past material! That will keep a little longer, since it has kept so long; utilize rather what I bring, of matters social & civic (even political & philosophical) for smaller publics, no doubt, but of more definite values to existing situations & incipient progress.

Let us rather keep in touch as much as possible. It may be I can retire with you – later when lecture season over – and people have dispersed for long vacation? Why should I go so far as California – that for another season perhaps. Chicago is quite far enough, or at most one or two of the State Universities in easy hail of it beyond.

I am writing to my old friend Ely at "Town Hall," as per enclosed, which read before posting. I am not sure of address. If you know him, look him up, & talk things over. Old mutual friend James Mavor of Toronto speaks highly of his achievement of organization & its educative variety & value, so he may give me a rostrum to the right sort?

As to your health, I *may* be able to help you to work out of this (post-war) condition so common, as I am pleased to have done with my son (here in the past year & half).

Yours cordially

P. Geddes,

1.   New York has not had a mayor named McEnerey.

Department of
Sociology and Civics
Royal Institute of
Science
University of
Bombay
Bombay

14 February 1923[1]

[Ts copy Oslo]

Dear Mumford,

CITIES AND TOWN PLANNING EXHIBITION

This occupies about ¼ mile of screens – in two galleries each 180 feet × 30 feet. This however affords, in half of one, the needful large tables for students to work at, etc., and space for our seminar-sized small classes. A limited amount of graphical and sociological illustration also is included – and usefully so.

The 117 big screens, set close along walls, are all 10 feet × 7 – and full from top to bottom, with little or no space between them. [. . .] The 75 minor screens in galleries are only 6 feet broad by 6.6 feet high – but covered on both sides. There are also tables with books of reference at various points. There are about 4150 sheets of exhibits in all, great and small, from long friezes 10 or 12 feet by 2 or 2½ feet to prints of this paper size. Many minor things are mounted together, and are counted as one each. Of course, the method of hanging frames on a central post, or at ends of screens, would admit of some condensation (though not so much as appears at first, since you then can't hang above or below).

The bulk of cases is thus considerable, about 200 cubic feet of boxes are required for shipping or railing it.

In my former Exhibition (at bottom of sea since October 1914) almost all were framed – this time not – too expensive!

I have also a lot of pictures in London and Edinburgh – but mostly left, as of less importance – though some are good, and used there. And I'll buy all I can find on journeys through Europe. I'm also getting in material from cities, planners, etc., as I can: so it's always growing.

All this is thus getting too big to carry about, or easily find gallery space for! But a representative Exhibition, such as I had in Paris in 1916, and in Jerusalem in 1920, can be packed in one or two cases, holding 60 cubic feet; and will fill one gallery say 100 feet by 30 feet – or still better, arrange into a suite of rooms of about the same area, and thus with more wall space, and less crowded arrangement of exhibits accordingly. I might send this smaller collection to the States some time – after this year's visit – if a group of cities cared to take shares in it – paying about ? a piece for a month's Exhibition each, this charge being inclusive of all charges and expenses on Exhibition's and expositor's part – but assuming given this gallery space, screens, lighting, cleaning, and ordinary guardianship. (It has been paid for at a rather higher rate sometimes in India, and at others rather lower.)

Catalogue will be provided, though at present out of print.

That I think is a general account of this show. I'll form my ideas more clearly during my visit to U.S. And I'll bring just a few bits from it to illustrate lectures (though not especially as samples, which would be too large a box).

Yours,

P. Geddes

1. This description of the Cities and Town Planning exhibition was apparently enclosed with Geddes' letter of 17 February.

Dear Mumford,

(Boni & Liveright) Enclosed explains itself. (Indian printing, at Madras especially, is cheap, and decently passable: all I can afford just now.)

The crafty Russell Sage, on hearing I was coming anyway, has dropped any fixing up with Unwin, but not idea altogether (as perhaps they will since my cloven feet, horns, etc., are not disguisable).

Still, more & more I am interested in learning what I can of your world – and thinking out things more clearly accordingly: so, though *some* potboiling is needed, you need not worry over this. My theses, of various kinds, will doubtless find some doors to be pinned on! I have not written to Dr. Adler or

University of Bombay

1 March 1923

[Ms UP]

Professor Boas, who will probably have forgotten me, but if you can unobtrusively let them know I'm coming that may lead to our meeting again.

A likely milieu would be the New School of Sociology & Economics – Dewey, Veblen, & Co. – which seems hopeful, & more akin to our world of ideas than most, is it not? I have not written them either, indeed do not know if Dewey is still in China or home? I wrote Ely however.

Has Veblen any constructive ideas to match his criticism of Universities? Is there any feeling yet that Universities are not what they need to become? My half-dozen new schemes & plans of that sort should at once gain by what I hope to see; but will they help to arouse some questioning spirit? I dream indeed some gradual organization of the University Militant, but no easy matter in so complicated a tangle of conservatisms! I'm trying again to plan this Department as more than Tower, Sociological Society, etc., a sort of definite clearing house for all kinds of studies & specialisms, & by enlisting aid from these in first place, & so showing them their place as clearly as may be in social whole, & so in pro-synthesis accordingly. Write me next to Sociological Society for 22/3 April – or c/o Mrs. Geddes Mears 14 Ramsay Garden Edinburgh from 26th.

Yours,

P.G.

Since preceding was writtten, a letter from Raymond Unwin has come in, suggesting that I cooperate with him (if Russell Sage Foundation, etc., agree, of course) in the survey/study of New York to which he lately started on short visit in October – & asking my terms – which of course will be his own – subject perhaps to a reduction for longer time.

Three months seems to me but brief – but much might be done with that in view of able collaboration available.

So *don't fix me up for engagements* till you hear how this matter is arranged. The Russell Sage Foundation, etc., will probably settle it within a fortnight of your getting this, & their office will then tell you what has been arranged (or not as case may be).

Of course best way of doing 3 months would be with intervals, during my stay – I leaving jobs for others. P.G.

P.S. Will it clarify matters if I outline main tasks, as I see them, beyond planning in Palestine & Jerusalem & new University, though much in relation with these, & adaptable more or less elsewhere, to other cities & Universities – & with *Cities* Books – & *Universities* Books.

In Sociology the long dreamed *opus* is nothing short of general (i.e. encyclopedic!) outline, & with such more developed chapters as may be; and with relations to other sciences, especially biology & psychology (ethics & esthetics too) made clearer than heretofore; as also with history & economics, with considerable criticism & I hope construction in these fields. With geography too of course.

I don't know how far you know or care for my diagrams; but all await fuller exposition & development, for which materials abound, & which will I think reward your working over. I quite agree that Mann makes them too difficult by over-elaboration & also by introduction of theosophic symbolisms

– to me irrelevant, however clever, & suggestive.

My feeling is that with you to work with – given not only your general interest & sympathy (which a few others share), but also your training & habit of definitely getting stuff to press (which the above save elders occupied don't share!) – we might get a good deal done! And without some such co-operation, I don't see how I am to get much done. For brain still teems daily, & thus interferes with writing, though always promising to make it better!

Still, with my unsettled plans, it is difficult to fix place & date. And with correspondingly unsettled pay too! But I'll clear up as soon as I can. P.G.

Dear Mumford,

University of
Bombay

9 March[1]

[Ms UP]

Little to say this week, my previous letters having been voluminous. But herewith a letter which explains itself. (I hope not too much liberty?)

I am also sending one of the packets to you direct – but not all, for I presume I shall be living first at least in Henry St. or other Settlement.

You may open it of course – but I warn you it, & all others are a mess to nth power! My morning meditation yields a copious mass of jottings, very desultory often, or in any case mixed up by the demons of chance & change, as they accumulate & I migrate. But there is some stuff in them, if & when extricable.

It is my dream that I shall find perhaps my long sought collaborator in you – who may be at once willing & able to try to work with me & over these Teufelsdröckhian paper bags – as did Thomson long ago for *Evolution* and *Sex*, etc. – & Branford re. various collaborations also.[2] Of course I have to guarantee you against loss in the matter – & to share authorship, if & when we get so far.

So if you & Mrs. Mumford are going for holiday during summer – can you let me come with you? The essential is to find space to lay out papers, on plenty of tables or desks – the ideal might be to find a village schoolhouse closed for vacation – or still better to be beside some college, in which some friendly colleague might let us work in his department, laying out papers upon laboratory tables, not in use. Think if you can fall in with this scheme, & if so arrange it. Might not rooms, etc., be obtainable in Yale, Harvard, Princeton, or other pleasant town in vacation & at not too great expense. Or at Clark University, Worcester? Or at Toronto? I think I could manage it at either of these. We might thus have change – peace – yet work too.

Yours ever *P.G.*

1. Apparently enclosed with Geddes' letters of 1 March 1923 to Mumford and Boni & Liveright.   2. A reference to Thomas Carlyle's *Sartor Resartus*: Teufelsdröckh sends to his editor "Six considerable Paper-Bags, carefully sealed, and marked successively, in gilt China-ink, with the symbols of the Six southern Zodiacal Signs, beginning at Libra; in the inside of which sealed Bags lie miscellaneous masses of Sheets, and oftener Shreds and Snips, written in Professor Teufelsdröckh's scarce legible *cursiv-schrift*.... Selection, order, appears to be unknown to the Professor.... Daily and nightly does the Editor sit (with green spectacles) deciphering these unimaginable Documents from their perplexed *cursiv-schrift*." "Professor" Geddes recorded his thoughts on unorganized "Shreds and Snips" of paper; he apparently hoped that Mumford would become his long-suffering "Editor."

University of
Bombay[1]

1 March 1923

Messrs. Boni &
Liveright

[Ms copy Oslo]

Dear Sirs,

You may remember writing to me a year or two ago, suggesting that I send you a book for your consideration. I have always been too busy – planning towns, colleges, universities (*six* on hand this year), botanical gardens, & "zoos"; & my "Reports" on these there are as yet mainly of local interest – though Mumford is quite right in seeing that a book of more general interest could be compiled from them – & so may perhaps do it.

But I have some past things, & my students have been at me to reprint two or three, especially – viz. my *Masque of Ancient Learning / Medieval & Modern Learning* as *Dramatisations of History*, which were the interpretations of big functions in connection with my Edinburgh Students' Halls of Residence in the first place, but which interested the public also, and next were performed in Hall of University of London. These booklets went to *12,000 or so* (my only large circulation at any time) but then got out of print before war, since which I have had no time to think of them, much less of going on. But these helped to stimulate demand for Wells' *Outlines of History*, & now this reacts in reopening interest in my *Dramatisations* – so that at length one of my masterful young assistants has made me send off my last surviving copy to the Madras printer of the Modern Publishing Co. here. I am thus having 1000 from him – & as there may be also some demand at home, & some even in U.S.A. I am getting another 1000 for each. But now, on way home & to U.S. (New York City) as I am next month, there is no time to correspond with London, let alone New York. (I have to correct proofs before sailing.)

So I take a liberty, & with each, as I never did before! I put down for London publishers Sociological Publications, Ltd. (who are Branford, Farquharson, etc., old friends, from whom I can freely take their name, even if in vain). But what to do for U.S. edition? With an audacity you can never have experienced, I put down your name on title-page also! No time to ask permission!

If you resent this (as you are quite entitled to!) I can paste your name over, with a strip, when copies arrive in America: but meantime the thing is being done, & you are too late to stop it, even by cable!

Per contra,[2] if you will accept this, I am in your hands: and I accept beforehand any terms for this small edition you see fit to define, when it comes into your hands. (Or I can give them away if you don't take it.)

There is even another paper (of which I can send sample of its later part, as last complete copy again has to be kept for printer) entitled – *Essentials of Sociology in Relation to Economics*. This expresses some of the views which, as a wandering student, I am coming over to your side to maintain as best I may. I am out for the watery blood of political economists (orthodox or Marxian matters little) and so again can give this away, as challenges, if you don't take it!

So I leave you as free as I can! despite my audacity as a total stranger to you. This is at any rate a very distinct overture towards acquaintance, & not without hopes of this becoming a friendly one. I trust you will see that as mitigating circumstance, & so remain

Yours cordially,

P. Geddes

On second thoughts, I send this through Mumford as he can perhaps help to explain, although despite long acquaintance, we have also never yet met! (He can perhaps show you *Masques*, etc.) P.G.

1.   Enclosed with the previous two letters. Mumford's note on the envelope dated 16 December 1956: "I was shocked and horrified by this letter and never gave it to Liveright – nor did I introduce PG to him."   2.   On the contrary.

Dear Professor Geddes,

7 Clinton Street
Brooklyn, New York

25 March 1923

[Ts copy UP; Ts, pp. 3–4, NLS]

Six weeks or so have passed since I last wrote you; and in the meantime I have, of course, received your various letters and reports. This last week the draft sent by your daughter came too, but I am holding that here against your arrival, since I have not yet begun work on the book, and since, moreover, I have no present need for it.[1] My hesitation in beginning on the book was due partly to the fact that I wanted to have all your reports on hand before I plunged in to them, partly to the necessity to earn current cash, and partly because I wished to canvass the publishing situation a little more carefully before beginning it. I wish to arrrange a tentative contract for its publication before actually beginning work, so that the book may be written with a particular audience in mind; since it is one thing to write for the city planners and architects, and another to open up the field for a wider public. Branford in the meanwhile has written to say that Arthur Thomson recommended his own publishers, Putnam's, and in addition offers to write a general introduction to the book; and I intend to lay the matter before them next week.

Now as to your own engagements here. The Russell Sage foundation has not communicated with me since my first conference with Flavel Shurtleff, the Secretary of the Committee, on a Regional Plan for New York; and since receiving your advice, I have let the matter rest, on the theory that if I pressed them for a decision at once they would probably say no, whereas if you come here, and if your arrival is duly heralded in the newspapers and the magazines – as I shall attempt to arrange it! – they may be tempted to engage you on the spot. I saw Thomas Adams the other day; and found that he was busy on the Russell Sage plans; but as to your prospects he was quite non-committal. I may be doing Adams a great injustice, but I fancy he is assiduously feathering his own nest; and if you are to obtain an entry with the Russell Sage committee it will have to be on your own merits, plus Unwin's recommendation. None of the other tentative offers or parleys about city planning work have come through; so that all that is at present on the calendar is:

1. A lecture before the school of Landscape Architecture, under James Sturgis Pray, at Harvard; by 25 May at latest.[2]

2. Three lectures, if the trip can be arranged, at the University of Michigan (May).

3. Course of lectures at the New School for Social Research, beginning June 24.

Besides this, there are the following possibilities:

1. Invitation to speak at the annual meeting of the American Institute of Architects (16, 17, 18 May).

2. Lectures before the Department of Sociology or the School of Landscape Architecture at Dartmouth College, New Hampshire.

It will be easier to get further lectures if you will give me a list of the subjects you will speak on, in city planning and sociology; and if possible, a synopsis of lectures. Before you leave America, if I am able to work up the right sort of "publicity" and if you lecture sufficiently at the colleges, we may be able to work up a demand for university planning by you, to say nothing of zoos and botanic gardens; but I am in doubt as to what to do in advance – if indeed anything can be done. I trust, by the way, that you will arrange to give a course of lectures – say three – on universities, past, present, and possible. Unfortunately, you know the university terms end in America with the first of June, and there will not be very much time between your arrival and their close to devote to lecturing.

The group that is most prepared to receive you here is a small body of architects, engineers, accountants, and journalists known as Whitaker's group, because they have been drawn together by Charles Harris Whitaker, the editor of the *Jounal of the Institute of Architects*. One of the members, Clarence Stein, an architect, is doing a book on the past, present, and future of New York City, with regard to townplanning and regional development; he has also started, or rather is in the act of starting, a Garden City Association. Another, Benton Mackaye, is a forester, who is bent on developing a eutopia in the Appalachian region, beginning as a recreation trail and developing through an exploration of resources into a comprehensive scheme not merely for reforesting but also for recivilizing the region. The first steps in this development, the laying out of the trail, have already been taken; and next October a meeting will take place for the purpose of projecting further steps. I am sending you herewith a rough outline of his project, which he drafted. Still another, Stewart Chase, is a public accountant who has followed Veblen's lead – with hints from Ruskin and others; his pamphlet on waste I am likewise sending you. Here is, in fact, a real university, which is doing some of the things that the New School for Social Research might have set out to do, had it been more wisely directed; and I think it will pay to give this group some of your time.[3]

Don't be alarmed by the discouraging letters I wrote you whilst the horrid streptococci, which took possession of my mouth last December, were still clogging my system: during the last few months I have gained twenty pounds; and am again in normal condition, so that all the precautions and cares which seemed imperative, in sheer defence, last January can now be forgotten. My getting run down was not so much post-war as post-utopia: I cannot stand intense work over a long period, and in writing, re-writing, and copying over *The Story of Utopias* in four months I got badly run down. If I am to do good work at all, I must be able to spread it over a considerable period, and do a little at a time, pretty regularly, after the fashion of Herbert Spencer. Your habits, Branford has told me, are quite different; and I daresay if you don't stay here too long I can adapt myself to them without serious damage! At any rate, it will be much easier to give this adjustment a trial in New York, than it would have been in India!

As things go at present, I intend to devote all of the first two months of your stay to you; and the greater part of the second two. This of course is subject to any changes which either of us may find it necessary to make in the meantime; I merely want to assure you, for the present, that I shall not "run off to the hills." (At most I will rent a little cottage to weekend in!) As to how my discipleship is to be financed – that is a question which can well be postponed till we meet: I shall do my best in the meanwhile to accumulate a little reserve, and by reducing my work to half-time I shall probably be able to scrape through somehow. I still stick to my notion of a biography as the biggest personal contribution I can make towards spreading your thought: a biography not in the sense of a mere tombstone or a monument to a life that has been lived, but as the essential explanation and raison-d'être of your philosophy and plans, seen as things which have had a life and evolution. The genetic approach to an idea is surely through the life of the person who has given it forth; and this is particularly true of your ideas, because of the fact that they are interrelated so completely that if one does not see the whole one does not, really, understand any particular part. Geddesianism is a thing that one must speak of as a whole, like Platonism or Aristotelianism: that is why, I think, you have usually met with complete acceptance or complete rejection: each particular point you have set forth being either blessed or damned by its implications! In order to get your separate theses into order, and have them published in articles and books, I can of course be of assistance as editor; and I cheerfully offer my services in this capacity to the limit of their use. I am dubious about genuine *collaboration* because of all that the word implies: Branford and you, Thomson and you, have labored together, and have interracted with each other whilst pursuing certain lines of thought and action. My life, on the other hand, has followed a different trajectory, as a result of being born in another part of the world, being almost a generation and a half apart in time, being affected by other circumstances, with its momentum not a little reduced – in common with all the other members of my generation – through spending adolescence in the shadows of a devastating war and a corrosive peace. So my relation to you can never be fully that of collaborator: it is rather that of pupilship, in which I absorb what I can of your thought and in turn make it over and revamp it to suit the particular life-experience that I have encountered. Branford and Thomson were working in the forge with you whilst the iron was still hot; at best I can do no more than use the emery cloth on the finished product. That part of your thought that I can assimilate – to make the metaphor a little more vital – will probably come out more successfully in a novel or a drama or a philosophic essay like *The Story of Utopias* than it would if I attempt directly to cast it into the framework of economics or sociology or what not. The chapter on Coketown and the Country House, and particularly Megalopolis is of course largely *you*; and I have pointed this out to everyone who has praised the book; and it is you much more effectively because it had become organically a part of me than it would have been had I set out, deliberately, to make an exposition of your ideas about our paperized civilization. The use of my practiced hand as editor is quite another question; and doubtless we shall be able to put a little more Geddes in print, this next year, than has ever appeared; although rest assured, I shall do nothing to take away the salt and savor of your personal style – with a Celtic crypticism it shares with Carlyle and Meredith.

I have followed your discussions with Swinny and Stanley Hall with intent pleasure; and before you come I shall go back to Comte, to refresh my mind – although I had a pretty thorough bout with him three years ago, and emerged from it full of admiration and respect.[4] I seriously doubt the existence of any sort of linear progression in a human society as the law of three stages presupposes; it seems rather to me that all three are implicit or potential in each society and that in the rythm of history one or the other may be stressed or subordinated, as society develops, only to return again; but this may be merely my own ignorance, and perhaps with as wide an historic knowledge as Comte I would find myself in the same position. Your elaboration of Comte discloses the possibility of one of n number of possible developments; but it is only after the actual historic process has taken place that one can confidently say that n is a or b or c. The reason that social predictions come to pass so frequently, or rather, one of the reasons, is that the mere laying down of a prophecy gives people the confidence to act in terms of it, and so, by accepting the truth of the hypothesis they in turn verify it. Surely not a little of the industrial development of the nineteenth century was due to the fact that the Smiles's and the Porter's so confidently predicted it; and no one had the hardihood to turn his back and say no to that which everyone else was laboring to achieve, with the fine assurance that Mammon and Moloch were on his side![5] We may do as much by predicting eutechnic, eugenic, and eutopian developments; and so far good; but we will win out not because truth was necessarily on our side from the beginning, but because faith was there and faith brought truth over.

The day is wearing; and I must close. A final work about arrangements. I shall take it for granted that you'll stop at the Henry Street Settlement the first few days at any rate; during which time you can decide for yourself about staying or moving. The settlement is highly inconvenient – on the extreme east side of the city, away from any of the main lines of transportation; duller and more depressing than the slums around Toynbee Hall. But you shall see and decide for yourself. There is another settlement, Greenwich, at the meeting of various transportation routes, where the old city left off and the new part, planned in 1811, began; and I shall make inquiries there.

Yours ever gratefully,

Lewis

1. Norah Geddes Mears sent Mumford a bank draft for $250 (6 March 1923, UP f. 3224).   2. James Sturgis Pray, author of *City Planning* (1913).   3. Founding members, along with Mumford, of the Regional Planning Association of America: Charles Harris Whitaker; Clarence Stein (1881–1975), author of *An Outline for Community Housing Procedure* (1932); Benton Mackaye (1879–1975), proposed and planned the Appalachian Trail (1921) and author of *The New Exploration: A Philosophy of Regional Planning* (1928); Stuart Chase, author of *The Tragedy of Waste* (1925) and *Men and Machines* (1929). Another founding member was the landscape architect and community planner Henry Wright (1876–1936), author of *Re-Housing Urban America* (1935).   4. Shapland Hugh Swinny (1857–1923), editor of the *Positivist Review*; member of the Sociological Society Council and frequent contributor to the *Sociological Review*.   5. Samuel Smiles, author of *Men of Inventions and Industry* (1885). George R. Porter, author of *Progress of the Nation* (1836–1843). Mammon: the devil of riches and covetousness; Moloch: Canaanite idol to whom child sacrifices were offered; both are depicted as devils in Milton's *Paradise Lost*.

Dear Professor Geddes,

7 Clinton Street
Brooklyn, New York

3 April 1923

[Ts NLS]

Your recent letter, with the enclosure to Liveright, caused me considerable perplexity. I laughed with gargantuan appreciation at your Napoleonic stroke in printing *The Masque of Learning* with the New York imprint: but in the end I decided to counter it with a stroke equally Napoleonic: and I have not given Liveright the letter! The reasons for this are numerous; but they are all reduceable to this: If Liveright is willing to publish your book he will do so without being forced into a Hobson's choice: if he is not willling, your maneuver will only make him irritated and resentful, and what is worse, the story would be sure to leak out – and there are enough difficulties in the way of getting an appreciative reception for you without adding that to it. Were Liveright your equal in any sense, were it really worth while to make such a bid for friendship, I should say, well, take the chance; but in all friendliness I must point out that he is not your equal, that he is likely not to understand you and that you have much more to lose by such a stroke than you have to gain. Save your audacity for a more important occasion.

But, you will say, the deed has already been done; and I am only spoiling its final execution. Well, let us see what we can salvage.

First: please observe that there is a heavy duty at the U.S. customs on bound copies of books. Therefore, before importing the *Masque* into America it will be better to decide whether they are to be sold or are to be given away. I suggest that you keep the American edition in London until this matter can be settled. If Liveright, when I show him my copy of the *Masque* and tell him he can have five hundred copies, printed in India, without any risk, accepts the offer, there will be plenty of time to ship the books over from England; since, one way or the other, they could not be published before the autumn. If no publisher can be found, there will still be time to have Liveright's name blotted off the cover and title page, and a little paster "With the compliments of Professor Geddes" put on the title page: this may enable it to enter the country duty-free. (On a thousand books the duty might be as high as a hundred dollars, or over.) If you import the books with a B. & L. imprint, without an authorization from Liveright, you may get into serious legal difficuties; and it would be foolish to waste time and energy on such matters.

There is still another reason for not forcing the issue with Liveright. Professor Thomson has advised that I try his own publisher, Putnam's, with respect to the book on your townplanning work; and I have already written them, but have not received any reply. If they come to an agreement, it would be advisable to have the *Masque* published by them; and it would not be difficult to get them, I think, to do that. In fact, I shall not approach Liveright about the *Masque* until I hear from Putnam. If this happens, it will be well to unbind the cover and the title page, and import the book to America in sheets; partly because the saving in duty is considerable; and partly because Putnam would doubtless prefer to give it their own imprint and cover. I have a notion it may be better to wait until you have been here a little while anyway; for the reason that the publicity notices and personal contacts will give you a hold with the publisher far more effective than any bolt that may be shot from the blue. I trust you've received copy of the *Survey* with an advance notice of your coming. There will be more of this sort of thing.

This matter has given me no end of anxiety; but I am quite sure as to my soundness of judgment. I don't want your visit here, or future visits here, jeopardized by any false steps; particularly upon what is, after all, a relatively unimportant matter. At any rate, your letter is locked securely in my desk; and it will stay there till I can give to you in person. As for the books: pray store them in London or Edinburgh until we can make some reasonable disposition of them.

The other day I met for the first time Percy Mackaye, whom you probably know as a playwright and fellow-masquer; and I found that he, too, was a great admirer of yours; and felt that you were the one person whose approach to art was sufficiently synthetic to understand his own.[1] He wants very kindly to talk over things and to learn more from you; and he hopes that you will be able to visit him during the summer at his home in Cornish, New Hampshire. I should add that he is a brother of Benton Mackaye, about whose book on an Appalachian utopia I told you: they are the intellectual and the emotional indeed, working in complete harmony, as Percy Mackaye is now writing a cycle of folk dramas, I believe, for the hill people of the Appalachians.

Your first lecture is scheduled at Harvard on May 11, unless you make some other disposition upon your arrival. Professor Pray suggests that its topic be: City Planning in India.

Cordially yours,

Lewis

Please give particulars of your steamship: name, line, date of sailing and probable arrival.

1.   Percy Mackaye (1874–1956), prolific American playwright; his "cycle of folk dramas" included *This Fine-Pretty World* (1923) and *Kentucky Mountain Fantasies* (1928).

---

7 Clinton Street
Brooklyn, N.Y.

11 April 1923

[Ts NLS]

Dear Professor Geddes,

A hasty note to greet you before sailing. You do not mention having received my letter of 18 January in which I announced the arrangement of a course of lectures to be given by you at the New School for Social Research, on the subject of City Planning and City Development; so unfortunately the catalog announcing the course had to go to the printer without an outline of what the course will provide. If you will make an outline of the course, and a short description – say a hundred or two hundred words – Mr. Johnson will specially advertise it. As scheduled now, it will call for two lectures a week for six weeks; at fifty dollars a lecture, beginning June 24. It is posssible that you yourself may be able to make arrangements for still another course, on sociology say, when you meet Mr. Johnson; but this awaits your arrival.

I inquired this week for rooms for you at Greenwich House, a central location, but find that the only ones available are small and noisy. I shall find out at the Henry Street Settlement what sort of accommodation has been reserved for you; and unless something better is available, I suppose that you will at first stay there. The best permanent address for mail for you would probably be in care of

The New School for Social Research
465 West 23 Street
New York City

I am enclosing a few items that may interest you. Here is an anticipatory clasp of the hand, and a cordial welcome.

Lewis

P.S. Don't be startled if you are interviewed by an American reporter before landing.

Dear Professor Geddes,

Instead of running in this morning for a few hours I have decided to sit back and take a reckoning of where we are, for you have now been here two months, and I have a woeful sense of having squandered precious hours without, as yet, being of the slightest use to you. In the meanwhile I have been learning much, of course; at times you've given me so much to think about that my brain buzzed and reverberated in sleep, and I awoke the next morning completely weary; but I don't like to have all the advantage on one side; and it is quite plain that in these occasional meetings we can get nothing done, nor yet, for that matter, planned.

In one sense, I have the feeling that we have yet to meet. We both have been aware of the obstacles to meeting: but it is rather hard to climb over them, partly because of the gap between our generations and our varieties of secular experience, and partly because my respect for you is so great that it reduces my mental reactions in your presence to those I used to feel in the presence of my teacher when I was twelve years old – that is, complete paralysis! Putting this last matter aside there is a real barrier to understanding between us in the fact that you grew to manhood in a period of hope, when people looked forward with confidence to the great world spinning forever down the ringing grooves of time; whereas I spent my whole adolescence with a generation which, in large part, had no future. Your pessimism about the existing state of civilization as portrayed in IX does not prevent you from still working eagerly at the problem of the transition to 9, because your own career still has a momentum acquired under an earlier period of hope and activity; and so, perhaps, you don't realize the paralyzing effect of that pessimism, which is inherent in the situation, upon those of us whose personal careers have not yet acquired any momentum.[1] Rationally speaking, there is as much chance of doing good work as there ever was; rationally speaking, a work that is worth doing is worth doing for itself without regard to the possible mischances of war, famine, or what not; rationally speaking, all the interests that we had acquired before the war are just as important and as valuable as they ever were. True enough: but something of the impulse has gone; whatever one's conscious mind accepts is not enough to stir the unconscious; our efforts are no longer, as the saying is, whole-souled. If I found this bitter sense of futility in myself

7 Clinton Street
Brooklyn, N.Y.

6 July 1923

[Ts NLS]

alone I should be tempted to attribute it to an unsatisfactory personal experience; quite to the contrary, however, my own career has on the whole been a happy and eventful one; and the forces which undermine its satisfaction are at work in almost every intelligent and sensitive person I know between the ages of twenty-five and forty. Those who are younger than I am differ from my generation in the sense that they are "realists" who have no hope for the morrow whatever and no faith or interest in the polity at large; whilst those who are over forty are still living, as it were, on the capital acquired during the days of hope, and if their store is rapidly running out they manage to scrape on from day to day. Our sense of a "calling," our sense of any one task to which we could profoundly dedicate ourselves, is gone; and until we can recover this sense a certain intensity of devotion to our professions is the only thing that prevents our lives from being altogether inconsecutive and dispersed. I have fought against this drift of things from the very moment I detected it: but it is like trying to relieve one's bosom of the pressure of the enveloping air; and I see no way of relieving the crippled psyche except by trusting to some slow and obscure process of cure. It is no use saying, be different! for we are like the sick man that Saadi mentions whose only desire was that he might be well enough to desire something.[2]

You came over to America without, I suppose, any sufficient awareness of this change which, apart from any mere difference of age, separates a large part of the younger from the elder generation; you came over, too, with a somewhat over-idealized portrait of me in your mind, as a vigorous young apprentice who might work at the same bench with you for a while, and keep on at the task when you had gone back from America. You are naturally disappointed to find that by natural bent and by training I am of the tribe of Euripides and Aristophanes rather than of Pythagoras and Aristotle; a fact which is, possibly, a little obscured by the fact that mere necessity and convenience obliges me to get my living from day to day, with the Sophists of journalism. Faced with an actual me, you have naturally tried to make me over a little into the idealized portrait, whose aims and interests and actions were more congruent with your own; and, instinctively, I find myself resisting these frontal attacks, although my defences have again and again fallen down before unpremeditated movements on my flank! In the light of this difficult adjustment between the Ideal and the Actual, it would not be at all surprising if the original portrait had turned into a Caricature – that of a clever young hack writer, rather sullen in temperament and unamenable to conversation, who has no other interest in life than that of turning out a certain number of sheafs of copy per diem! The inability of this creature to follow your talk for more than a couple of hours at a sitting you would, in the light of caricature, attribute to a lack of interest or worse still! to a general lack of synthetic intelligence, whereas it is only the obvious reaction of a thorough visual to auditive method of presentation. And so on.

Plainly, neither the ideal nor the caricature corresponds to the real creature; and one of the things that has hindered our work together is, perhaps, that you began with one and shifted to the other without our ever having (except in chance moments, quickly forgotten) the chance to meet. If instead of thinking of me as a quack journalist you'd conceive of me rather as a young scholar who publishes his notes and lectures instead of speaking to a class; and if you'd see that I have chosen to get a living in this manner because it is for me the one means by which I can work at my own pace and keep at least a third of my time free for thinking and studying of a different sort, there might still be a little exaggeration in the picture, but it would be an exaggeration towards the truth. Eutopitects build in exaggeration towards the truth. Eutopitects build in vain unless they prepare the mind as well as the ground for the New Jerusalem; and nothing you have said has shaken in me the belief that the best part of my work must be in the first field rather than the second, although it may be true that I shall do the first task more sanely and adroitly if I have had a little direct experience of the second; and I have so far admitted this as to go ahead with the plans of the Mohegan Colony....

All this has been a mere clearing of the ground; the real purpose of this letter is to discover, with time pressing upon us, what we can do together; or rather, perhaps, what I can do towards helping you do your own work. 1. Do you wish this year to embody your opus in a book, or do you wish to prepare the material so that, with a little exposition between pages, it might became a book? 2. Can I be of help in assorting existing diagrams, and copying them out on uniform paper, to make their handling in a book easier? 3. Can you hope to get even part of your opus into shape as long as you keep on elaborating its parts instead of, at some given moment, cutting the flow of thought and taking account of what already exists? 4. If you are determined to elaborate, should you not, at any rate, lay down the main outlines, so that others might get the essentials of your method, and to some extent elaborate for themselves? 5. Do these represent, roughly, the essential parts of the problem:

I.   Graphics. Personal account of use; their elaboration; their utility in charting relations. The theory of Graphics.
II. The Vision of Life. 1. Cosmos Space, E.T. 2. The thirty-six and its various parts. 3. Olympus and Parnassus. 4. IX to 9 Etc. etc.

6. What disposition shall you make of sub-theses, like the bibliographica Synthetica? History of Universities, and so on?

Let me strenuously plead that, for the moment, you take all the sub-theses, which obscure the main problem, and cast them aside; and devote your attention to the architectonic whole, and to the problem of how that whole is to be presented. Is it to be a book with diagrams, or is it to be diagrams, with running explanations. And is it not high time now to give a shape to the broad outlines of you work, letting particular refinements and details take care of themselves, or working them out at a later date when you are satisfied with them. Your morning's pile of paper has become almost a vicious habit, and the pile will be so big presently that it will defy almost anyone's patience and organizing power – your own included. If you leave others to do this work it will be done badly, or not at all. Besides being a general philosophy, your biosophy represents a personal experience and vision: given the same counters other people would derive different values; and such fragments of your thought as Ben or Victor Branford, or Mann, or Farquharson, or myself might present are necessarily altered in the course of their transmutation from mind to mind. At best, such assistance as I can give outside of sympathy and understanding, both imponderables, must be largely of a manual nature, writing out diagrams or editing a stenographic mss., if you choose to dictate it.

Can we arrange a definite job? A more regular plan of work and intercourse for the next month and a half or two than we have yet achieved? Please command me.

Lewis

1. The "IX–9" chart, "The Ways of Transition – Towards Constructive Peace," preoccupied Geddes during the 1920s. See Geddes' "Wardom and Peacedom: Suggestions Towards an Interpretation," *Sociological Review*, 8 (January 1915): 15–25, and "Ways of Transition – Towards Constructive Peace," *Sociological Review*, 12 (January 1930): 1–31 (April 1930): 136–141. The "IX to 9" chart is reproduced in Boardman (480–481).   2. Saadi (c. 1200), Persian poet of the Sufi tradition; author of *Gulistan* and *Bustan*.

Saturday 6 July
1923[1]

[Ms copy Oslo]

Dear Lewis,

Good letter!

Yes, we have yet to meet. I feel that.

But do believe I understand, only too sadly, the tragic situation of your generation (or rather ¼ of generation) as compared with pre-war. Let alone wider experience (as from students, in Scotland, before & after war, etc., etc.). I have had this experience in the sharpest clearness – e.g. my full partnership with the joyous Alasdair – my eight anxious years, during & since war about Arthur – who is only now gaining something of equilibrium, of joy in life, of vigour towards tasks with renewing life – & this only thanks to leaving home & Europe for India & its fresh environments which have relatively escaped this western desolation. True, I can, & do, live by my earlier capital of pro-synthesis, experimental sowings, grains of reapings, & hopes accordingly – but he has had & is having to reconstruct a shattered youth, as you all more or less have to do – or fail in.

You are not one of those who will fail: that is manifest. There is no caricature of you in my mind, whatever be sometimes said in half-jest, or warning from others – nor yet any *excessive* ideal either, I think. I claim *some* little "understanding" of the real creature! – since always studying men as they move.

I quite admit, & even appreciate, the value of your writing, & (though I have still to see your cullings books) I am convinced you are largely at least on right lines. I have done some reviewing & leader-writing too in past – and though once & again renouncing it, to go on thinking instead, I am often very conscious of the loss of periodic stimulus to organize one's ideas on any matter, such as they are, which writing gives. And I may well feel, at this age especially, but long before, the endlessness of dispersive thinking.

But you have felt the advantage of doing a *Book*, and now you will have it preparing for a fresh edition by & by – & with improvements & developments which may also take form in articles, etc., as you can write them. And other books are doubtless coming. What are they? And, while I rejoice to see you also started on Eutopitecture, & doubt not it will lead to more, I do loyally agree that your forte may be, nay almost certainly is, more towards literary presentment. But, I stick strongly & clearly to my view of life ("Mr.

Squar's" as Huxley used to say) – *Vivendo discimus*! Living is an experimental science, more or less full of applied theory; but fresh thoughts & theoretic visions come with & from each task. I might, & no doubt would, have written more had I stuck to my professing, like my colleagues, but I'd have had far less to write. With all of my failures, & disappointments, in thought & action alike, I hold by alternating & associating these. See *word & blow*, see *sword & horn*, in the old ballads.

But while we are clearing up, let me explain too that instead of the active old fellow, hustling the East as in these past years, for whom you thought of preparing an active time here, I have found myself here on holiday & free, again a spectator & of new and urging things. Thus am stirred by them, new ones come, & old ones get fresh lights – & so I have been having my best holiday since first seeing Indian cities in 1914 before work began with them. It is just because I know I have to brace up again before long to focus my ideas for Bergson, for last year of Indian teaching, for publication too I hope, for development of Tower, Sociological Society, etc., etc., that I have let my mind run as it will – & even encourage it to! (It has not been lost time, if I live a few years longer.)[2] Time for bracing up however is now coming on.

Best thanks then for your final questions – quite to the point! But since I don't want mere secretaryship (though a little of that is needed & gratefully received!) – but your maturer development above all, and thus as no mere exponent of mine, I ask you to go on thinking over what you see in my middens of papers & my discursive talks, as well as in the (really more ordered) mind they come from. I'll turn over matters quietly too, tomorrow, as quiet Sunday. Come in on *Monday* morning if free.

Yours cordially – affectionately – & not ungratefully,

P.G.

1. Mumford's annotation: "After my protest." 2. Henri Bergson (1859–1941), French philosopher; founder of the International Organization of Intellectual Activities in which Geddes was interested.

Dear Professor Geddes,

I didn't want to break in on you and Lasker yesterday; so I ran away silently. I am enclosing the result of today's work; it has served to clarify my own mind a little, and might, if it meet your approval in whole or in part, serve either as a separate essay in the *Soso* or the *Journal of Philosophy*, or form part of an introduction to a book of graphs.[1] You will see that I have tried to make explicit certain points which occur in the mind of a listener, and which, in the haste of exposition you usually take for granted. Most of this has probably seemed too obvious for you to waste time over; on the other hand, the point of view is so foreign to the ordinary reader, even the ordinary academic reader, trained in science, philosophy, or what not, that it seemed wise to spend a little time in harrowing the ground and in general loosening up the soil; indeed, I am not sure but that the first section should be extended, by introducing your account of its genesis.

7 Clinton Street
Brooklyn, N.Y.

26 July 1923

[Ts NLS]

I am grieved at having to miss your first lecture today; but I shall be on hand for the rest of them.

I shall drop in after lunch tomorrow; if you're not in just leave a note on your door as to whether or not you expect to be back.

Lewis

1.   A short essay entitled "Graphics" is attached to the letter. It discusses Geddes' graphic method. Mumford writes: "By substituting a visual scheme for an auditive one we can, as it were, lay our cards on the table, and see them, not merely as separate cards but as related members of the pack.... The graph ... permits one to organize and to bring out and to relate items which seem otherwise scattered and insignificant."

Department of
Sociology and Civics
University of
Bombay

6 November 1923

[Ms UP]

Dear Lewis,

I had an 11 days voyage – turbines out of order & feebly working (one day not at all) so was short of time for London & Edinburgh but got some writing done, as also on Genoa–Bombay voyage (18 days) – some 6500 words in all – the *Regional & City Surveys* book thus sent off to Mr. Kellogg in instalments, and another one are begun – not bad for me, was it![1] In fact, although on arrival here I am naturally smothered in arrears & in unpackings of notes brought back from U.S., etc., I feel I have got a release by my long rest and change in N.Y., etc., and can get to writing at last. I feel too my old dream of a fifty years course of studentship now long enough over passed to compel some ("growing up," "coming out," etc.) to a more active life, of expression and action.

I had a warm reception at Edinburgh & at London – so adding these to Ely's send-off, here were three stimuli in rapid succession towards above. Next too at Brussels – where I found friend Otlet also suffering from over-study period, & delayed expression. Then on to Bergson, etc., with long discussions, & next a week nearly in Geneva at League of Nations & Labour Office. Both strangely mingled – of truly progressive spirits like Nitobe, Thomas, Miss Wilson (Librarian), etc., and mere bureaucrats of bureaucrats: so pulling in opposite directions: & I think I see various ways of aiding the former – of which more later.[2] I am sorry now I did not get in touch with League of Nations sympathizers in New York, etc.: for imperfect though these two organisations are, they are better than nothing, and have hope & promise of growth. And even at their woodenest, they are angels of light compared with the mass of my fellow passengers to India!

So – can you refer me to the League of Nations organisations in U.S. – or better, to save time, ask them on my behalf, & from me, as preparing a report for M. Bergson & his committee on International Intellectual Relations – to send me here any of their printed mattter that they can spare? (I hope this won't too much strain your critical conscience!)

Branford was cheered by your writing him a line about the Ely luncheon. He was fairly well, for him, but still very sleepless, etc. It is a great thing to have *Science & Sanctity* off his chest. A very little of either is enough for most of us: no wonder with so much of both, he has nearly burst!

Write too, at leisure, yet before long, of how far you are *helped*, & how far *discouraged*, by our active acquaintance of these past months, and what sort of cooperation, if such you still think of in general ways, seems visible. You understand, do you not, that I quite appreciate *Utopias*, etc., as such very real co-operation. But I must stop now – for post. I hope Mrs. Mumford & you have found your country house? Let me have any news of friends, & give my remembrances.

Yours ever,

P.G.

1.   Paul U. Kellogg, editor of *Survey Graphic*, which published the series "Talks from My Outlook Tower" (February–September 1925), the lectures Geddes delivered during his 1923 New York visit.   2.   Inazo Ota Nitobe (1862–1933), author of *What the League of Nations has Done and is Doing* (1920). M. Albert Thomas served as Director of the International Labour Office of the League of Nations.

Dear Master,

7 Clinton Street
Brooklyn, New York

26 November 1923

[Ts NLS]

This letter has been long delayed, and I fear now that it may even be too late to give you Christmas greetings on the appropriate day; in which case, perhaps, you will accept them for the New Year. I have had a letter or two from Branford in the meanwhile; a word from Mabel Barker to say in what good form you were when you passed through London; and from your silence I conjecture that you have had few vacant moments on any part of the trip. For my own part, I have been busy, but to no purpose: first a month spent in preparing the lectures on architecture, and then another month uneasily spent in attempting to salvage the material left on my hands when the course was removed from the New School program for lack of students. Four people came to the first lecture; and no one registered: this fact undermined my whole program of work for the winter, and it is only now that I have gotten on my feet again. Lacking a permanent foundation you will see from this how difficult it is for Johnson to turn those empty rooms into a genuine school for social research: the school is attended by people who have no serious interest in study as such or in obtaining any general grasp of the world they live in or in the world of culture; at the present moment their dominant interest is psychology with the result that any course which was either on psychology or had the word psychology in its announcement has been popular this fall, whilst Hart's course on education, Johnson's on economics, and my own on Architecture have fallen by the boards.[1] A more militant spirit might of course attract real students; for they did indeed come when the school was first announced; but lacking this – and it is either there or it isn't there – the school is dedicated to the snapper-up of unconsidered trifles. I had at first planned to do a comprehensive book on architecture in America, which would in the final chapter relate the possibilities in architecture and city development to the transition from IX to 9; but the criticism I have received on the first couple of chapters makes me see that this attempt is a little premature, since my knowledge has not yet been sufficiently seasoned and salted, and I shall postpone any systematic attempt until I have time both to gather more material and to "forget" it – that is to say, in the indefinite future, whilst I content myself with touching on certain aspects where my knowledge is already sufficiently deep and authentic. My chief task for the present is to gather material for a report on Housing in Europe for the Housing and Regional Planning Commission; there is a great deal of education to be done before the task of housing can be seriously undertaken, for the impression still is common that the housing shortage is an accidental, temporary thing, instead of what it really is, a chronic inadequacy of a mechanized and mammonized civilization.

I offered *The Masque of Learning* to Boni & Liveright on any terms they might care to handle it for; and they refused altogether for purely business reasons. There are scarcely any publishers now who feel any responsibility as publishers apart from their interests as business men; and they turn down perfectly worthy books which might sell a few thousand copies in order to concentrate their attentions upon best-sellers. In the case of the *Masque* they probably figured that it would not sell more than a few hundred copies at best, and that even if the books were given to them for nothing it would not pay them to place the thing with the booksellers. This is merely one of the vicious elements in a wretched system of distribution: they do things much better in Germany, for there, despite the complete demoralization of the market, books of all classes continue to be printed and distributed. I am still sending the *Masque* on its rounds amid the minor publishers; with the faint hope that something may finally be done with it.

Mid all the diagrams that I have been puzzling over from time to time there is one that causes a halt in my mind each time that I come to it; and perhaps by now I can formulate the difficulty. The objective squares of p:w:f are each comprehended separately by geography, economics, and anthropology, and as a working unison by sociology. So far clear and good. When I turn to the subjective correlations of S:E:F I can't find any sciences which deal with aspects either separately or as a whole.[2] True: there is esthetics, psychology, and ethics; but esthetics does not deal with sense but with the objects of sense, psychology does not touch experience by itself but the whole psyche or personality in its manifold aspects, and ethics does not deal with feeling itself but with those *modulations* of feeling which work out for the good of the community; so that if there is a connection between S:E:F and the subjective disciplines it seems to be incomplete. Moreover the categories of esthetics, psychology, and ethics do not agree with the beautiful, the true, and the good; at least there is a pretty stiff discrepancy between psychology and the true whereas the correlations are perfect and self-evident between p:w:f and Geography, Economics, Anthropology.[3] One of the things which would prevent misunderstandings in presenting these diagrams, it seems to me, is to emphasize that they are an organon of the sciences rather than a description, and that the actual sciences only more or less fulfill the requirements laid down in this logical frame, esthetics and ethics for example being for the present, as cultivated, only the simulation of such disciplines as have eventually to be formulated; and so, too, economics and anthropology.

Without having read it yet, I am sending Jung's *Psychological Types*. Apparently he has reached a classification which follows the lines laid down by Comte and developed by you; and the generalization is all the more striking because, I suspect, he has never read Comte and is quite ignorant of systematic sociology. At any rate, he distinguishes four types, compounded out of the introvert and the extrovert, and I have no doubt that they can be assimilated to the historical classification. Henry Osborn Taylor has just announced a volume on *Freedom of the Will in History*; an essay to show the influence of ideas and motives independent of economic influences, it is an attempt to redress the balance of EFO's school, and it comes appropriately at the moment when the influence of the OFE's was never so low.[4] The most interesting piece of research that's come out recently is a *History of Witchcraft* which shows its origin and continuation as a priapic cult. Apparently witches not merely existed up to the seventeenth century but believed in their powers and boasted of them; institutions like Witches' Sabbath were the periodic orgies of the cult, and in certain parts of the European countryside during the late Middle Age children grew up without knowing any other religion! It explains what seemed otherwise a perfectly idiotic persecution mania; for all the decent members of society had good reason to be alarmed at the appearance of this cult, particularly in New England where in Puritan times – despite the blue laws and the connotations of "puritanism" – sexuality was pretty lax. Please tell me if you want me to send you Lynn Thorndike's *History of Magic and Experimental Science in the Middle Age*: a ponderous work that shows the activity of the scientific tradition, independent of Aristotle and the Greeks, throughout the Middle Age.

Hurwitz wrote me to say that he has asked you to contribute to the *Menorah Journal*; and I am still wondering what has happened to the Survey lectures![5] Branford tells me that you now plan to take the Cities Exhibition to Geneva: does this still hold? Benton Mackaye asked me to send you his best respects when I wrote you, and Sophie adds her deep good wishes to mine.

Lewis

1. Joseph K. Hart (1876–1949), author of *Democracy in Education* (1923).   2. "S:E:F": sense, experience, feeling.   3. "p:w:f": place, work, folk – the fundamental Geddesian triad derived from Le Play.   4. The title of Taylor's book was in fact *Freedom of the Mind in History* (1923). "EFO": environment, function, organism.   5. Henry Hurwitz (1886–1961), founder and editor of the *Menorah Journal*.

Lewis Mumford AND Patrick Geddes

THE CORRESPONDENCE

1924

Dear Master,

7 Clinton Street
Brooklyn, New York

5 January 1924

[Ts NLS]

Your letters of the sixth and twenty-fourth of November have come; and in the meanwhile I've written you; to no very great purpose, I am afraid, since I was still laboring under an unusually long and painful succession of disappointments. For the present, however, I am troubled by nothing more than a cold; and I can again lay plans to snare the moments as they go. I have already sent you one article from the series on Architecture; and will send you the others as they come out.[1] As you will note, they are painfully incomplete; but without illustrations it seemed to me worthless merely to talk about the architecture itself, and so I resigned myself to analyzing some of the forces that had produced it. The P.W.F. motif appears, of course; but rather by implication than by direct statement. Unexpectedly, the first two articles have attracted a little attention: Mr. Nock has suggested that I write a history of civilization in the United States along the same lines, and has gone so far as to hint that a way might be found of subsidizing the work; at the same time, out of a blue sky, Charles Beard, the historian, whom I have never met, asked me to dinner! I shall eat the dinner, and sleep over the first proposal.[2]

There's a distinct drift towards history at the present time, and as I said the other day in writing to Branford, if we attracted the attention of people to our outlook and methods by demonstrating their use in analyzing the Making of the Past, they might be a little more willing, perhaps, to see the place of our sociology in the Making of the Future! The formulation of sociology has gone almost as far as it can without undertaking a specific task: so it is now time to work out the plan of campaign by a series of moves in the field. All of us, Farquharson, Dawson, Ramsey, and the rest, would profit a good deal if you and Branford would put your heads together for a turn to discuss the *strategy* of our sociological attack.[3] You will remember the series that Branford projected in 1910; the series that finally became the Making of the Future? Well, it seems to me that without any definite announcement, all the books of our group might be centered in more or less the same way, simply by an exchange of understandings as to what each of us was working on and what he purposed to do. This in turn would be helped if you would point out the subjects and the points that seemed to you most in need of attention. I pleaded for this orientation of research and effort when I was withdrawing from the *Sociological Review*; and I still feel that it would be economical and productive. Although you will doubtless find it necessary to elaborate your own contribution, you have given us an *organon* of sociological investigation whose main lines are firmly established; even those who do not consciously work with the organon have assimilated the point of view – men like Wellbye and Ramsey.[4] What we must do now is to *occupy* the field. So, to begin with the concrete, please tell me how you think a history of civilization in the United States is at present an important undertaking? Since all the things that have entered Western civilization during the last hundred years, and that threaten eastern civilization during the next hundred, have had a clean sweep in America it seems to me that a diagnosis and a new regimen would profit by an historic analysis. I am not sure of my capacity to do the thing properly; but it seems to me that it should be done. Articles Bertrand Russell has written about China and the West would be more incisive and penetrating if some such sociological survey lay before him.[5] Or do I weigh too heavily the thing that lies nearest me? Incidentally, I have suggested to Branford that he publish his historic *Survey of Westminster*, since the appetite for history is now raging lustily; as I remember it, the mss. was almost complete. *Science and Sanctity* is filled with good things; but one must bring much to the book before one can draw much away from it.[6] I have already suggested to Branford various people who would be interested in it; and am waiting for news as to whether he will seek an American edition before I undertake to place or stimulate any reviews. The enclosed letter from Harcourt, Brace and Co. speaks for itself: the reason is, as I found, that no publisher will undertake to handle any books unless he has convinced himself of a market: with the limited number of books he can place with the booksellers he will undertake nothing unless there is a definite prospect of profit – the mere absence of risk is not sufficiently attractive. The fact that the *Masque* is a reprint; the fact that it is a small book, and the fact, above all others, that it is done in the unusual *form* of a masque causes him to shy away. I shall keep on sending it out. Incidentally, I am now reviewing Rachel Taylor's *Aspects of the Renaissance* for the *Journal of the American Institute of Architects*. It is, as you said, a fine, passionate piece of scholarship; there is nothing to compare with it in spirit on the period except Eli Faure's *History of Art*: in the midst of my debilitation, it was as good as a cordial.[7] Do not think me churlish for having taken so long to read the book. It follows from a pretty solitary childhood and adolescence that suggestions and stimuli from the outside do not immediately pass over the threshhold; I rather keep the door ajar and examine them cautiously before even entering into conversation with them. This was, I think the chief barrier to our intercourse; for in proportion to my respect for your slightest opinion, the door became a gate and the threshhold a moat; and it is only now that your precepts and practices are beginning to take effect. In the long run, the door *does* open; and meanwhile much that is silly and noxious stays out, or tends to. . . . I send you herewith the latest project for our tank in Central Park; but alas! I have been able to find no photo. – Running through Samuel Butler's biography the other day I came upon the story of the Perfect Cockney. Butler, who was a kindly old duffer, took his clerk, "Alfred" to Switzerland. They ascended one of the big mountains and Butler pointed out with enthusiasm the various peaks and passes that they could see from their vantage point and laid out the course of their journey. Intoxicated with the view, Butler turned to Alfred and exclaimed: "Isn't it splendid?" "Yes sir," said Alfred, "and now, sir, if you don't mind I'll just throw myself down on the ground and have a read of *Tit-Bits*."[8]

I am looking forward to the *Survey* articles and am eager to see what you have done to them. I have written to the Peace House and the Carnegie Endowment asking them to send you all their publications. This is all for the present.

Lewis

P.S. Sophie and I exchanged rings at Christmas.

1. "American Architecture: The Medieval Tradition," *Freeman*, 8 (19 December 1932): 344–346. This and other articles in the series were incorporated into *Sticks and Stones* (1924).   2. Albert Jay Nock (1872–1945), editor of the *Freeman*. Charles A. Beard

(1874–1948), author of *Toward Civilization* (1930); Mumford criticized Beard in the late 1930s for his isolationist views.   3.   Christopher Dawson (1889–1970), historian at the University of Exeter; author of *The Age of the Gods* (1928) and *The Making of Europe* (1932). Stanley C. Ramsey, member of the Sociological Society and author of *Small Houses of the Late Georgian Period, 1750–1820* (1924).   4.   Reginald Wellbye, member of the Sociological Society. 5.   Bertrand Russell (1872–1970), renowned philosopher and social critic.   6.   Victor Branford eventually published "A Sociological View of Westminster," *Sociological Review* (1930); *Science and Sanctity* (1923).   7.   Elie Faure, *History of Art*, 5 vols. (1922). Rachael Annand Taylor, *Aspects of the Italian Renaissance* (1923).   8.   *Tit-Bits* was a popular "saucy" penny magazine founded in 1881.

Dear Master,

7 Clinton Street
Brooklyn, New York

9 January 1924

[Ts NLS]

A.E. Orage is in New York now, the advance agent of a cult that has established itself at Fontainebleau; and some of the things that he told me the other night about its doctrines – among much which was mysterious – made me think that you would be interested in it.[1] The head of the cult is a Greek with the Russian name of Gurdjieff. According to the apocrypha, he spent more than a decade in the interior of Asia – Tibet, the Desert of Gobi, and other places – seeking for the fountains of mystic lore which according to the theosophic tradition were still welling forth there. In various monasteries and temples he made himself acquainted with the sacred teachings, and noted down some four thousand sacred dances; and now he has come back to the west to inculcate these teachings into those who are capable of following them. His discipline consists in shaking free, by means of the dance, the physical centers of habit: this physical dissociation has a similar effect, according to him, upon the intellectual and emotional centers; and the result is an expansion of all one's potentialities, which can be increased, at a later stage, by appropriate intellectual and emotional disciplines. Ouspensky is a follower of Gurdjieff, but according to Orage his *Tertium Organum* is worthless, since it contains doctrines he had formulated before meeting the master, and that he has since renounced.[2] Gurdjieff is undoubtedly a man of insight: he put Orage on a regimen of sixteen hours of physical work per diem, chiefly digging in a garden; and each pupil he treats according to his capacities. His psychological doctrine of course recalled your diagram; and I thought you would be particularly interested in the fact that he permits no manual work without accompaniment of continuous intellectual and emotional activity. The estate at Fontainebleau was one that had been given to Maitre Labori by Dreyfus; and since its death it had fallen into ruins. For the visitor it is a sort of abbey of Thelema; for the permanent disciples, it is a Benedictine order, whose austerities are tempered from time to time by the master. Men and women live in the Château itself to the number of about 100: it is a college, however, rather than a cloister, for the pupils pass through it and go out into the world again. A demonstration of the sacred dances is being given in Paris now; and Orage is paving the way for a similar demonstration in New York. There is a certain amount of cryptic and unmentionable theosophic hocus pocus connected with the cult which Orage with consummate skill minimized in conversation; but there is also a valuable discipline which Orage's complete poise and self-sufficiency testified to more eloquently than any words of his. He endows Gurdjieff with godlike insight into man's powers and possibilities; and altogether he seems a far bigger man than Steiner.[3] (Have you by the way heard of the school that Count Keysersling, author of the *Reisetagebuch eines Philosophen*, has established in Germany? The number of these schools seems to be increasing.)[4]

So much for the present. Orage can be reached at Fontainebleau, or in care of the *New Age*, to which he still occasionally contributes.

Lewis

1. Alfred Richard Orage (1873–1934), British Fabian; editor of the *New Age* (1907–1922); raised funds in America for Gurdjieff's institute at Fontainebleau.   2. Peter D. Ouspensky (1878–1947), follower of Gurdjieff; author of *Tertium Organum: The Third Canon of Thought: A Key to the Enigmas of the World* (1923).   3. Rudolf Steiner (1861–1927), founded the Anthroposophical Society in 1912; established the Goetheanum, a "school of spiritual science," at Dornach, Switzerland.   4. Alexander Keysersling (1815–1891), *Aus den Tagebuchblatten des Grafen Alexander Keyserling: philosophische–religiose Gedanken* (1894).

--- --- ---

Bombay
(Reply to Le Play
House not Tower)

3 February 1924

[Ms UP]

Dear Lewis,

Is it possible I have not written since I returned? I thought I had! However I sometimes think letters – & then by & by come to imagine them written – so if you have not heard that is a likely alternative hypothesis to blaming the P.O. system.

I stayed only a weekend with Branford on return, & a day or two also at Edinburgh: went on to Brussels, to old friend Otlet then to Paris, for discussions with Bergson over International Intellectual Relations – & with Paul Desjardins on things in general: cheered up too by old friend Schrader – the grand old man of Geography, who has completed his great atlas at last – with tremendous spectacles replacing both lenses removed for cataract – & also is settling down to other portions of his opus.[1] Then Geneva for a week – looking into League of Nations – & Labor office. On the whole very favorably impressed – as per summary of lectures I am sending you. What do you think of them as the outline of a book? Give me your impressions – why should reviewers always wait till they're too late to be of any use!

Very busy here when I got back – then attack of colitis – neglected until very serious – packed off to nursing home for 3 weeks then near fortnight in hills. Getting on all right till chill the other day – now on my back with touch of fever (malaria) which makes one very weak. Sailing homewards – on 15 March – so don't reply here – but use *Le Play House* address, from which letters will be forwarded.

I'll probably go again to above-named places – perhaps also to Berlin, then home – perhaps settling for a time in Edinburgh as I have more room in Tower than I can get elsewhere – & am really minded to get to writing – though as always fresh speculation intervenes.

So much then for myself. My son took up my Lucknow Zoo at Christmas, & will remain to keep the planning practice moving – though I don't think he should or is likely to remain long. My chair tenure is over with this term – so I may not return to India next winter: but plans are not yet defined.[2]

Very sorry indeed to hear your lectures do not come off! The deadly mythologies of economics – classical & socialistic alike futilitarian – have no place for you or me in them – & we remain in dark ages, on both sides of Atlantic until (when?) we can give our idealotopias & Eutopias some appeal instead! I have been thinking of an onslaught – of which more by & by – but it can have little effect – of course. Still, things are moving, criticism beginning anew, as in Carlyle's & Ruskin's day – this time I trust more scientific & effective.

Greatly pleased to find in library here an American thinker I never heard of before – have you? Samuel Johnson, *Oriental Religions* – India 1872, China 1877, Persia '85, 3 vols. – extraordinarily good – the best I have ever seen of the general history & sociology of the East – & in full perspectives – full of insight – but, as unconventional & unorthodox, doubtless ignored. See at least Frothingham's introduction to last volume. (By the by what did Frothingham do? was he a divine? – if so, it does him credit!) Was Johnson a disciple of Emerson – who did Emerson especially influence?[3]

There is a memoir by Samuel Longfellow.[4] Is it good? Who is publisher – I suppose long ago out of print. But to my mind another striking glory of Boston. (How & why did not this continue? Why does it not revive? The contrast of Concord & Cambridge is a striking light upon this.)

I must stop – very limp.

Yours cordially

P.G.

Have you yet got your country home & settled down to domesticity? Best wishes. P.G.

1. Paul Desjardins (1859–1940), French philosopher, attended Summer Meetings in Edinburgh; author of *The Present Duty* (1893). Franz Schrader (1844–1924), author of *Atlas classique de géographie ancienne et moderne* (1916) and *The Foundations of Geography in the Twentieth Century* (1919).   2.   The Lucknow Zoo was one of Geddes' planning projects. Geddes was completing his appointment as Chair of the Department of Civics and Sociology at the University of Bombay.   3.   Samuel Johnson (1822–1882), author of *Oriental Religions and Their Relation to Universal Religion*, 3 volumes. Octavius Brooks Frothingham (1822–1895), Unitarian minister, author of *Transcendentalism in New England* (1874).   4.   Samuel Longfellow (1819–1892), a Transcendentalist; his *Memoirs and Letters* was published in 1894.

Dear Lewis,

Bombay
Use Le Play House
address till further
notice (marked
Please forward)

9 February 1924

[Ms UP]

Yours of 5 & 9 January just in, & I had better reply at once before getting absorbed in clearing up, as I now must be for sailing 15/3. Unluckily set back from touch of malaria, but not severe, getting over it all right; the only trouble is that the two illnesses have weakened heart action a bit; however Dr. (a first rate man) is confident of clearing these poisons out; but recommends a little time in Switzerland (on low levels) to recruit before going into the press of Paris, London, etc. I think too of a short stay in Berlin to see things from that side; the more since Dr. Hardy and family will be cheered up by a visit, as well as make mine interesting for me.

Very interesting this new development of Orage! The connection of dance with life I've always believed in – not having it myself but seeing it in our children, who were all early encouraged, & trained. Connection too with mathematics per Ouspensky is also as per Life. There are indications all round here & there of its being at last in the air. (See Farquharson's coloured printing of my paper in *Sociological Review* – Part 1 – simpler side of schema.)

V.B.'s book not going fast – since of course unintelligible to most of the scientific & radical public – but rejoiced over by a leading Catholic reviewer, Dr. William Barry – who of course thinks as better Christian than we are![1]

Very glad you are recovering from your disappointments. Vive la jeunesse! After all, as things go, what other writer of your age has got a better reputation than you have managed to make!

The trouble about attempting a *History of Civilization in U.S.A.* is the magnitude of the task – one of years & years perhaps, as you get absorbed in it – in fact an *Opus*! Even Mrs. Taylor's small but ambitious volume (send copy of your review of Banabhard) was practically completed years before the war – but she kept it by her, & I daresay improved it as the time went on.[2] You will feel the same sort of difficulty – as I too in my tasks have always done – never good enough!

Still, plan it out! Keep it before you for a year or so; and see how planning needs to be altered? If it survives your own criticism, and that of others, write it in instalments as ordinary articles when possible – each thus a sketch of a chapter – & eliciting criticism & discussion – preparing others for the coming book. Then too would not this plan make a *course of lectures* – in New School, etc., with more audience than possible for architecture? (What *alternative* books have you in mind of less vast scope?)

Do not bother any more over my *Dramatisations*. At most this – jot down on a postcard the names of a few bookshops to which copies might be sent, 1 or 2 dozen each, "on sale or return" by *Le Play House Press*.

Thanks regarding Peace Movement publications. What is this new scheme of *Filene's*? Who got the Bok Prize?[3] Can you give reference to any publication of the pick of schemes, which I suppose the trustees would or will bring out? That might be suggestive. (I'd like to have *your criticism of my IX–9 scheme – Wardom to Peacedom* – which still seems to me clearly woven, and gets bits of additional detail, etc., now and then. What seem to you its difficulties, limitations, etc.? Pray tell me!)[4]

I hope to get to writing, when this removal is over, perhaps at my Tower in Edinburgh in late summer & autumn (and more room to spread out there than I can hope elsewhere!) But plans not yet clear.

I feel as if there may be a possibility of increasing demand for sociological ideas of many kinds. There is surely in progress a wide discouragement with politicians – and Labour may help in this. Will its influence be felt among your labourers & intellectuals? Towards more criticism of the powers that be, no doubt – but with any growth & organisation?

One criticism I wish you'd write is of London attitude of mind – what revelation of fatuity, as bad as your anecdote of Butler's servant – in the mass of its novels, plays, talk, etc.! Your "Main Street" is Athens to most of it! I seriously think the foolishest mass psychology in the world – & tragically illuminative of how great cities & empires have decayed to death.

Always yours *P.G.*

(Congratulations on your rings! & all good wishes for golden wedding by & by!)

1   The reference is to Victor Branford's *Science and Sanctity* (1923). William F. Barry (1849–1930), writer on religion and literature; author of *The Two Standards* (1898) and *The Wizard's Knot* (1901).   2.   "Banabhard" was Geddes' nickname for Rachael Annand Taylor.   3.   Edward A. Filene (1860–1937), retailer and founder of Filene's store in Boston, offered prizes for "practical" European peace plans during the 1920s. Edward W. Bok (1863–1930), successful editor of the *Ladies Home Journal* and philanthropist, in 1923 established the "American Peace Award" of $100,000 for "the best practicable plan by which the United States may cooperate with other nations to achieve and preserve the peace of the world."   4.   Mumford's note of February 1971: "How could I tell him that this seemed a mere marshalling of abstract categories?"

Dear Mumford,

Bombay
(but address Le Play
House till further
notice)

18 February 1924

[Ms UP]

I am now about all right again – getting the last germs of fever (now quite over) dislodged from their last lurking places in spleen & spinal cord, etc., by a new drug – pernidine I think it's called – said to be of greater efficiency than quinine, which does not penetrate there to sufficient extent. Also being massaged daily by the powerful barber of the Nursing Home I was at, who is an Indian expert answering curiously to your osteopaths, & who similarly unstiffens the old back, etc. – & has made me with a month's treatment quite keen for exercise, & even exercises of Müller![1] So there's life in the old dog yet!

I'm busy clearing up to sail 15 March, but leaving son to wind up Exhibition, Town Planning of Lucknow Zoo, Tagore's University, etc., & return in autumn. He has meantime shaped into a keen writer – & after these three winters with me, has got hold of main ideas, & is convinced he can put them better than I do – into less unreadable form than mine – in which view he seems to be right. So it looks as if I were getting my long-needed collaborator in getting more stuff out.

Tell me what you made of my paper in current number of *Socological Review*? Will any of your sociologists read it, I wonder – & with what impression?[2] Anyway I am following it up with another chapter.

I am a little concerned about £.S.D. ($), as both my sources of income now come to an end very shortly (chair, & Indian Town Planning) so may have to sell my Exhibition. If you think there is any chance of its being bought in America – as by Brooklyn Museum, Russell Sage or other group – the price, to repay its heavy outlays at all, would have to be $15000 (or *at least $12500*) out of which I should pay you $1000 for your time & trouble. Do you care to enter that speculation? For *complete* Exhibition of it very considerable wall space here is required, about 440 yards length of wall-space – close filled from floor level to average of 9 ft – say then *1300 sq. yds.* or thereby. (You have photos giving some idea of its extent.)

But in three cases – Paris 1916, Jerusalem 1920 especially – I have made a smaller Exhibition – showing about ⅓ or less of that and leaving the rest in its cases for reference as required. That is really the best way, for on the full scale the visitor sees too much detail. Another possible idea is for organizing an itinerant Exhibition – through a succession of cities, as with London, Edinburgh, Dublin, etc., before war Ghent Exhibition (1913) & since then the five main Indian capitals, etc. In the last case, the provincial govts. syndicated & paid about £1000 each = $5000 in round numbers, which was all needed to meet expenses of transit & tour & assistants' time in hanging – which alone needed a month before the Exhibition – & then of three weeks or *a month's Exhibition*, with very frequent, indeed regular explanations to visitors – itinerant lecturing with pointer – by self & assistant daily. I don't know that *I* should face this now – but my son can now explain it as well as I, & more popularly; & can easily train an understudy from the available town-planning assistants of a little experience.

But for this a group of four or five cities would be needed, alike to make it worth while to meet such heavy transit expenses to U.S. & over long distances there. Indeed perhaps that fare is too low? – unless they were the Eastern Cities, not too remote.

So that is that. (You yourself would find it worthwhile to take a turn with Exhibition of course paid for time.)

Now however for the intellectually more interesting matter – of general exposition on the theoretic side – of long dreamed *opus* – to which this recent illness has roused me, as nothing before. Do you still hold the view of this being a very remote affair? For one's contemporaries & the mature generally, it is not simply difficult, but generally impossible. I have not produced much impression on University & City here, as you may well believe; yet there is a certain influence, & as long as young men in moderate number take to such presentment – as even here enough not to leave me quite discouraged – there is no ultimate fear.

But now suppose we accelerate matters? – as by "syndicating" further than heretofore of possible Le Play House, Tower, Otlet, etc., & with a group actively pushing International Intellectual Relations at Geneva, & from these places more generally – should we not all have a much better chance of hearing? I might come over, as Ely was good enough to suggest, & lecture at his Town Hall (by the way what of that for initial (smaller) Town Planning Exhibition – could you talk that over with him?).

A good test which would help me is if you'll answer how far that graphic presentment of the transition from Wardom to Peacedom "(IX to 9)" remains in your memory & mind as effective, or how far it has faded? Broadly speaking, all existing Peace Plans are broadly within IX, as limitations of Wardom, & only use 9 as practically but Utopia. Whereas for me it is *Eutopia* & "practical politics" accordingly. (Please send reference to book of best Bok Prize schemes.)

And what other Schemata, etc., appeal to you as serviceable? – *Which as not yet so?* How far have even you taken in main thesis – that *all science – each & every science* – is working towards graphic presentment – & that those who don't see this are so far fixed amid verbal *abstractions*, & not freed for clear *generalizations?*

Yours,

P.G.

1. Perhaps a reference to Johannes Peter Müller (1801–1858), the German physiologist and "father of experimental physiology" who wrote on the relationship between sensing and thinking.   2. "A Proposed Co-ordination of the Social Sciences," *Sociological Review*, 16 (January 1924): 54–65.

7 Clinton Street
Brooklyn, N.Y.

25 February 1924

[Ts NLS]

Dear Master,

I learn from Branford that you have been ill; and I am sorry and trust that you are well recovered. On the other hand, he tells me that you have decided to return to Europe; and this is good news; for I fear that in too many ways India has been desert air, and it is time that you were settled long enough to gather a school for the fruition and reaping of your thoughts – peripatetic of course, but within limits! The ripples occasioned by your plunge in New York are just beginning to appear, like those which reach shore long after the boat has passed: Waldemar Kaempffert, for example, has written an article on power in which he used the terms neotechnic and paleotechnic to distinguish between the two ages.[1] In my own life, the effects are beginning to appear also: I recently acquired Whitehead's *Introduction to Mathematics*, for example, and to top this, Sophie and I are going out to the country for three months, beginning 1 March, to look after Whitaker's farm while he is in Europe. Both of us have weathered the winter poorly, and need a rest; more than this, we want to find out how we take to living alone in the country, for both of us, as you know, are city bred, and are acquainted with the country, in so far as we are acquainted at all, only in the summer. I shall take *Sex* and *Evolution* out with me; and since we will be starting the vegetables for the garden, first under glass and then in the open, I shall seize the opportunity to refreshen my biological knowledge. All these plans for renewing contact with the earth call me back to six years ago, to the spring when I joined the navy, for my going away broke into a course in advanced biology I was then taking at City College, and upset the plans that I had made for tramping up the Hudson Valley and spending a little while in one of the quarries or brickworks. Instead of this, the navy gave me its sterile routine; and I have waited six years for another opportunity. Both of us are full of pleased anticipations. So far I have no definite work planned for the interval; on the contrary, it seemed better to resign myself to short rations for awhile and give a chance for the old habits and associations to break up, and for newer ones to come to the surface.

I think I mentioned in my last letter the possibility of getting a subsidy to write a "History of American Civilization"; but this mirage has now disappeared. The *Freeman* will suspend publication with the present issue, and with this will go not merely the weekly outlet but my prospective subsidy, too! It's too early yet to find out how my income will be affected by all this. Since his cancellation of my course at the New School, I have heard nothing from Johnson: I suspect he has a bad conscience, and, since he has treated me somewhat shabbily, has transferred the reproach to me. There are times when I regret the impossibility of an academic career; but I have only to talk with a University professor or instructor to clear away such regrets; for the position is an almost impossible one in America, even under the best of circumstances. Even men of great capacity and strength, like Stanley Hall, have been curbed into conformity against the pressure of their own interests and enthusiasms; and I haven't enough energy to fritter away so much of it on a hostile academic environment. I have been reading, or rather skipping through, James Mavor's autobiography; and have been delighted with all the glimpses he gives of your old life in Edinburgh. That was a vivid and educative situation: the glimmer that appeared over thought in life in the eighties was almost dawn itself by comparison with the dullness and misery of the environment out of which it grew; and I can't help thinking that the generation which came to maturity then, the generation that included you and Kropotkin and Mavor, had the best of it, by far, over any that had gone before or have come since, since Goethe's time. With your great speed of dictation, I should think you might write a book of your own reminiscences in less than a month, and it would be a precious thing to have; for it could show your pursuits and your ideas in their genesis, and thus it would supply a needed element for their understanding and acceptance. The story about the fall from the trapeze and the sudden resolution of the first synthesis, for example; and you told me many others that were equally illuminating. If history, in Croce's sense, is criticism, so is autobiography; for what one does for the community the other does for the individual.[2]

My chief intellectual experience these last few months has been the finding of W.H. Riehl, the German historian. Gooch mentions him in his *History and Historians of the Nineteenth Century* as one of the main culture historians; and his *Natural History of the German People*, writtten in the forties and fifties is a masterly application of the regional method to History. Treitschke once dismissed Riehl as a "historian of the salon," and the reason is plain, for with Riehl the place, the people, the work, the home, the industries, the arts, the folk-music, and so forth are in the foreground, and the political organzation enters no more into his history than it does in the life of anyone who is not a functionary of the State.[3] Without the Comtean clues to the four orders, Riehl nevertheless combines instinctively the methods of Le Play and Comte. He has written a chapter or two on method which might serve for the regionalist's handbook of history; and I have half-hinted to Branford that this might make an interesting little item for *Discovery*. As early as 1850 Riehl saw that the effect of the railroad was to join city to city, and to depress the countryside by draining it into the railway capitals, whereas the old system of roads enabled the city to get out into the country, and preserved an economic balance. He also predicted that freedom would be gone in America when the forests were destroyed; and the beginnings of imperialism here do indeed date from the passing of the frontier, in 1890, and the exhaustion of the Appalachian forests.

I have been delinquent in answering Arthur's letter; but I trust he will not think too hard of me; and I will make amends when the country gives me a little more leisure. Tell me what your plans and moves are. Sophie joins me in sending you affectionate greetings; she has already decided that when she has a baby, it must be named after you, if you will consent. (Incidentally, as another proof that your words weren't lost, we exchanged rings last Christmas.)

Lewis

1. Waldemar Kaempffert (1877–1956), science writer and editor; first director of the Chicago Museum of Science and Industry; the article is "Era of Super-Power," *Forum*, 71 (January 1924): 62–76.   2. Benedetto Croce (1866–1952), Italian philosopher, historian, and critic. 3. Wilhelm Heinrich Riehl, *Die Naturgeschichte des Volkes als Grundlage einen deutschen Social-Politik*, 4 vols. (1866–1882). George P. Gooch (1873–1968), *History and Historians in the Nineteenth Century* (1913). Heinrich von Treitschke (1834–1896), German historian.

Dear Master,

This is the hastiest of answers to your most recent note; and another will shortly follow.

The editor's name was *Glenn* Frank. His magazine is *The Century*. New York is sufficient address.[2] I should write him by all means, and if possible send him the first chapter, typed, and explain the scope of the rest of it.[3] The only manuscripts that ever remain on hand for a long time are those that have been accepted half-heartedly; no well run magazine keeps such a large amount of literary capital idle.

As for literary agents, they are all bad – even worse than I am, so you may as well let me take care of the manuscript if Frank doesn't accept it. There are three or four other magazines that may – *Scribner's*, *Harpers*, *The Forum*, the *Atlantic*, the *North American Review*. I trust this note finds you better.

Affectionately,

Lewis

Twelve Opossums Farm
Hackettstown, N.J.[1]

27 March 1924

[Ts NLS]

1. The Mumfords lived in the New Jersey farmhouse owned by Clarence and Gene Whitaker while the Whitakers traveled for three months in Europe. 2. Geddes' "recent note" has apparently been lost. Glenn Frank (1887–1940), editor of the *Century*, later became president of the University of Wisconsin; see Mumford's letter of 11 December 1928. 3. Probably Geddes' "reminiscences" of his work in India; see Mumford's letter of 9 April 1924.

———————

Dear Master,

I am within a chapter of having finished *Sticks and Stones* and I am at last free to answer at some length the last few letters that have come from India, dated the 3, 9, and 18 February. The news of your relapse grieves me, but I trust that this will find you in Switzerland or Scotland, well-recovered and rested. Spring is coming tardily out here, one warm day forward and two snowy ones backward: we mind our little garden under glass, I chop wood in the afternoon and write in the morning, and so, at slow rhythm, but all too quickly the days pass. This little valley is a miniature of America: its rocky hillsides are full of iron, and a century ago a mine was worked not quarter of a mile away from our house; except for two bleak churches, whose only congregations are a handful of old women and children, the mine left no trace but its own pits and debris. On the farm above us the land has been tilled for five generations; the soil was originally stony, and is now worse; that and all the farms around are running down, and the houses and barns are rotting away, since there is no use in "improving" land that waits only to be sold. "Mine and move" is the caption for all this effort. As if man had not done his worst here, Nature, in final irony blighted all the chestnut trees and the naked whiteness of their trunks stands out against the hillside like a forest of birches, beautiful now in the distant haze, but bleak, when the leaves begin to appear. All this sounds rather miserable; but the esthetic effect is really much better: the country rolls into upland, like the Cotswold, and many of the scenes are genuinely beautiful. It is a happy change.

What you say of the difficulties in writing a history of civilization in the

7 Clinton St.
Brooklyn[1]

9 April 1924

[Ts NLS]

U.S. is all too true. What I had in mind was a synoptic interpretation which might pave the way for the history proper; but even that would be colossal; and since the fund I had anticipated has failed, as they so often fail, to come forth I have done what I could in a small way, in *Sticks and Stones*, whose subtitle is: *An Interpretation of American Architecture and Civilization*. I have just sketched out a fairly lengthy, expository bibliography in which, (1) I rehabilitate Viollet-le-Duc's *History of the Habitations of Man in All Ages*, obsolete in its ethnology and archaeology, but magnificent in its humanistic grasp of architecture, (2) point out the outstanding contribution of Ruskin as critic and prophet, and (3) draw elaborate attention to your and Branford's contributions, including the analysis in the January *Sociological Review* and the *Principles of Sociology*.[2] (2) will make amends for the over-hasty treatment in the *Story of Utopias*, until a new edition gives me the opportunity to rectify the impression there.

As for the IX to 9 diagram. Once it is laid out I have a difficulty in presenting it to others as a picture of the existing order and of the possibility of its antithetical alternative. The difficulty lies in the manner in which it is built up, particularly with the initial terms.

Clarence Stein was out here last weekend, and I consulted with him about the possibility of placing the Cities Exhibition. On his recommendation I am writing to a colleague in Pittsburgh, named Bigger, who is secretary for an active committee on City Planning.[3] The main thing to do is to circulate the news that the exhibition is for sale. As soon as I get back to the city I shall put together a little memorandum describing the exhibition to serve for answer to correspondents. The real difficulty is not to get the exhibition bought, but housed. I have more hope that this may be done in a provincial museum than in New York. Stein thinks the same. I shall also write to Pray at Harvard. As to your reminiscences in India, there is still another New York magazine that might welcome them: *Asia*. I hope that you not merely get this particular bit done, but that it sets you on the more general task of writing an autobiography of your ideas. If once you started this, using no more than the material of your conversations, you should get the whole thing written in a few months. It would be enormously valuable; for it would be a key which would open up a great many diagrams which now, to the outsider, seem obscure. Do not leave this to anyone else. I had hopes of undertaking it myself before we met; but it did not take me long to realize that the essence would evaporate in the course of transference. Miss Defries' book is misleading; anyone else's would be the same.

As for the IX to 9 diagram. Once it is laid out I have a difficulty in presenting it to others as a picture of the existing order and of the possibility of its antithetical alternative. The difficulty lies in the manner in which it is built up, particularly with the initial terms.

Military/Theological > Political/Abstractional > Mechanical/Physical

Comte generalized these terms from history: the ordinary student can grasp them in sequence, but fails, for the most part, to grasp their interaction and cumulative effect. Even if IX is not challenged on the grounds of failing to represent the historic process, there is a further difficulty with the 9. Are the three corresponding terms also to follow in order? Or are the general elements of the 9 to develop more or less co-ordinately? If we roughly date the Military Order at 1299, the Political at 1600 and the Mechanical at 1800, are we to date the Biological at 19–, the Geotechnic at 20– and the Eupsychic at 22–: or are they all to be resolved at an indefinite point in the future, y. This last point has never been clear in my own mind. How would you answer it? I hold entirely with your thesis that if we could work out the *logical* antithesis we should have a key to the *pragmatic* synthesis and by *acting upon our hypothesis* would ensure its success. It is on this very point however that most of our dispersed modern minds will be in rebellion; for they do not recognize any inherent connection between logical order and the world of fact. This point occurs in the initial explanation of the diagram, too. One can either say, taking one's three terms as granted, that theology reacted upon abstraction and created the myth of the Powers; or one can say, logically, theologize your *politics* and you get Theologized Abstraction, i.e. the State. Similarly one can say that the hunter's military tradition, reinforced by machine industry, causes war; or that if you mechanized the military order you get mechanized militarism, i.e. modern war. The first set of statements is historical, the second is logical. In verbal explanation you get the benefit of alternatively using one or the other, as seems more profitable; but in a rational description one cannot slip so easily from one category to another. Quite apart from this, I have only one or two suggestions to, tentatively, make. The first is that the spiritual element of the financial order is not $, the theory of money, but the Prospectus or advertisement, and that the antithesis of this, under social finance, is the Policy. Likewise the antithesis to the Ballot seems to me not to be the Transition, but Group-Direction. These are of course minor points; possibly inevitable ones, for if the prime elements are given, the values which will be substituted for their combinations will be different ones for different thinkers until divergent interpretations are brought together and reconciled. Thus, using Good, True, and Beautiful, I got an entirely different set of institutions and states for Good-truth, True-good, etc., than you had given. This is obviously because Good, True, and Beautiful are only counters, or tokens, for a whole variety of things. How can this be made more rigorous? That is, I think, your capital problem. Otherwise the diagrams tend to remain as personal as the more chaotic philosophies they replace. Is not this one of the reasons you sometimes lose adherents? They are not convinced of the impersonality of the logical method: it seems neutral, but what comes out of it is Geddes! Some way must be found to show that there are not strings, that anyone who takes pains can manipulate the same instrument, to his own advantage!

Tell me of your movements and plans. I keep in fairly close touch with Branford.

Affectionately,

Lewis

P.S. Fagg has just sent me an interesting paper on Freud's contribution to the Evolution Doctrine.[4] Have you seen it? Fagg is an excellent fellow.

1. The letter was obviously written while Mumford was at the Whitakers' Twelve Opossums farm in New Jersey. 2. "A Proposed Co-ordination of the Social Sciences," *Sociological Review*, 16 (January 1924): 54–65. Probably referring to "Civics: as Applied Sociology, Part I" and "Part II," *Sociological Papers* (1905 and 1906). 3. Frederick Bigger (d. 1963), Pittsburgh architect and city planner and Executive Director of the Citizens Committee on the City Plan of Pittsburgh; member of the Regional Planning Association of America. In 1919 he had written Mumford to

inquire about the possibility of bringing Geddes' Cities Exhibition to the United States.
4.   Christopher C. Fagg, follower of Geddes; leader of regional survey efforts in England; with C.E.
Hutchins authored *Introduction to Regional Surveying* (1930).

Montpellier (but
reply as usual to Le
Play House)

26 April 1924

[Ms, incomplete and
unposted, NLS]

Dear Lewis,

(Yours of 27 March (as literary agent!) & of 9 April, with personal news,
& criticism of schemata.)

Glad you are in such interesting environment (albeit of "La Terre qui
meurt"),[1] and getting on with *Sticks & Stones* (an ingenious & effective title!)
which I hope will be available soon.

For myself, a very poor time since sailing 15 March. On back mostly all
the 18 days to Genoa – only in last days up on deck for few hours but never
at dinner in cabin – mild diet. No illness but utter weakness & lassitude – a full
experience of advanced old age, such as never before! Then from landing on
2d till last week, cold wet or windy weather – very late season in Europe – thus
depressing, & continuing weakness. Still with injections of neurasthenic serum
& then of strychnine, I am picking up, & "summer is y comen in." But I must
take it easy. No sustained energy even for correspondence, till these last days.
I begin; & I take yours, partly to avoid delay, & partly since stimulating. I have
read a good deal, & had some ideas all along day by day, so the long dreamed
synthesis still feels *growing*, & clearing in main features, but I don't think at
this rate I can face Paris & London for a month & more yet. I think of going
next to the swim cure which has so benefited the Branfords. Then perhaps the
needed energy for writing may return. (Indeed I must try, since now I have
neither chair nor practice!) And also more calls for unpaid work than ever –
say four chairs if I could fill them! – at Paris, Brussels, London, Edinburgh, and
even here I could stay on a long while, if life were not now of such shortened
perspective!

Bad news from India of my remaining practice, e.g. court intrigue around
Nizam has (temporarily I trust) knocked out university, & probably myself (&
Mears too) permanently. Other worries & losses too. But as sign of con-
valescence, these troubles invigorate.

Well now your criticisms (or rather, as I think, *difficulties*) with
(consideration of) IX–9. Such highly condensed generality needs endlesss
exposition for its appreciation & use: & it needs long talks beyond compass
of a letter. Yet see (& say) if this does not help towards clearness.

The main point, I take it, for you & for others, is that all this sort of thing
seems so personal to me – idiomatic, individual & so on.

But what was at first more personal than this notation of *Mercator*,[2] at
first the exclusive patent of that particular chapman* of ideas! – and so
uninteresting and abstract to his contemporaries, fascinated by the concrete
interests of new geographical discoveries! At first sight his blank diagram of
squares was mere dullness & emptiness & lacking in all the concrete
knowledge of the old charts, let alone the vividness of the old pictures &
descriptions of many lands. But by & by people understood how it showed all
the world & for the first time with the relations, the respective positions, of
such land & sea appropriately clear.

*Hence (as example of conversational & personal thought-stream) a possible excursus, on the Latinisation (or Grecisation) of surnames so general among the intellectuals of the Renaissance, & its explanation too, as expressing their passage from PWF to FWP, Town & School to Cloister. But here too, the next idea which happens to rise is the larger one of Hegel's *Aufhebung*[3] – as his intuition & version of this process – PWF, FWP. These 2 ideas, one so detailed, the other so general – just happen to come to me as I write: but to any one else, may well appear queer digressions, & not what they are, chance suggested illustrations – of which every one will make his own, as the notation, etc., become familiar.

Yet even now, how few can really read a map? Not merely to recognise places, but to see, & each place in its position – & with its regions & cities, thus evocable, to their detail.

But now to your specific difficulty. Comte did indeed generalise his "three states" from history, but also justified them in terms of his scientific philosophy, his sociological pro-synthesis. But here on IX I elaborate them afresh, & in > 9 the corresponding three others (of which Comte leapt the first two, so the third (for him fourth practically), of Humanity, but for me sixth).

Every step, from the earlier agricultural order – is also here worked out & explained afresh, and at once historically & rationally. It is an admitted fact that the peasant world was ever beset by the predatory & military onset of the barbarised hunter, utilising the nomadised shepherd & utilising him often as guide, leader, strategist; & utilising too the woodman & miner, with their fire, their tools, & weapons, their *powder*. So too often the fisher as pirate & conqueror.

The peasant defence is weak; his Kraal, his wall and ditch, are not enough. He admits, attracts, bribes, tames, civilizes, a hunter, a nomad, a pirate, etc., – a warrior, in short, compounded variously of these. The barbarian is fairly true to his salt, & keeps back his tribal kindred like any other. Hence Fioun, hence the barbarians – generals & more in service of Rome, the Varings of Constantinople, & so on – to Napoleon or Von Moltke[4] – or to the (gipsy) Lee, the (Highland) Grant! So Wallace = "Welsh," & Bruce is Anglo-Norman. So again British generals, etc., so largely Irish or Scots (Wellington to Haig but conspicuous instances).

(Again conversational elaborations – instances just as they happen to come, from teeming multitudes of history. Any number of times these illustrations might (indeed would) be different, since the Masque of History is broad as well as long.)

Agriculture – indeed peaceful industry of all kinds – is dispersive: each man to his own field or job. War *must* concentrate, the warrior *must* mobilise, control, command. The old folks of authority & influence (male & female) in the peaceful peasant world have no choice but to accept this dictatorial command. (Pity it took our old fogeys in power, or old wives in palaver-houses & in press, from '14 to '18 to come to the single command, with its corresponding victory!)

But war needs weapons, defensive & offensive, shields, spears & swords, bows, by & by guns! Thus a new impulse to industry, the Smith in power! Tubal Cain, Wayland Smith, "Hal o' the Wynd" to Shaw's Armorer (= Krupp, Schneider & Co.) are again a fresh pageant stream of history. (And so the Vikings – Olaf to Nelson – & to Mahan's understanding of their historic significance.)[5]

What the historians however are too gentlemanly & patriotic to point out is that all these smiths are profiteers! What will not a man give for his life & pay for [...] Brain-lister! War-Industries thus not only supremely skilled (from Regin's forge of old to Alwar's in recent memory, or in our own time to each & every national experience) but also supremely costly. And the more armament we need, the more must industrial workshops be organized – they grow to arsenals of State accordingly. So alike from military authority on one side, & its need of technical equipment on the other, we see the increasing development between them of the military & armed State! (For this fundamental origin of State, from militarism (authority & equipment alike) & its development for & from conquest also – see much historic literature – e.g. Oppenheim on State)[6] but thus we have Military Development creating Political Development as Industrial Development too, with fertile reaction on this – indeed interaction all through (to IX).

Here then a sharp difference from Comte's approach & presentment (a) from their spiritual sides, & (b) from their respective historic ascendency, their successive preponderance.

Yet these are not excluded or forgotten: on the contray, more or less freshly presented & illustrated. Thus with change from peaceful agriculture or pastoralism to wars of defence or offence, the old mother-goddesses fall out of sight; & the gentle partriarchal Father, who walked with Abraham (his prototype) becomes replaced by the God of Battles: "the Lord of Hosts is his name!" Thus in short, the Theology appropriate to militarism, authoritative & formidable, of gods or God, now literally to be "*feared*"! ("*La crainté du gendarme est le commencement de la sagesse*!"[7] And even "*Alles verboten*"[8] is but the perfect re-temporalisation of this sort of spiritual power!)

The other peace-world's Theologies were each an ideal-ology with idealizations alike of life phases & their ideals; but now cause and develop their contamination, increasingly indeed their replacement, by Theologisms (as Comte calls it) which have more & more of Militarism. *Caesar divin!*[9]

How "the people" becomes "the Kingdom," and thus "the State" need hardly here be elaborated. But these ideas are increasingly intellectual and abstract, & not theological, however partly theologised, though this as far as emotionalised.

With military discipline (& its needs of men to discipline) the old simple village council, the Indian Panchayat, are increasingly dominated by a very different conception of justice, in which "customised tradition" gives place to "authority" & "law." These two latter diverge – the first remains & intensifies as personal & is strengthened by theologism to veritable "majesty" with its outspoken will ("*Sic volo, sic jubeo*"[10] says – & must say – every commander, not simply the poor Kaiserling). But thus we have not only "Right," but "Right Divine," consecrating the Force which assures it, applies it, justifies it in all eyes; and thus evoking submission, even devotion, i.e. "loyalty" in full old sense, from people: so far justified by results of course, when war (be it of defence or offence) is successful: even taking heroic & pathetic forms when the opposite.

1. The earth that is dying.   2. Geradus Mercator (1512–1594), Flemish geographer who created the meridians of latitude and longitude for his map projection.   3. "Hegel's *Aufhe-*

bung": to supersede or sublimate; the process of change, incorporating the good from the past into an improved new stage.   4.   Helmuth von Moltke (1800–1891), Prussian field marshal who worked to make the Prussian army a formidable one.   5.   Tubal Cain: the first forger of iron and brass (Genesis 4.22). Wayland Smith: blacksmith and armor-maker of English folklore. "Shaw's Armourer": Andrew Undershaft, the weapons manufacturer in George Bernard Shaw's *Major Barbara*. Alfred Krupp (1812–1887), German armaments manufacturer known as the "cannon king." Joseph Schneider (1805–1875), French industrialist whose firm, Schneider et Cie, manufactured armaments, ships, and locomotives. Olaf I (*c.* 963–1000), king of Norway. Horatio, Viscount Nelson (1758–1805), famous British admiral. Alfred Thayer Mahan (1840–1914), American naval historian; author of *The Influence of Sea Power Upon History, 1660–1783* (1892).   6.   Lassa Francis Lawrence Oppenheim (1858–1919), German jurist who emphasized national sovereignty over international law; author of *International Law: A Treatise* (two volumes, 1905–1906).   7.   The fear of the police is the beginning of wisdom.   8.   All is forbidden.   9.   Caesar divine.   10.   As I wish, so I order.

Dear Lewis,

You have been long & much on my conscience! & it is only because there is so much to say that I never say it!

I fancy my varied restaurant life in New York was not very good for me – and early in winter in Bombay I got run down – then dangerously ill – with colitis, but so severe that good specialist told me afterwards he thought he might lose me. However, three weeks of his nursing home & several in country brought me round – but soon with relapse & fever, touch of malaria, & three weeks more – after which he turned me out of Bombay for Europe. Very weak in voyage – long rest two months here at Montpellier – then month in Switzerland with V.B. at Sanatorium – thus again seeming o.k. – but after spurt of writing a book (*Biology*) with J.A. Thomson, Home University Series January issue – I had a minor relapse in September.[1] Then better – Congress of Religions in London – & busy time with Otlet reguarding his Geneva schemes – then here, where I am in my natural element (as not in New York City last year!) as builder of old into new – cottage into châteaux college, with unusually interesting grounds, to be made mostly out of two old quarries – one gardens & now wild shrubbery, & other of last century, precipitous field, with garden terraces to renew or build – also orchard to plant, etc., etc., ad lib – a far healthier job than writing books! (Vivendo discimus! Managing so much & varied work is stimulating too.)

Still I am building here – with small, but I hope not ineffective, Outlook Tower (photos by & by) for purposes we talked over. My son Arthur is with me, but busy for his geographical Thesis on Village – comparative village-ology! – but carving & decorating a bit in spare time (& so far "working his way").

I hope you got your new book well received. V.B. writes that he is much pleased with it; & will no doubt review it appreciatively, & then send it on to me here. What next?

If & when you have next one clear in view, why not write it here! This is College des Etrangers in general, pending an American house! And take your doctorate in nine months (D.Litt.) thereby, all at small expense? It might come in useful. And I think you & Mrs. Mumford, too, would be better here than in a great city – for a nine months change! And France is a better market for

Collège des
Ecossais
(Plan des Quatre
Seigneurs)
Montpellier (Hérault)
France

14 November 1924

[Ms copy Oslo]

ideas than most (although I am always chaffing Montpellier people, regarding City & University as each a "Belle aux Bois Dormant!").[2]

It was very kind of you to suggest to V.B. the writing up of my old notes – & it is true that there must be a good many thousands of pages! – most of which will/may never get written. Yet these are mostly jottings by the way – products of sleep on waking – & though I don't re-read them & fail to arrange, their writing fixes a good many ideas for further incorporation into the ever-growing web. To get that written is the main problem – & largely why I am making a winter home here.

What of these *Survey* articles – in which I've more than before "let myself go"! P. Kellogg is an excellent editor of them – & I am almost always improved by his treatment.

What of such articles on *India* for one of your magazines in that line? (I'd rather be invited than offer!) I have however a sea of arrears – nearly a year's! – and longer – of which this to you is but the briefest of samples! Still, once my big study ready, I'll do my best, with such time & strength as may remain, & with advantage of country life & work – though only 10″ beyond the tramway limit to the town – seen from this hill top – rising finely on its hill (*Mont*-pellier) projected against the Mediterranean & so an excellent outlook; such as I need more than I can say!

Alas I must stop, & go on to others: e.g. There is trouble at Jerusalem University – danger of lapse from our unified scheme into confusion of building by different architects piecemeal on different purchasers plots! – but Mears & I hope to get over it. It will be a real calamity to them if we don't! Mrs. Joseph Fels is to help, & may succeed in her way – but it is a fight.[3]

By the way I remember you write for *Menorah*. What of your writing Mrs. Fels a line? I'll tell her I am suggesting this, & that she show you my letter to her explaining situation! You might then be able to give a note to that journal? – *if she thinks it expedient*? Or to advise regarding Exhibition of our plans? Her address is *1 West 81st Street, New York City*. Call & see her. (Don't think this letter is for that reason – this point only occurred to my mind at last paragraph above!)

Cordial remembrances to Mrs. Mumford (whom I have seen so little – & I think of often with regret & even shame for failing her hospitality that time!)[4]

Always yours,

P. Geddes

1. Geddes and Thomson, *Biology* (1924).   2. Sleeping beauty.   3. Mrs. Josephine Fels, the widow of an American millionaire soap manufacturer, was active in Zionist causes. 4. During Geddes' 1923 visit to New York, Mumford writes, "after promising to spend a whole Sunday afternoon with my wife and me in our little apartment on Brooklyn Heights, visiting our home for the first time, he entirely forgot the engagement and never even phoned" (SL 325–326).

What would Mr. Kellogg say to "Talks" from this Outlook Tower?

Collège des
Ecossais
Plan des Quatre
Seigneurs
Montpellier (Hérault)

3 December 1924

[Ts copy Oslo]

Dear Lewis & Missis!

Here now these seven weeks, hard at work building and gardening, and with constant supervision of both – half-a dozen gardeners at it – not to speak of numerous masons, joiners, tilers, etc. I've not had such a time since making Dundee garden, and building Ramsay Garden – for though of course this is a much smaller affair, it is more complex, as by enclosing this old two storey cottage – and on all four sides curiously enough – e.g. the old kitchen (now for dining-room) has had to have eight holes (doors, windows, etc. great and small) cut through its walls!

Imagine this ragged moor I failed to buy – then its steep corner *here* – I pounced on; with medieval quarry overgrown with wild shrubbery above and grassy theatre-like slope below to old stone walls, now beautifully blue with time; below this again a little olive yard (there are also olives above): below again, a small vineyard opening to a village road: and below this again our future tennis court, between high walls and fine pine wood of my present home – the little château (not *castle* of course) of Les Brusses, of which my place was the *garrigue*,[1] all pretty well lapsed from its reclamation many years past; whence the five Italians needed to supplement the permanent Frenchman, whom I am training, I hope, from very decent labourer to gardener proper. My word! how these Italians astonish the French, and me too! Never before did I understand so clearly how the Romans conquered, and with spade as well as sword: one in particular I cannot but call Hercules! And if we had such men for and in education, things would "hum" and become Promethean as well.

Imagine then the old terraces broken or fallen – new ones neeeded also – terrace fields partly unreclaimed, many outcrops of rock to be blasted or cut out to get such land as may be ploughed for use! Big stones also utilised as "garde-fous"[2] for my *other* quarry – a XIX Century one, now being developed as Rock Garden, above on edges and below cliffs; which I'm altering by blasting here and there to improve picturesqueness, and to get space for planting fig trees in crevices, and stones, great for foreground-heightening and smaller for Rock Garden pockets, dry walls, etc.

Imagine too "Le Volcan" – a local name and fame I at first supposed to apply to the crater-like aspect of the quarry: but no – a rock opening, from which rises warm vapour-laden air from depths (and apparently with much CO_2), and condensing its invisible vapour on cool mornings. Hope soon to turn on geologist and chemist and get it explained and analysed. I see also how to warm *Salon* besides, and to run a small greenhouse – without injury of course to nature or amenity of quarry effects, etc. And even to get a "grotto" in the hole below Salon door-window.

By and by too the indispensable Bois de Pins, which every proper farm or country house, small or great, in these parts has by this time at full height, as admirably around Les Brusses below – but which has here to be planted (soon)! I have only as yet one good tree (and that is not very big), the best of its kind in the district, and greatly admired accordingly – an old *Chêne-vert*,[3] usually mere brushwood. But of course thereby all the more sun – the essential of this region in winter, as shade in summer.

So you see I'm greatly pleased with place, and with myself, as again happily at real *work* – open air all day long – and bed after bread and mild supper – say 8 p.m. – to begin next day with a spate of thinking, from 5, 4, 3 (or 2) a.m. as the case may be, and before going up to work, at 7 or 7.30 or 8. In short, I am again realizing *vivendo discimus*! (and even *Creando pensamus*).[4] I am only writing today – and the first non-business letter for long – because it is pouring and no one can work! (We have had the edge of the great N.W. European storm here – and beyond local memory they say.)

Another matter which pleases me is to be making (and to the joy of my dear old friend Flahault) a not inconsiderable succursale to the Jardin des Plantes (long ago over-crowded and not able to extend, since now suburb-blocked) especially as regards my wild Flora and shrubberies (of course respected), my incipient Bois – as afforestation experiment on small demonstration-scale, and my rocky quarry for succulents and *dry* plants generally.[5]

Also I hope here to illustrate and develop my theory of variation vegetable/floral in planting out – indeed I dream anew of getting at my long-labored botanic *opus*, entombed since I left Dundee and indeed never cleared up there!

So I might ramble on, for here are dreams emerging towards deeds, and some of the high Gestes I've failed to do opening out for a new try at each! e.g. University Hall, Edinburgh and Crosby (each in its way failure, though I hope towards new avatars) improved on here; and as experimental type for new Universities and old ones; and Collège des Ecossais for renewed siege of that at Paris.

Four studios too, on plans, of which first two will soon be ready. And a studio cottage, not too far off, forbye.

Forty nooks also, or more (in-nooks to balance outlooks) for dreamers, poets, etc. (not to obtrude biosophers!). So when you personally conduct Banabhard to Italy, pause ye here awhile.[6] Though of course most is still in the rough, or in germ, she will I trust see with inward (and partly outward) eye, since bringing both, in the words of one of the great future poets she inspires –

"Broad space of blossoms 'mid my garden closes,
Long lines of terrace walls afoam with roses!
(E'en now, spring bulbs are popping up wee noses)"

Arthur is carving the blazon on the Tower, already at land-mark height, since up to his study-oriel above mine, and with Salon as world-room below, but going up higher – for fullest possible view, like that of Edinburgh again to sea and Braids, etc., behind, from Blackford Hill; for we are well outside town – on its nearest adjacent garrigue. Camera at once, and I hope telescope dome later. For here every science has to come back in ordinary life.

The essential idea is this:

But at present Universities distil out all that is most vital in Region and City. Thus past achievements, but also too much now – into

i.   *"Religion,"* rancidising, fossilising, etc.
ii. *Learning and Science*, rancidising, fossilising, & dis-specialising also – as into dried plants of herbalists, and their distillates & extracts and alkaloids as per bottles on chemists' shelves (= University Departments and "Faculties").
iii. *Art* of all kinds thus excluded or paralysed.

(Now is the winter of our discontent!)

Yet how unite, synthetize all 3 (& 9) in cloister – Emotion-Ideation-Imagery – and re-incorporate all these into life of Individual & Group, in Place, Work and Folk, Sense, Experience, and Feeling anew; and by and by of Region and City alike? Whence in turn to fresh cycle of Cloister and City, dream and deed.

Hence an experiment towards Jerusalem, Tagore, etc., as well as to take back to Edinburgh if all's well. And a way of stirring up University here I trust! This is better than most, but still, as I tell them, "Belle au Bois Dormant"! The Studio Cottage is over the first ridge and thus out of sight only a quarter hour away. A new inland landscape, towards vines on hill-sides, moors and forests beyond, and lovely hill-village crowning valley. I offered it to Donnachie, but his School of Art duties keep him prisoner (and he is depressed as well!). Who'll accept the use of it? Tell me!

And now I've got (since paid for) a quainter place than either (which I perhaps told you of, as in prospect) at Domme (a picturesque little old hill-town of Dordogne) and next to my old friend Paul Reclus ("G.G.").[7] It is the *col* connecting to the town-ridge at its precipitous apex-level with old Domme Castle – now quarried away, save for vast souterrains.[8] I have thus on my small plateau-space two wide views – over Vezere Valley to North and another smaller valley to South – and on top of ridge the most massively built old windmill you ever saw = alike appropriate to situation in past and in future my particular line of quixotism, that of reconstructing after capture. Damp castle-moat for garden: old terrace walls for fruit, and for lean-to camp-shelters, and a medieval tower-stump, perhaps not past utilization. Then from a steep path up from public road sloping below, you'll by and by enter by old *"Porte des Poissoniers"*[9] as they came up to town with their catch from river deep below. This Gate Tower of town has four storeys ruined, since wide-wall fallen in, but still complete on two sides, cliff forming fourth. And thus you get up to terrace levels around and below wind mill hillock. In short all this affords the material for 2, 3 or 4 *châteaux* (term not necessarily so imposing as *"Castles"* in English).

Arthur is to spend Christmas vacation in study of possibilities, and to draft plans.

Finally, I'm after two or three ruined cavern-cottages at Les Eyzies.

You'll think my ship has come in. But no: my studio cottage (four rooms studio and an acre) over the ridge hard by, cost only £50 and this Domme place say £100 (though with taxes and expenses say £60 and £120) while the Les Eyzies cottages will be dear at £10 apiece! say thus £200 in all. (Of course, I can't afford to do them up very much, or almost at all, till ship does come in again – somehow!)

There are enough of these rambles! but there is method in my present madness, as in previous ones.

P.G.

P.S. I don't think I told you that I arranged lately with Tagore, whom I met lately in Paris (on his way to South America and Mexico) for his presiding at a *Conference on University Progress* in Edinburgh next summer vacation while I act as Organizing Secretary. So pray *send me names of all the appropriate live people you can think of*! In my ten years' absence In India I have missed new contacts, and most valued old ones are gone or going fast. So don't imagine I know whom to ask.

1. Waste land.   2. Parapet.   3. Evergreen oak.   4. We think by creating. 5. Charles Flahault (1852–1935), French bontanist at Montpellier and long-time friend of Geddes.   6. This was apparently a form letter that Geddes sent to more than one person. Mumford's copy has some handwritten additions. Boardman identifies "you" as Mabel Barker. 7. Paul Reclus, nephew of the French geographer Elisée Reclus (who was an important influence on Geddes' concept of the "Valley Section"); Paul had to flee France in 1893 because of his anarchist activities, became affiliated with the Outlook Tower in Edinburgh, and wrote under the pseudonym "George Guyou" (Meller, 119).   8. Underground.   9. Fisherman's gate.

New Address:
135 Hicks Street
Brooklyn, N.Y.

4 December 1924

[Ts NLS]

Dear Master,

I had been musing over a letter to you when yours came in and broke in on, or rather precipitated, my intention: and here it is, with, first of all Christmas greetings, and my hopes that next year will find you working at your usual vigorous level – although, heaven knows I cannot imagine you so depleted in energies as not to put in a bigger day's work than the vast mass of your younger contemporaries, who groan more perhaps but work less! I have followed this last year of yours eagerly through Branford; and although I have missed your letters, I could not bring myself to beg for a word from you until you were ready, since I know the burden of your correspondence, and know too how quickly with you the deed follows the decision. Your suggestion that I might spend a winter in Montpellier met half-way an impulse which was forming in my own mind: the chief condition is that I must have a book under way (with an advance form Liveright!). At the moment this seems pretty far away: for I am deluged under a horde of trivial tasks – the usual round of reviewing, plus two periods a week teaching English literature at the Walden School, plus the editorship of the Regional Planning number of the *Survey*; and as a result my time at present is too scattered to contemplate or brood over any consecutive piece of work. By next spring however the load will lift, I think; and inspiration may come. There is one other matter that is both an incentive and a difficulty. Sophie expects a baby in June. If travel isn't impossible with a nursing baby, and if no other complications arise, a winter in Montpellier would temporarily solve our housing problem, would give Sophie a certain amount of isolation and rest in a good climate – in short, would be quite admirable. But what about practical matters? Are houses or cottages available? What price? What would be the probable expenses? I know how difficult this sort of inquiry is: but it's impossible to make plans until one gets it answered. If you have time any day before next March to jot down a few answers, as well as any other information which I have not been keen enough to inquire about, please let me have them. The presence of the medical school is reassuring on the score of medical attendance: but it raises doubts about housing conditions! Sophie, by the way, is just as eager to go as I am: the brief glimpses she had of you have given her a respect for your Thoughts and Powers which does credit to her good sense and insight: so that the mere thought of you gives her a feeling of satisfaction.

We had an "Appalachian revival" meeting at the Hudson Guild Farm in October: we danced and walked over part of the Trail and spent long hours threshing out the contents of the Regional Planning number; and again and again some memory connected with you and your visit there would fall from our lips: so you remained with us and were among us. Let me describe briefly the number as it stands – or rather, as it is projected, for only a few of the articles are ready. It is to be a special number of the *Survey*: thirty thousand words; numerous diagrams and illustrations. About ten articles. The first one is on the Fourth Migration. Each great migration in America has spelled a new kind of opportunity: first, the covering of the continent and seizing the land: second, the migration into the industrial town; third, into the financial centers, New York and the ten sub-metropolises. Now, we point out, the community is on the eve of a fourth migration. The occasion is electric power and auto transportation, which, plus the radio and the telephone, tend to equalize advantages over a great area and thus rob the centralized city of much of its "attraction." On top of this is the fact that industry, housing, transportation, and so forth no longer operate automatically: their automatic growth tends to pile up embarrassing conditions, so that no industry for example can afford to pay for the urban housing of its unskilled workers. Since planning is necessary, why should we not plan so as to reap advantages from the Fourth Migration. The next three articles paint a picture of the impossibility of life under metropolitan conditions; show how the machine is clogged and stalled, how it is falling under its own weight. The city planning movement which attempts to rectify the conditions by treating their results has not rigorously examined the problem: regional planning means, not greater city planning, but facing the problems of the city by relating them to the regions that support it. It leads to urban conservation. Then comes the constructive section: an article by Mackaye on recreation (he has gone far since you talked with him), an article on transportation (its progressive elimination!), on the garden city, on education, and on social life as a whole. Here is an attempt to tie up the physical garden city, of which we know a great deal, with a program of civilization-building, in which we are, relatively, duffers. The issue is timed to come out in April: the same time as the Russell Sage report, the International Garden Cities Congress, etc., etc. If it stimulates any thoughts on your part, please jot them down, and I will try to squeeze them in, or, one way or the other, will make good use of them!

Stein, by the way, has been working ever since you came over here on the Garden City, and he and Wright have just made an interesting discovery: they are quite confident of being able to plan a beautiful shell: they are completely at sea as to what sort of *community* to provide for. I quoted to them Branford's notion that the townplanner neeeds the aid of the poet; and they agreed; and having succeeded so far, I told them a little about regionalism in Europe, and suggested regionalism must be made the cultural motive of regional planning, if it isn't to relapse into an arid technological scheme. Stein pretty well saw the point: I had waited patiently these last three years for an opportunity to make it. I think that in one way or another I shall be able to inject a little regionalism into the Regional Planning number; and that may give it a strange distinction.

As for *Sticks and Stones*, I sent a copy at the same time as Branford's to you via Le Play house; and I trust that by this time it is in your hands. It is provoking both enthusiasm and bitter enmity: this seems to me a good sign, much better than the tepid cordiality that *Utopias* met with. There will probably be an English edition; and the book is already being translated into German. (The chapter on the Imperial Age appeared in *Neubau* in August.) The criticism of the Mechanical Age is much more definite and drastic than it would have been but for our talks in New York, in the midst of the howling slums of Chelsea; but except for an occasional bit of historical interpretation the book has little that I value. Too much has been left out of it. Life is too short to find out all there is about Antimony! Branford's appreciation of it was more than generous.

I shall write to Mrs. Fels, as you suggest: it would be a pity to let the vultures get at your plans for Jerusalem. It is possible that one of the more militant Jewish papers would be an even better place for a letter or a note than the Menorah; but I shall abide by her suggestion. Don't hesitate to command me in any such emergency. It would be a sufficient reason for writing me.

I owe Arthur a letter; but I shall carry the debt on my ledger until sometime next spring; and meanwhile this carries news for both of you. Happy greetings to you both from Sophie and Lewis.

Collège des
Ecossais
Montpellier

December 1924

[Ms UP]

Dear Lewis,

This time I write by return of post – for fear of delays, & because your idea of coming here so warmly appeals to me, & seems so practicable.

Your own increasing *Regionalism*, & the idea of coming here are one idea – for it is from this South French region as Provence & *Langue d'Oc* that the happy appearance of a vital poet (Mistral) started the movement most of all, & fifty years ago or more.[1] I am only to a limited extent in touch with these regionalists – but when the constant pressure all day & every day of supervising building & garden abates (both more curiously full of problems than any I have tackled before) I know how to set about interesting them. Thus to the little tween-room theatre, indoors, I outlined in my first week here, the addition of a quite big open air theatre in a salient corner into the heath well walled by thickets of evergreen, oak, etc., & well away from road. (Of course this poetico-dramatic business is only one side of the movement – yet its flower, so far.)

Yes this has always been one of the *medical* capitals; & your only housing difficulty, I hope, will be of choice – between this house, ten minutes from Suburb Tramway, & my little guest house half a mile further off or so (though as that has a passable studio, it may be occupied by artist friends).

Your visit – your new book? – why not on *European* (or French) *Regionalism*? Here you'd be in one centre, & you are also in easy reach of Toulouse, Bordeaux, etc. – on one side, & of Grenoble & Lyon & Geneva, etc., northward, on the other – Marseilles too & Aix – & of course also near to Spain where Barcelona is trying to revive as Catalonia, & set example to "all the Spains." (Arthur's history Professor here protests against term "Provinces of France" – & insists on "*all the Frances.*" The movement in fact is latent everywhere, despite Paris, etc.)

Then too your regional or (kindred) book will be an excellent *D.Litt. Thesis* & that would be an additional string to your bow. It only involves 9 months in France – if & when you produce a passable academic record, as you can do. & as for this "Doctorat Etranger," they have accepted Arthur's irregular record without a previous degree & so will they yours. So bring (a) old college certificates of studies. (b) evidence of having been or being a Lecturer in various institutions. (c) any proof of assistantships? You were certainly a real one to me in New York City last summer semester – & they accept such informal & unacademic records all right, as I know by repeated experience. (d) of course you'll give a few copies of your books judiciously here – & a set to University Library (*but one at a time!*). So all this involves no mere return to student life, or *I would not suggest it* – though fresh & widening contacts with some of the best over here – quite worthwhile. A French translation of your *Utopias* would also (*perhaps*) do: but I rather think they require a *new* work.

Here, instead of the holiday observer (lying low) you knew in New York, I'll be really at work – and such collaboration as you chose to give me will relieve you of all expenses – & why not if all goes well, leave something over, & for each of us? (J.A. Thomson & I wrote a little *Biology* together for Home University Series last September & have now our big one on the stocks.) (Honestly, what do you think of my papers for *Survey*? I've not seen one yet nor got your book; but am writing for both to Edinburgh & London, as they have been only sending on letters, but not accompanying printed matter, a rule I gave them in India – but must now modify here – can such papers on *India* find a market?)

For first time for many years – several before these ten in India – my life has been so migratory a camping, that I have never had time or space to clear up in – & the space is for me peculiarly essential – now built here – & this garden environment most valuable. Here I've Tower & House going up together, so *can* tackle the Teufelsdröckian paper bags, without leaving them to your generous (yet too despairing!) scheme of having other decipherers alone!

A very valuable aid to potboiling, & to collaborative developments also would be the sale of Town Planning collections to cities & Libraries. If you see any opening let me know. I know the ropes of this job – & Arthur has an eye for it too. He has also done a good deal in the past year with my Civilisation Lectures in Bombay – but has now to go on with his thesis for "Villages" (I am sure you & he will get on together).

Another potboiling also interests us. Though the low franc rapidly adjusts on financial levels, & in city life, the old values change *far more slowly* in land & houses – and even in building & gardening wages, etc. (though in some measure of course). But for this I could not be building my *château* at all (or running 6 gardeners!) The *guest house* above offered you – 4 rooms with cowhouse easily converted into studio, & an acre of good land with fine views on 3 sides, though high cypress fence on one – has only cost me £50! This was exceptional – but this hectare & half = 3 acres, & 4 or 5 roomed cottage which I'm building round, a place of exceptional beauty & variety & possibility – cost only fr. 23000 with expenses, etc. – less than £300 – say £260 at time. While I've just bought, for £120, including legal expenses & taxes, one of the most picturesque places you ever saw – a windmill and ridge & château looking over two valleys on neck between old Domme & its castle – slopes, cliffs, caverns too I hope – & looking down into those from stoney city gate Tower (one side fallen certainly!), a medieval tower – & many terraces for fruit & for penthouse dwellings against walls – quite dry when cemented! also the damp old castle moat – a short one of course, for vegetable & flower garden. So I can camp my Archaeological & Regional School & make 2 (if not 3) *châteaux* out of the rest, if ship with a little capital comes in.

Here then are examples & materials for your "Fourth Migration" – an excellent phrase! – which I'll not fail to appropriate (with due acknowledgments!). Here too are experiments towards the *culture-villages* (& with ecomomic side too) you are discussing with Stein, etc.

Again this "Collège des Ecossais" (also a move towards recovery of our huge old Paris one) is provisionally also "Collège des Etrangers" – & an experiment in Hall (Hostel-Dormitory) conditions for students, which advances on my Edinburgh, London, etc. ventures – & will, I hope, tell on Jerusalem, Indian, & other Universities. Lots of Americans were here after war – but none practically now – mainly I believe because these old fashioned houses are not *warmed*, as your standard requires! But with improvement in this respect (much needed these frosty nights!) Montpellier might develop its rare advantages – as the only well known & eminent University in Mediterranean climate – & so attract foreigners from all North & mid-European ones. And so your visit may initiate an American house beside this!

Did I tell you I hatched with Tagore lately in Paris the project of a Conference on University Progress, in Edinburgh next autumn? (As yet all academic functions have been retrospective – time for a prospective one!) *Who should we invite from U.S.A.? etc.? R.S.V.P.*

Or of my Archaeological Station in Dordogne at Easter '25 (Domme affair is an element of this.)? I think so, in my last.

Indeed I wish you & Mrs. Mumford would come along now! For what if I've to go to Jerusalem next winter? – or back to India? For I am too deeply committed in both to be able to refuse – or able to afford to refuse! (Still I may be able to send Mears instead – if either ship comes in – though he is busy too at home, in Edinburgh & in Dublin also.)

Did your plan for the advanced Colony (to north of New York was it not?) (who were they exactly?) materialise – or is it waiting, like so many of mine?

Of course you read French, & possibly speak? With a little study you will find it the most brain-clearing of languages: & with all the defects, deteriorations, fossilisations too, so abundant, there is no doubt that this is still the best of countries for ideas & ideals, & more open to us than are our own – as I have lifelong experience of! So go on with your scheme – & believe me expectantly yours,

P.G.

1. Frédéric Mistral (1830–1914), French poet of the Provence region.

Lewis Mumford AND *Patrick Geddes*

THE CORRESPONDENCE

1925

Who is Herbert Crowley with whom Miss Alice Lewisohn sends me notice of marriage? I have seen name, but in what connection I don't remember![1] Did you write Mrs. Fels – & if so with what result or any?

Collège des
Ecossais
Plan des 4
Seigneurs
Montpellier (Hérault)

16 February 1925

[Ms UP]

Dear Lewis,

I've been hoping reply to my recent letter as to your possible visit, but just in case you may be free this season, I had better write to tell you of the upset of my present plans (& especially for Easter at initiative of a School of Archaeology in Dordogne). For the University of Jersualem is to be opened – in its (my) first small laboratories & temporary use of old mansion – April 1 by Earl Balfour – I declined invitation to come & speak – but the Zionists repeat it so strongly that I see between their lines, & indeed in them too – that I have no choice – but else to lose touch, & have all fall into a confusion of rival schemes! Thus I start from here about 1st of week of March, & shall be lucky if I get back here by last week of April.

There is to be a big World Congress of Education at Edinburgh next July, & I have to get Tower, etc., into better order before that. It also indicates the expediency of a smaller start of the Conference on University Progress, planned with Tagore – as this big gathering would otherwise too much overshadow ours – while they may give no facilities for making our project known for a better start next year.

Of course if you can come this summer, all this will interest you, I trust. I should like you to be in touch with what we have of Scottish atmosphere, as distinct from London's, Oxford's, etc. (Of course all atmospheres are very mingled! and our oxygen is well antidoted by soporific elements.)

I have never yet succeeded in extracting *Sticks & Stones* from Tower Circle, but I hope that is a good omen!

V.B. is coming here shortly for his Swiss cure (interrupted by visit to London, unluckily). I wish he'd give up business! – that life & the sociology don't agree, & make a muck of his health between them. Alas, he has just written he can't come – Dr. Jacot is keeping him – but Mrs. B. will be here for a short visit tomorrow – as are at present Mrs. Fraser Davies & Gladys Mayer – whom you probably knew in London.[2]

Our little *Biology* (Thomson's & mine) is now out, though I have no copy. But our leading Scots Librarian (Dickson of National Library) writes glowingly regarding its Bibliography – quite converted – & V.B. is rubbing it into Bergson Committee (International Relations) at Geneva & Paris.

Do you see how to get that element of this book reviewed by librarians in U.S.A.? Can you call their attention to it when you've looked at it?

Kellogg has just sent copies of my Talks, with your amazing preface![3] How am I to live up to that – unless you two come & help me!

Yours ever,

P.G.

1. Geddes may have been thinking of Herbert D. Croly (1869–1930), first editor of the *New Republic* and author of *The Promise of American Life* (1909). Alice Lewisohn married the British artist Herbert E. Crowley in London on 15 December 1925.   2. Mrs. Fraser Davies served as

Secretary of the Sociological Society.   3.   "Who *Is* Patrick Geddes?," *Survey Graphic*, 52 (1 February 1925): 523–524.

Collège des
Ecossais
Plan des 4
Seigneurs
Montpellier (Hérault)

18 February 1925

[Ms UP]

Dear Mumford,

As mentioned in my last, I have to leave for Jerusalem in little over a fortnight – so address not here, but *c/o Zionist Organisation Jerusalem (Palestine)*.

There is an active endeavour among various important Zionists to have separate competitions for University Library & other departments among *Jewish* architects & thus get rid of me – and Mears too – though thus converting our unified plan into a confused medley & even muddle. And Mrs. Fels & other help may thus be defeated.

One way of countering them is by compromise. Neither Mears nor I desire to make long stays in Palestine, to carry out plans there – so I should like to find some bright young Jewish architect whom we could associate with us, perhaps for trial as an assistant, but partner as soon as may be, or from the first if possible.

I am asking Unwin in London – but it seems to me far more likely to find such a man in America. Could you ask Clarence Stein, for instance (whose address I have not here) & other likely friends, if they can put me on the track? An early letter will much oblige.

Of course one would need to know all one could of his experience & capacity – alike for ourselves & for the Zionist Commission, who would be asked to engage him on trial in the first place, though as acting for & with us.

I'll be glad to hear from you too as to your plans. (I expect to be back here late April or early May.)

Yours ever,

P.G.

Jerusalem (but reply
to Collège des
Ecossais,
Montpellier)

21 April 1925

[Ms copy Oslo]

Dear Lewis,

I have now been here a month for the opening of the University & in the endeavour to retrieve an imperilled situation. For in my $4\frac{1}{2}$ years absence our planning of '19 and '20 had too much fallen into oblivion, and a block of biochemical buildings, etc., erected, extending my plan greatly with more money, but less simplicity & beauty & economy of result.

Fortunately however they had used our plan for world-wide advertisement of University scheme, & thus came to realise how they were committed to it. As also I think to some extent at least, that our "sublimation" of old Jerusalem was more appropriate & pleasing than the last Baroquerie of Berlin, etc. That block is happily out of sight – behind ours.

So now it is fixed that "the Geddes–Mears plan" is to be proceeded with; & with Library & Einstein mathematical–physical Institute, & even "as soon as may be, the Great Hall," as everyman says. Thus to some extent our ship is coming in (unless new dangers & difficulties arise!).

One is that their delay in purchasing sites is costing them dear. What was only £10,000 or so five years ago will now cost them five times as much. Zionism is deeply taxed by exorbitant land prices.

Now another matter. Pray buy a copy of the *B'nai B'rith Magazine* (National Jewish Monthly for March 1925) and see illustrations on page 209 with explanatory text of the most amazing character.[1]

I think this may be described as the worst thing ever built in New York or out of it! Is not that so? It would secure the dismissal of any practice from any architect's office as hopeless, would it not?

Is it not deplorable when architecture is improving in America, and notably in New York – and so much too for Libraries, that this great sum should be spent on establishing the record for ugliness. The Jewish cause is strengthened by such seemly buildings as it can erect, as I trust here – but that building is a vast advertisement to invoke the derision of every Gentile passer-by.

To describe thus in text below as of "*style used in ancient Palestine*" & "*in part after a section of King Solomon's Temple*" is again tragi-comic – & clairvoyance at its wildest! How could any editor pass such nonsense, & admit such building – refuge of lies?

As a friend points out, the dome is a parody of ours – a melancholy thought!

Can't you, as architectural critic, say a word of weight & as citizen, do something to warn the simple folk thus deluded, & put them in a way of a better design. Are there not surely better Jewish architects in New York than this awful duffer – said to have been here for a time, & thus bluffing them as above? Stir up your Jewish friends for their own sake, as well as for New York's.

Yours cordially,

P. Geddes

1. Illustrated are architectural drawings for the proposed Yeshiva of America on Amsterdam Avenue in New York City; the buildings are designed "in the style of architecture used in Ancient Palestine"; the design does in fact resemble Geddes' plan for the Hebrew University in Jerusalem – *B'nai B'rith Magazine*, 39 (March 1925): 209.

Dear Lewis,

I forget if I sent you enclosed in recent letter? If so pass it on to Stein, MacKay or other friend!

Sorry to leave beginnings here for a new fight in Jerusalem for Unity versus Babel as all proper Universities tend to be! If you can review Thomson's & my *Biology* (Holt – or Wms. & Norgate anyway) pray explain our Bibliography! (Encouraging that our Scots National Librarian falls in with it – interested as never before – so good omen for you perhaps. He sees it means organising the (generalising sort of) specialists to make such bibliographies for the librarians to help their intelligent customers!)

To go on recapitulating, I talked to you with pen anyway, about possibilities of selling the Cities & Town Planning Exhibition – but have since

Collège des Ecossais, Montpellier
– but please reply (for March & April anyway) c/o Dr. Van Vriesland, Zionist Organization, Jerusalem, Palestine

21 April 1925

[Ms NLS]

done nothing – & it lies stored & packed in Bombay, ready for dispatch. If you can do anything to help with that – (for may not the approaching (late April is it not) International Town Planning Congress, in New York, give an impulse to Libraries, etc. throughout the country?) A commission would be natural & proper – for you need pot-boiling too – while I must realise, to go on with my enterprises here – the more since now too old to go on touring with it – save perhaps to bring it up to date (& set it up) & put in order – I would like $20,000 – but should spend $5000 of that in completing it more adequately. Of course I know at Bombay it is a pig in a poke.

Now, what think you of next winter here – as per previous exchange of letters? A new point has arisen – of which I may now tell you with certitude, since after competent consultation of physicians (although of course this is private) as well as from my own experience, there is no longer any doubt that senescence has begun. Heart weakness definite. No serious or immediate danger. On the contrary, reasonable assurance, by both physicians (& own feelings too, as far as may be trusted) that with care one may prolong existence on fairly reasonable terms – with lacto-vegetarian diet, & giving up of tobacco, coffee & tea (alas great helps to thought!). But no longer for me the till lately apparently reasonable hope of longevity, encouraged by that of parents (91 & 84), though not by two brothers (70 & 70) or others of ancestry much beyond that.

So here I am, and as an old man making his will, to what legatees, & with what executors? To some extent my son, and daughter too – other friends here & there, for various particulars (e.g. perhaps Miss Barker for opus botanicum). Naturally, I have all along looked to Thomson (7 years younger) for the *Sociology*. But alas their life expectation is, in each case, *more* damaged than mine, their cases graver; as J.A.T. knows & takes care fairly, but V.B. forgets & runs down – much worse than last year.

Hence I ask *you*. Come next winter; & see what legacies you can take over from my mingled heaps – & what executry here or there appeals to you. Without of course expecting to be quite so far or so soon gone as all that – that is broadly the situation in broadest outline!

With cordial regards to Mrs. Mumford, yours ever,

P.G.

---

135 Hicks St.
Brooklyn, N.Y.

5 May 1925

[Ts NLS]

Dear Master,

My days have been too crowded for letters: this has been my first breathing space. By this time you've doubtless received the regional Planning number of the *Survey*: it was part of a general campaign of strategy against the megalopolitans, and by a little adroit maneuvering on Stein's part we almost captured the International Town, City and Regional Planning Congress. The State Housing and Regional Planning Commission had a great exhibit – some of which is reproduced in the *Survey* – which showed the diametrical opposite of the Sage plans for increasing concentration and growth; and with the help of Unwin, Purdom and some of the Germans we kept the human side of city planning pretty well to the fore. The week closed with a weekend at the

Hudson Guild Farm, which recalled the jolly one we'd had with you two years ago: it gave us a good chance to get on intimate terms with Howard, Bruggeman, Keppler, Schmidt, Heiligenthal, and Ernst May, and a number of other interesting and helpful people: besides, it gave them a chance to wipe away the taste of New York! If you have the opportunity, please criticize the *Survey* number.[1] Everyone wrote so much that a number of the pictures got crowded out, particularly my own drawings and cartoons, including the Advertisement for Blasto, the Constipation remedy, with the legend: An army marches on its stomach: the big city commutes on its pills. But it was the best that we could do with our limited means and time; and it represents a great deal of cooperative thinking.

Sophie expects her baby in July: and if all goes well I shall leave for Geneva on 29 July! Just before leaving New York, about a month ago, Alfred Zimmern invited me to give twelve lectures on American civilization at his international students' summer school: he had gotten some American million-aires I believe to underwrite the work fairly heavily, and by some happy chance he picked upon me to give these lectures.[2] I had never met him or corresponded with him until the day I got his offer: so it had the effect of coming from heaven! I expect to be in Geneva from 10 August until the Assembly meets in September. I am sorry I cannot get there earlier, so that I might attend your universities meeting in August; but the baby of course makes this impossible; and as it is I have to be more or less prepared to cancel all my engagements should some emergency arise. Please tell me what your plans are for the summer and early autumn: I expect to return to America at the end of September; but I should like very much to meet you somewhere in Europe. When do you return to Montpellier? Branford has just sent me your encouraging reports from Jerusalem; so I trust things are going smoothly in that quarter; and above all, that you will not return to India! I lectured at Wesleyan College the other week and was entertained by Dean Ladd of the Berkeley Theological Seminary: I found that both he and his wife were acquainted with your work; and he had so far been touched by our regional planning number as to wonder whether the contemplated removal of the Theological Seminary to New Haven were not a step against, rather than towards, a more desirable future! Congratulations on *Biology*! I imported a copy and was delighted to find that you had incorporated so many essential things in it: I worry less now about the midden-heaps since, between this and the Principles of Sociology, plus the *Sociological Review* schemata, you have managed to put in definite and clear form a good part of the Summa Sociologiae. I have not been able as yet to read it through; but I have already started to recommend it. If I am not mistaken it will make many converts. It's unfortunate that Paul Kellogg has been forced to run your papers in such a desultory way: both he and I wanted badly to have one of them for our number: it was in fact already set up in type; but it would have needed another form to put it in, and we couldn't afford it.

Our garden has been a great delight this spring: there is a magnolia tree and a tree of heaven in the rear; we've had a row of tulips in front of our window all April and this part of May; and the calendulas, zinias, and marigolds we've planted are already pushing through the soil. In spite of the fact that I've kept three or four jobs running at a time (editing, lecturing, writing) I managed to carve out nine days and wrote a play which had long been on my mind – a veritable nine days wonder. It is not a bad play: it stands up pretty well under my none too tolerant critical eye; and I am already looking around for a producer! The time of the play is that of our Civil War; but the problems and interests are those of our own day; and through it all I have attempted to weave the threads of the outside world, the seasons, the city, the whole milieu. The first act is called "Goldenrod," the second, "Sumach," and the third "Morning Glory"; and the whole play is called *Asters and Golden Rod: an American Idyll*. As soon as it is sure of production and I have a spare copy I'll send it to you. The play rises from disillusion – it is a bitter idyll – to reintegration; and I trust it will invigorate and cheer my contemporaries as the writing of it has invigorated and cheered me. It has been a very happy release; so much so that I came near calling it *Renascence*. I was writing it in my mind all winter; and I grudged every minute I gave to school or to city planning because it seemed to threaten to stifle the play. But spring brought a rush of sap, and all these numerous engagements and entanglements were swept away.

Sophie sends you her affectionate respects: if it's a boy it's to be called Geddes! And here's to our meeting.

Lewis

1.   Charles B. Purdom (b. 1883), British urban planner and Shakespeare scholar; author of *The Garden City* (1913) and *The Building of Satellite Towns* (1925). Auguste Bruggeman, Director of the Urban Institute at the University of Paris (see Mumford's letter of 17 May 1925 below). Keppler: perhaps Frederick P. Keppel who worked on the *Regional Plan of New York and Its Environs*. Probably Alfred Schmidt (b. 1880) who developed a land zoning plan for the Ruhr district in Germany. Ernst May (1866–1970), distinguished German architect; served as city architect of Frankfurt; worked in Kenya during the Nazi era and returned to Hamburg after World War II. Roman Friedrich Heiligenthal, author of *Deutscher Stadtebau* (1921). The reference is to the "Regional Planning Number" of the *Survey Graphic*, 54 (1 May 1925).   2.   Alfred E. Zimmern (1879–1957), British political scientist; author of *The Greek Commonwealth* (1911); founder and director of the School of International Studies, Geneva, 1925–1939.

---

135 Hicks Street
Brooklyn

17 May 1925

[Ts NLS]

Dear Master,

Your letter of 21 April reached here the day after I sent you my last; and so, a week later, I'm sending this appendix. I grieve that the wretched Indian winter took so much out of you, after America had done what it could in an equally fatiguing summer: but once you settle down to a more even pace, instead of the super-youthful and to me almost terrifying regime that you were accustomed to, you'll find yourself knitting together again, I'm sure, on a firm basis; and with a good part of your career and work still in front of you. One of my grand-uncles went through a similar experience at about 67; but today, at 83, in spite of two surgical injuries, broken kneecap and broken hip (both healed) sustained in the meanwhile, is more hale and capable than sixteen years ago. If there are no complications at home I'll arrange to spend an extra time in Europe so that we may consult more fully together and see what particulars I can help in and carry on. Sophie herself is anxious that I should do this; even if she is not quite prepared to take the young infant a-travelling, or quite reconciled to my remaining a whole winter away!

As for the wretched Jewish architecture that still crops up, I'm doing what I can: I made a long and careful analysis of the Jewish tradition for the *Menorah Journal* – they haven't printed it yet; and I have held up your Jerusalem plans as examples of the genuine line of advance. I have done the same thing with the Catholics by the way; and, as you'll see in the enclosed clipping, mentioned the Dublin Cathedral plans. There are plenty of capable Jewish architects; but I suspect that the inveterate nepotism of the Jews is responsible for some of the atrocities. The Deutsche Werkbund (union of manufacturers and artists) is going to start a new magazine next fall called *Die Form*, which will deal with form throughout our civilization, from kettles and pots and stoves right through to buildings and paintings and the modelling of the earth's surface by all man's works – what the Germans call Kulturlandschaft. The editor who has just been over here, Dr. W.C. Behrendt, has asked me to be the American contributing editor.[1] He told me that after the war there was an active movement in Germany to do away with the manufacture of the cheap and nasty, and to produce nothing but quality products, but the ecomomic disorganization caused it to break down. The Werkbund however is using skilled artists to "give to barrels, trays, and pans grace and glimmer of Romance!"[2]

I haven't been able to do anything about the City Exhibition. Have you thought of approaching Auguste Bruggeman, Directeur de l'Institut de l'Urbanisme, Université de Paris? I met him but hadn't any opportunity to broach the matter to him; I learnt, however, that his school has plenty of funds and is gathering civic materials of all sorts.

The amount of work I must do to prepare the Geneva lectures overwhelms me, even in prospect – and I have done little: I was foolish enough to engage to do twelve.

Pray take care of yourself; and let me know of your whereabouts and plans.

Lewis

1. Walter Curt Behrendt (1884–1945), German architect who became a close friend of Mumford; Mumford later helped him come to the United States to flee Nazi Germany; author of *Modern Building: Its Nature, Problems, and Forms* (1937).   2. From Ralph Waldo Emerson's poem prefacing the essay "Art"; quoted by Geddes at the end of *City Development*.

Dear Lewis – and Sophie –,

Congratulations & all good wishes to you both – and above all to the mother – and the coming babe – one of the fortunate (little but increasing) scattered group of the rising generation who will have a proper upbringing. I accept with pleasure the proposed naming – if boy – indeed, if girl, *granddaughter* for me – a no less welcome hope, I may petition for second name (if Patricia be unsuitable? or prempted by higher claims, which there are no doubt already!).

Very bad that you have to leave so soon after the event. But that is the ill-fortune of fathers: and the errand is a good one. But why should not Mrs. Mumford & babe come over later, & spend the delightful autumn, & if

Tel Aviv, Palestine
but write Montpellier

25 May 1925

[Ms UP]

possible winter also, in Montpellier with me, and you together? You will be charmed with my home, I think, though I say it. And you need not hesitate. For I can now afford to keep the house practically or completely my private home next winter – thanks to "pot-boiler" of town-planning Tel Aviv & Jaffa (& of *probable* start of Jerusalem University building too). I shall not be excessive in my demands of collaboration! But you too will do well as my essential heir, it seems, of my Teufelsdröckhian "paper bags" = to realize that the long dreamed *Summa*, as you so generously call it, is by no means so fully outlined as you think – but has many more developments, alike for thought & action than you yet have seen.

Moreover, you will thus enter more fully into the main heritage of civilization – that held & continued by France. And thus increase your efficiency for your own work & life in America thenceforward.

University Progress Conference in Edinburgh never got started! – owing to Tagore's serious illness; but I have to hold forth there on University Progress scheme at *International Conference on Education* about 28 July, & shall thence be in Edinburgh or Aberdeen (both of) which you might visit – especially if you take boat to Glasgow or Liverpool & before you go to Geneva. Of course that is only possible by fast boat – & with short time after all!

Congratulations on your play! Great joy to break out in a new place! I look forward to reading it when you come, & seeing it, if fortune allow! I too have such dreams – but more in dumb show – with bits of dialogue. (Perhaps we might do one together!)

My plannings here are coming out with more definite results I think than heretofore – you can have no idea, till you take to such work, in many towns, how superlatively they waste & bungle! I'm trying to save hundreds of thousands of £ in these small towns – & they are more than usually willing to let me – partly because Jews are intelligent – in appreciably higher proportion than gentiles! & partly because their difficulties are so great. I am adjusting all new city blocks to large ones, with *interior* bit of garden village. If you have such examples handy in U.S. pray send me what you can, to strengthen case.

Must close for post! Yours always,

P.G.

Still Jaffa, Tel Aviv
– but RSVP to
Montpellier

3 June 1925

[Ms UP]

Dear Lewis,

J.A. Thomson tells me the American publishers are making trouble over the difficult style of the little "Biology." This is a pity. It is needed as an introduction to the longer *Essentials of Biology* we are now at work on.

I once met New York editor of Home University Series – Prof. Brewster – of purely literary interests, he seemed to me – & then not at vital levels. (Holt refused my little *Dramatisations of History*.)

Can you review our *Biology* – & see that Holt's paper & Brewster see it? Can you say that it is not *merely* difficulty of style – but unfamiliarity with *life-view* on readers' part that makes the trouble? (That is of course my whole lifelong experience.) Children, working people, knowledgeable women, artists, etc., I've always found understand me, & like my ways of putting things! It is "the practical man," the accountant, the lawyer, the verbalist, literalist, etc., who find me so obscure, impractical, etc.

Very pressed with Town Planning here & home in ventures in that line which will interest you. Cordial regards to you both & renewed best wishes & hopes of your coming here. It may be my last winter – *Anno Domini*[1] is shattering me a bit – though I may readjust – & *opus* progress would help that more than you can believe!

P.G.

1. Old age.

Dear Lewis,

I'm now well enough on with Tel Aviv & Jaffa (q.v.) and Tolkowsky's *Gateway to Palestine: A History of Jaffa* – worth review see Atlas – to see I've made some good new advances on art of Town Planning & city-making – & am on the brink of more.[1]

(2) I was interested by your saying in recent letter that Clarence Stein, etc. – in short I suppose the *live* group of recent International Town Planning Conference, etc. – were waking up to ask – *what's this all about? What for? What use – what hope* – so here a little paper on *A Town Planner's Day*. (Alas no time to write it!)

(3) All this raises question also of your coming to me & in further capacity of a future possible partner – on Town Planning as city-making – Eutopitects as E. Politotects – evolving "*Main* Street" towards *High* Street wider in all lands alike. You'll find this won't choke your literary work & life – but evoke it {'Vivendo discimus!/Pensando creamus!'} – and thus *Creando pensamus; Agendo parlamus*! (though I also recognise the fact/need of *Dubitando agrimus*! & want your help & stimulus for that too).[2]

(4) But even to relieve anxieties of life & needs of funds for your own as well as my "opus" – mingled of would-be *Summa* & of *Seminal Initiatives* as with old-world creative "artists"/"painters" – and anyway *endeavourers* – we need creative work as well as literary *productivity*. To think & write about cities is of great interest (witness my own Exhibition you've never seen – my books & teachings as Professor of Civics, etc.) but shaping them is yet better. "Let us be impartial & take both!" I believe that something should be done with my Town Planning Exhibition – & I wish I could hear of some American outlet. I have some indications from South Africa–but *that's* less attractive!

This in hopes of extra earlier mail.

P.G.

Write Outlook Tower, Edinburgh

7 June 1925

[Ms UP]

1. Samuel Tolkowsky, *Gateway to Palestine: A History of Jaffa* (1924).   2. "We learn by living; we create by thinking." "We judge by creating; we speak by doing." The third Latin phrase is obscure, perhaps: "we go on by doubting."

---

135 Hicks Street
Brooklyn, N.Y.

20 June 1925

[Ts NLS]

Dear Master,

As soon as I got your letter from Tel-Aviv I asked Clarence Stein to send you the report of the Community Planning Committee: but for the rush and the bother of the last few months I should have sent it to you myself long before. You will note that the Diagrams at the end of the report deal with your very problem; and give statistical proof of the ecomomy of the large block. It was a customary way of planning in America around Boston, at Longwood and Brookline, and because of it those two suburbs have kept their rural character much better than most of the other adjoining towns. Moreover, the big block gives the opportunity to produce charming little dead end streets, as in the diagram. It is interesting that the very careful investigations of Stein and Wright should have brought them back to the same solution: the current age of Transportation sacrifices to its Road-God the money and space that should go into gardens.

How much have I told you about Sunnyside, the housing development in Long Island City for which Mr. Bing formed a limited dividend company?[1] Stein and Wright and Bing have been working steadily at it, ever since the meeting we held in your honor at the Hudson Guild Farm, and the houses are being sold and occupied faster than they can be built. Unfortunately, Stein has to work within the rigid street pattern laid down by the municipal engineer; but through economies in design, and through limitation of the interest rate, which is very high here, he is producing houses and gardens and playgrounds at prices that the skilled worker and the clerk can meet. With a drop of three percent in the interest rate – six is now the very lowest – he could begin to plan for the unskilled worker; but with money and wages what they are now it is impossible to provide decent dwellings for two-thirds of the people of New York and so it will remain unless the first goes down or the second up!

I have just heard from Zimmern that I will have only six lectures to give; this makes my job a little easier, but I have still a great deal of work to do on it. I had to put my sailing at the latest possible date; and I shall have to go directly to Geneva. I expect to stay there until September eighth: after that I shall be relatively free. But you can tell me of your plans in August: My address will be: c/o Fédération Universitaire Internationale / 6 Rond-point de Plain-Palais/Geneva.

It will be impossible for Sophie to come over to Europe this winter; but if my course at the New School prospers there is a real possibility of my coming over with her early the following spring. My ability to help on the opus is, alas! small out of all proportion to my interest and sympathy: it is due partly to my temperamental inability to do anything with things or ideas that I haven't assimilated and thoroughly worked over; and the worst of it is, in the act of assimilation and working over the material becomes very different! That was what made me such a duffer when we met in 1923: I simply found my internal processes paralyzed, and in that state was much less effectual than the worst dunce would be who could somehow react directly to you and keep moving. I have a notion that this happens in all discipleships; it is the real tragedy of the relation; certainly it has happened again and again in the historic cases; for Plato undoubtedly betrayed Socrates, and Aristotle turned away from Plato; and Paul deformed the doctrines of Jesus; and each did this in perfect good faith and loyalty. You may say that this is all you expect; but at bottom no one is really satisfied with this: what one wants is a promulgation of one's activity *in the same form*. I am fully conscious of all that I have absorbed from you: conscious how different, how much weaker my ideas would have been had I not taken hold of yours: but somehow all that I have absorbed undergoes a metamorphosis, so that at times you yourself perhaps don't recognize it, and in spite of all my valiant acknowledgements people refuse to see the obviousness of the connection. I am still learning from you; still taking in more; and I welcome each fresh opportunity for this. But it is another matter from taking over something that you have achieved or built. Had I a more extraverted temperament I might be able to do this: as I am, it would only lead to frustration and disappointment for both of us. Still, of course, I can do much; and I hope to do much; more at any rate than in 1923!

Sophie has been very normal all through her pregnancy; and she has enjoyed the change from what had become an irksome routine. From my own short observation, few girls are happy in the non-biological occupations; or if they make themselves happy, they do so at a sacrifice and become hard; which is to say, that they are not happy after all, but blind themselves to their underlying misery! We are taking an apartment over in Sunnyside; which we will keep permanently, we hope, the time we have to spend in the city; and as soon as opportunity offers, we shall get a farmhouse in the country, far enough off the beaten track to be cheap. Sunnyside lies in the midst of a wretched waste of land; but it is built for children and mothers; and as long as both must remain fairly near the house it is the most desirable environment in New York. Sandpiles, wading pools, gardens, etc; and small well-arranged houses and apartments, none over three stories high. I'll send you photos later.

Here's to your bettering health.

Affectionately,

Lewis

1. Alexander M. Bing, the developer of Sunnyside.

Dear Godfather,

135 Hicks Street
Brooklyn, New York

8 July 1925

[Ts NLS]

Your namesake came into the world on Sunday, in a rash, hasty, impetuous way unusual for first babies! which makes us think that perhaps some of your spirit got into him at the beginning![1] We are both radiantly happy; and Sophie is feeling very well indeed: a wide pelvis and a fairly normal existence rob childbirth of a few of its terrors. He weighs seven pounds, has large hands, and big bones generally; with a nose already as Napoleonic as his father's. His early arrival has relieved both our minds: we dreaded long delays and anxieties. A truly nice child; and although we had hoped secretly for a girl, we are more than compensated by the fact that another Geddes has come into the world!

I leave on the 29 July. Please write me in care of Fédération Universitaire Internationale / 6, Rond-point de Plain-Palais / Geneva and tell me what your plans and movements are. Branford tells me you are tied up with J.A. Thomson for August and September.

Hastily but affectionately –

Lewis

1. Geddes Mumford was born on 5 July 1925.

Outlook Tower
Castlehill
Edinburgh

27 July 1925

[Ms UP]

Dear Mrs. Mumford,

Hearty congratulations on your happy event, and all good wishes for the new and maturing start in life it opens – good wishes which include of course my wee godson, for so I interpret his naming. And thus not without such responsibilities as may be, though in the shortness of life of septuaginarians these may not be great, I lose no time in offering him all manner of good wishes; & through the varied stages of his life-career, from his present *Eros* phase onwards, & throughout. Children in conditions such as you & Lewis will make for him have a great advantage from the start; and I doubt not that both you and he will make the best of advancing and developing conditions (and even of difficulties too, as they arise, since one must never forget that these are also *opportunities*!).

I am sorry you can't all come to Montpellier this winter – but put it in your plans for another (& as soon as may be, since my time may be short).

I am writing to Lewis at Geneva – & in hopes of a visit from him here before I have to go south again.

With renewed good wishes to you both, believe me always yours very cordially,

Pat. Geddes

Geneva

26 August 1925

[Ts NLS]

Dear Master,

I'm sorry I forgot to acknowledge your letter, which was lying here with one from Branford, which outlined his project for a grand symposium on the transition in Western Civilization. I'll discuss both with you in detail when we meet, and will try to see Branford and have a talk with him before that. Here is the schedule of plans at present. I will leave here on September fourth, and, since you won't want me the first week, will go to England via Basel and Paris, perhaps picking up Gladys Mayer on the way. I should arrive in London by Wednesday, the ninth, and will go up to Edinburgh on Friday the eleventh, unless you drop me a word to Le Play house in the meanwhile to alter the arrangement. I'll spend a week with you – if you can afford that much time. My passage home is scheduled for the 30: but Sophie and the baby are both calling for me loudly, and I shall try to put it ahead a week, as soon as I can make arrangements in London. The baby thrives: he is unusually strong, and has big bones, although he weighed only 6¾ lbs. when born. I genuinely regret missing two whole months of his development!

The lectures here have come off pretty well so far. Zimmern has gathered about a hundred students from all over the world; or rather, there are about a hundred for each fortnight's session, some old ones remaining over and some new ones coming. While there is a great deal of emphasis on the legal and political side of internationalism, he has also managed to get together a number of interesting courses on the cultural aspects of nationalism; and from the way the Zimmerns have responded to my own lectures, I feel that you are very close together – although his approach and background is so different. He is the Secretary of the General Section of the Committee on Intellectual Relations, and is in a strategic position.

Spencer Miller came in the other day; and reassured me with the news that you were in splendid form at Edinburgh.[1] But alas! I wasn't cheered by his reports over the attitude of my countrymen in the Educational Congress. We are a boorish and insular people: I'm afraid that more than one *Civilization in the United States* will have to be written before we change perceptibly!

Two young students of philosophy from Columbia told me that there was a young biologist there – but they didn't know his name – whose work you and Thomson surely ought to be in touch with. If their report is correct he has experimentally confirmed your theory of sex, by causing the sex to change through an altered rate of metabolism. Have you heard about this?

I am looking forward keenly to seeing you.

Lewis

1. Spencer Miller (1891–1949), author of *Labor and Education* (1932).

Dear Master,

4112 Gosman
Avenue
Long Island City, L.I.

17 October 1925

[Ts NLS]

I've been back here little more than two weeks; and still I am scarcely settled. The baby is well: strong, big, and good tempered; and Sophie keeps him company in that: they send you very hearty greetings, and I add to them my own on your birthday, which has meanwhile passed. The sea trip was too miserable for me to write to you; and the time since has been crowded with packing and unpacking and carpentering, for I arrived in time to complete the moving which my mother had partly superintended. Sunnyside is, on the whole, very good; and I'll send you some photos of it soon. I've been into Manhattan only twice, and have had no chance as yet to see Kellogg or Johnson; but I am going to have lunch with Kellogg on Monday. Have you hit on a name for the proposed federation yet? Why not something that could be framed in initials if necessary: Civic and Intellectual Alliance. Or Civic and Collegiate Federation. They sound clumsy, perhaps; but C.I.A. would presently have a meaning of its own, as symbol.[1] As far as work is concerned my homecoming is a little bewildering; although only two students turned up at the New School for my course – and so it's not given – I have received all sorts of invitations to write and lecture, principally upon architecture and city planning; including one for a short course at the University of Michigan. Ça va!² Meanwhile, I have things drumming through my mind which have nothing to do with architecture; and feel a little that I may be burdened with a reputation which I have no desire to keep up. All my European impressions are still floating in my mind; the marvel, for one thing, of all that you've done in Edinburgh – quite enough to occupy any one ordinary life. I still think that the drift is setting towards all the things you've stood for; and that we've now only to begin to act on the assumption that the stage of criticism is past, and that we can consider our opponents as prostrate or quite negligible, to take possession of the field. My great regret about *Sticks and Stones* is that I finished it in a mood of intense physical depression; and that I did not give it just the final, confident ring that might have worked a change. I had no notion that the public mind, at least the intelligent public mind, was mature enough to accept half the criticisms I flung out and then say eagerly: "Yes: but what can we do about it: what would you have?"

I'll write again as soon as I've seen Kellogg. Need I say how deeply I felt and appreciated your hospitality?³

Our love always –

Lewis

1. Mumford's 1971 annotation: "! ! !"   2. It goes well.   3. Their second, and last, meeting in Edinburgh in September 1925 was not a happy one for Mumford.

4112 Gosman
Avenue
Long Island City, L.I.

21 October 1925

[Ts NLS]

I saw Kellogg the day before yesterday. He was genuinely apologetic: his delay in answering your letters was partly due to a four months' absence from his office, and partly to the fact that he hoped from week to week to have enough money in the treasury to send you a cheque; but he ended his financial year without anything to spare, and lost one of his chief supporters; so that he is now busy campaigning for contributions. He has published all six of your revised articles by now. There are still three that, I think, you did not revise: he will try to publish these next year, but is chary of making a definite promise. If he does publish them, he will of course pay you for them. He really means well; and he shared your feeling about Van Loon's work.¹ He had hoped, of course, that you'd be able to supply some of the illustrations yourself; but he understood this was impossible. He is sending me a complete file of the articles; and I'll see if there's enough material to submit to a publisher; and if there is I'll submit it. Accept his hearty good wishes and apologies; and meanwhile, I shall keep in touch with him, since he wants me to take Van Loon's place, to the extent of doing occasional cartoons for frontispieces!

I returned to find that Stein was suffering from an overdose of Regional Planning and Housing work last spring, and now has his nose glued to his draughting board, swearing that he will do nothing, but play with his architecture from now on! A temporary reaction I trust. Mackaye is giving a seminar at the Civic Club: a further development of the New Exploration. He has not yet got anyone to finance his world atlas of *potential* commodity flow; so he is elaborating the more general and social parts of his thought. I have scarcely had time yet, however, to see anyone.

Have you or Thomson, by the way, seen a new book by Alfred Lotka, *The Elements of Physical Biology* (Baltimore: William and Wilkins). It is along the same lines as the book you showed me in Edinburgh.

Van Wyck Brooks, our chief literary critic, is now engaged on a work on Emerson which will recreate the whole life of Concord: he is tracing out the psychological genesis of the community and has discovered all sorts of astounding things. From Emerson's time onwards Concord's chief export to the rest of the country has been teachers and artists; and to this day, although the great figures are gone, Brooks found that there was a much higher level of intelligence and culture among the old Concord families than he has found anywhere else in America.

My own plans for the winter are not yet definitely formed. In November I am going down to Baltimore to write a series of articles on the city and its architecture for the *Evening Sun* – a newspaper on about the same level as the *Yorkshire Post* – and I am beset with a number of lecture engagements, although my New School lectures only attracted two students, and so are indefinitely postponed.

Sophie joins me in sending you our hearty and affectionate greetings. Remember us to Mabel.[2]

Lewis

1. Hendrik Willem Van Loon (1882–1944), historian and illustrator; author of *The Fall of the Dutch Republic* (1913) and *The Story of Mankind* (1921).   2. Mabel Barker.

Dear Master,

At last the hurly-burly of the last two months has quieted down; and I can send you these wishes for a happy Christmas, out of the happiness and health which we three, Sophie, young Geddes, and I, are now enjoying. I spent October and November madly feathering my nest for the rest of the winter: a long article on the evils of unlimited city growth, for *Harper's Magazine*, a series of articles on the architecture of Baltimore, made after a flying visit, for the *Baltimore Sun*, and numerous other little excursions, including two cartoons for the *Survey*.[1] The last takes a little too much of my time, for I haven't Van Loon's glibness and facility; but the net result of these efforts is that, with the addition of the advance royalty that waits me on my next book, I can settle down to a winter's uninterrupted work, and need write only when I feel like it. I am now ploughing ahead merrily; I have already advanced far beyond my Geneva lectures, and if I can manage to say half the things I am already feeling and sensing, I shall make a fairly interesting American coda – with enough new things to justify it – to Branford's *Science and Sanctity*. There is a certain advantage in writing about the intellectual and ideal development of the modern world, from America; for here one is brought face to face with a more naked reality, here the bottom has dropped more completely out of the old traditions; and one gets a fuller sense of what is missing for a complete life.

I had no opportunity to see Alvin Johnson, but I wrote him about your

4112 Gosman Avenue
Long Island City

11 December 1925

[Ts NLS]

scheme of sympathetic federation; and in answer he said that he "didn't see his way clear to such a cooperation at present." He is a shameless opportunist, and there is very little to hope for the New School as long as he is in command of it. In the meanwhile, I have kept in touch with Branford, and learnt about the more limited scheme of publications he now has under way. The new Mayor of New York, who is very close politically to Governor Smith, who in turn is fairly close to Clarence Stein, has just announced his intention to institute a comprehensive policy for distributing the population of New York and instituting a better system of housing; and the Park Board has recommended the expenditure of 30,000,000 dollars to create spaces and playgrounds in the unsettled areas of the greater city; so there is a tiny spark of hope here and there.[2] The enclosed article, by my old teacher Dr. Slosson, will show how a more synthetic notion of the sciences is gaining ground.[3] As far as I have been able to observe the younger people in America, they are entirely disillusioned with the abstractions of politics, and though the abstractions of finance have still a greater hold on them, they nevertheless, as opportunity offers, are plunging more concretely into the arts; and this accounts for the growing interest in architecture, even when it is presented – as I always present it – not as an isolated esthetic phenomenon but as a capital social art, whose successful practice depends upon the well-being of the community. I have a notion that the students who would be attracted to you abroad would welcome most easily the more concrete parts of your work, and that this would be an introduction to all its other ramifications. I know that was the way my own initiation began; and I am still more at home in your synthesis of attitudes than in your synthesis of categories, for some of which there exists no corresponding elements in my own experience. I can, for example, use and see the application of every part of the 36, except the nine squares that stand for sense–feeling–experience. If I substitute Geography, Economics and Anthropology for them the parallel becomes satisfactory, for these represent the sum of the positive sciences, complementary to the ideal sciences on the right hand side of the diagram. Sense–feeling–experience however simply won't stand in my mind as the subjective aspect of PWF; they remain aspects of individual psychology, that is, they correspond to part of me, and not to parts of PWF. I have patiently turned this matter over in my mind this last five years, feeling that perhaps I would hit on the clue which would make it plain to me, as the righthand squares are all plain, now that I see that they put more definitely the Greek notion of "dialectic" as distinct from the "physics" of the lefthand side, physics representing what the world is, and dialectics representing its potentialities, when taken up by the mind and reacted upon. The mistake of so much of the science of the last hundred years has been, has it not? to regard all dialectics as worn-out and outmoded and sterile and superstitious; whereas the only part of dialectics which won't stand is that which attempts to apply its special method to problems which can be solved only by science. Thus the theological account of Heaven and Hell, the two ideal points of reference, has still a certain authority; whereas the the biblical account of the earth or the creation of animal species is an attempt to settle, by a method no longer legitimate, a pure matter of fact. The 36 clarifies all this immensely; but, as I said, the "s–f–e" remains a blank spot to me; and if it does this to one who has had the advantage of sitting at your feet, and has made some effort to master it by himself, it must be a much greater obstacle to anyone who encounters it casually from the outside. . . .

As to getting the co-operation of Keyser, Dewey, Robinson and company for your scheme – I am going through your letter of 26 October again to see if I have neglected anything – it is impossible for me to say.[4] You know all of these men, I fancy, better than I do. Robinson is now editing a series of short books which will attempt the "humanizing of science": the other day he asked me to write a book for him on pragmatism. Spencer Miller has told him about my Geneva lectures. Unfortunately, I had to say no; for pragmatism is only a half-way station on the road to a Weltanschauung (I daresay there is a square for it on the 36) and the sort of essay I should write would not meet Robinson's expectations. My American elders almost all seem to me to have spent the greater part of their lives poking down blind alleys; and I don't feel for them, intellectually, the sort of respect I should like to have. Dewey, Robinson, Veblen, each of these has given me valuable fragments; and they could give me no more because their work itself was fragmentary and limited, and had never escaped, in spite of all their acquired knowledges, from the naivetes of the pioneer.

Is Mabel with you? Please tell her I'd like a word from her, if she can spare it. I still remember with awful delight her goatlike antics on Arthur's Seat.[5]

If only I knew a millionaire I think I might dispose of the Cities Exhibition to advantage! All the city planners to whom I've mentioned the possibility of acquiring the exhibition are wary, not of any initial expense, but of the difficulties of housing and continuous exhibition, and, since they are not directly acquainted with it, they are not tempted into buttonholing the proper benefactor. Alas! I know no millionaires. I should think that the sound way to dispose of the exhibition would be to print a little descriptive leaflet, describing the collection in general, and announcing that it was for sale; and to send this to city planning bodies and city planners all over the world. Then the offer might really get a little attention.

Again our hearty greetings and good wishes for the new Year.

Lewis

P.S. How goes the biological opus?

1. "The Intolerable City: Must it Keep on Growing?," *Harpers Magazine*, 152 (February 1926): 283–293. In the *Baltimore Evening Sun* 1925: "The Bricks of Baltimore" (1 December): 23; "Deserts versus Gardens" (4 December): 25; "Modern Public Buildings" (8 December): 27; "How to Ruin Baltimore" (10 December): 27.   2. Jimmy Walker served as Mayor of New York, 1926–1932.   3. Edward E. Slosson (1865–1929), wrote on science for a non-technical audience; author of *Keeping up with Science* (1924) and *Sermons of a Chemist* (1925); the reference is probably to his article "Philosophy of General Science," *School and Society*, 20 (27 December 1924): 799–806; Mumford had taken a course under Slosson entitled "The Modern World" at Columbia University, *circa* 1917.   4. Cassius J. Keyser (1862–1945), philosopher and author of *The Human Worth of Rigorous Thinking* (1916) and *Mathematical Philosophy: A Study of Fate and Freedom* (1922). Geddes' letter of 26 October has apparently been lost.   5. Hill in Edinburgh.

Lewis Mumford AND *Patrick Geddes*

THE CORRESPONDENCE

1926

Dear Lewis,

Collège des
Ecossais
Plan des 4
Seigneurs
Montpellier, France

15 February 1926

[Ts UP]

Herewith copy of letter to Kellogg; for though I esteem him personally, and *The Survey* too, I think it well, after this long lapse of time, to be clearly *off*, or *on*, with both.

Do you think anything can be done with these papers? Will any publisher look at them? Or must the whole thing just be dropped for good?

Frankly, if it should go on, I think most of Mr. Van Loon's illustrations would be better dropped, for he really missed one's points, or took them too lightly – though I suppose his name is now one to conjure with!

Despite some symptoms of incipient age – inevitable to a fairly hard-worked and worried septuaginarian – I am getting a good deal done which will interest you. As you know, I incline to other sorts of expression more than to writing; and so here classification of sciences and arts; Biology, Sociology, Life Theory, even to Olympus and Muses, are all taking concrete form in symbolic gardens – none the less exuberant for that.

Then my masses of papers, of which you wrote so kindly to Branford, and of which I've half-ton arranging in new studies here, are really getting towards order; and though I make many new ones every early morning, as ever, I am none the less getting more of the old ones classified, or incorporated, and filling the waste-basket accordingly.

Returning to practical work, I forget if I told you that we have now both an American and an Indian College on plan. Towards starting the latter, two Indians are already here, and a third is coming in shortly, so as to be able to tell Indian universities that the latter is begun. *Tres faciunt collegium*![1] I have also another house on the hill, which the Zionists think of using, especially as towards training for Palestine in the neighbouring School of Agriculture – the leading Mediterranean one; but also as helping the many Jews who leave Eastern Europe for education, since excluded, beyond limited percentage, from their own universities.

Returning to the *Collège des Americains*, the active promoter is Professor MacKenzie now of Juniata College, Pa., who hopes to return in summer with an excursion of about fifty American students for short visit; but to make this also the basis of propaganda for his large hope of residential hall of fifty students here. Other national groups will also doubtless arise; and the proprietor of the adjacent heath has now relented, and is willing to consider a fairly large concession.

The five students here will thus present five theses this year – a good initial record, and helpful towards the scheme. But all the same I of course want something much better than individual theses, of usual scattered university type. In Bombay, although Research Professor, I disappointed many by not encouraging theses; for I felt it necessary rather to direct students towards fuller general views of sociology than any young man's thesis can attempt to deal with. And a thesis has the difficulty of premature specialism. Now my problem here is to synthetize these two methods. One who has practically finished his thesis is now beginning as an assistant and is helping me towards clearing up over the general field. But I would fain make (a) such assistantship a part-time job for others, leading to (b) personal selection of thesis, and as his other part-time job. So that his special investigation may be continuously seen as part of his larger view, and his contribution as towards greater clearness of that.

Recall the industrial training of the middle ages – prentice, journeyman, master. The prentice assisted the journeyman; and when ready to become one had his wander-year, with its vivid experiences at once technical and general. He then returned to his town as journeyman – i.e. skilled assistant to master; but meantime also prepared his masterpiece for due and full admission to the Guild. The conception of education as deeply *occupational*, and that of *Vivendo Discimus* in general clearly suggest the corresponding method in study, as above indicated. And I now see that this is how I learned my own biology – first assisting Huxley, etc., and yet also doing my own research as well, in spare time and on vacations.

Pray criticize above, as you will see its educational importance, its possibilities – and difficulties.

Again I see this growing place as a definitely possible academic league of nations – a junior Geneva, and suggestive towards the like elsewhere. See enclosed copy of letter to Dr. Magnes, University Jerusalem.[2] Further it should work towards definite treatment of problems of Bergson Committee, International Intellectual Relations Committee of League of Nations, as also with Otlet's large endeavours – Bibliography, International Association, etc.; and as I have probably already mentioned, Otlet and I had our resolutions adopted by International Education Conference, emanating from America, which had its first European meeting in Edinburgh last year. These resolutions committed their association and ourselves to report (a) on Universities past and present and possible: (b) Bibliography: (c) Sciences and their organisation – all main theses! And now the president of said association (Mr. Augustus Thomas, State of Maine Dept. of Education, Augusta) writes me that he hopes to arrange an appropriation of 1000 dollars, or more if possible, towards giving us some help for the needful payment of assistants.

Enough to indicate that the long-dreamed synthesis is tending to emerge from its "Teufelsdröckh" paper bags!

Though the making of this garden has taken much time these two past winters, and again will next, it is also very helpful. For though there are bigger ones of every possible kind, and I have made some myself, never before, so far as I know, has there been an attempt to express the essentials of so *many kinds* of garden – e.g. from nature-reserve for formal garden, and evolutionary botanic garden; from ordinary cultivation to desert reclamation, and foresting sample; and so on concretely. Furthermore, beyond ordinary aspects, scientific, aesthetic and practical, there is the attempt to work out the expression of *philosophy peripatetically readable* – as from the classification of sciences, laid out in squares of pavement on the house terrace, to gardens of Olympus, the Muses, etc.; and all in such forms as may be intelligible to the Boy Scouts, yet also stimulating to the professor.

In your varied reading for *Utopias*, etc. have you come across any such gardens? Bunyan of course dreamed them largely; and I am getting Samuel Butler's "Sacromonti of Piedmont." But I should like to know more, and I would be very glad if you can help me, for I have here a rare opportunity, and one which might be made a suggestive addition to University planning, as from Jerusalem, and round the rest! That for instance is what I want to lead Dr. Magnes to – the laying out of Scopas – in various ways the most magnificent of all Acropolis hills – towards *intellectual* as well as other use and beauty.

I hope it is settled you are coming back to Geneva for Zimmern, etc.? If so, you must really come here this time after its close. Give me your dates, and especially of closing, as soon as you can, so that I may arrange to return from Scotland in time for your visit. And if you can put in the September fortnight at Les Eyzies and Domme, our Summer Schools (peripatetic), you will find these two very vivid experiences of archeology and regional survey respectively.

Remember me very cordially to Mrs. Mumford; and convey the like, in such symbolic fashion as may be, to young Hopeful!

Yours always,

P. Geddes

P.S. I am sorry to say both the Branfords have been very ill. Mrs. Branford is still down, and V.B. should be. But he is returning to a new plunge into the City vortex. Can we not get him out of that? P.G.

1. Three make a college. 2. Judah Leon Magnes (1877–1948), American Zionist and first president of the Hebrew University in Jerusalem.

Dear Master,

Branford has been good enough to keep me informed of your thinkings and doings; and this last mail brought me both a letter from him, a number of copies of your letters to him and Ross, and finally your own long and ample letter to me, for all of which I am grateful. The winter has been pretty severe here; and the news of your gardenings in Montpellier, coming as it did in the midst of an attack of colds in which all three of us have shared, raised more than a little envy in our hearts. Your namesake has grown lustily; and up to now has been without fault or flaw, except for the fact that his diet has been troublesome: that is, he has a deep physiological repugnance to cow's milk – ten drops diluted in half a pint of water, and administered to the extent of one teaspoonful acts like a poison – and so the business of weaning him has been, and still is, something of a problem, since the vegetables and cereals, though nourishing enough, leave a little, dietetically, to be desired. However with the spring days and direct sunlight, both of which will presently be here, we hope for a better regime.

The Zimmerns were here in December and January to collect money for their School of International Studies, as it is now called; and, except for the fact that we were both too delicate to discuss finances, my going abroad again has been arranged for – or I expect will be, as soon as the problem of subsistence is settled. At any rate, they wanted me. If I go, I shall reach Geneva by 10 July, and leave it around the third or fourth of September; unless in the meanwhile some other arrrangement is made. I of course want to see you; and one way or another we will arrange to; if at Montpellier, all the better of course. And if any new plans turn up, I'll write you forthwith.

I am at present in the midst of my book, and I hope, but am not quite sure

4112 Gosman Avenue
Long Island City

7 March 1926

[Ts NLS]

yet, to have it finished by the first of June.[1] Parts of it go over old ground: the effect of pioneering on the American mind, the naïveté of pragmatism, and so on: but the interpretation of the period from 1840 to 1860, the Elizabethan age of American literature, and perhaps some of the things I'll bring to light in the later period, 1870 to 1910, will be relatively new, and I think fruitful. Dewey is only now beginning to be conscious of the lack of a place for religion and art in his philosophy; and he is making a brave effort to redeem this. Did I tell you that James Harvey Robinson wanted me to write a book on (in praise of) Pragmatism for his Humanizing of Knowledge series? I couldn't do it; for I am treating the subject in my present book, as well as I can; and I knew Robinson would be disappointed by it. The Carnegie Foundation of New York, which exists, among other reasons, to "promote art," asked me whether I would make an inspection of the various centers of art-teaching in America and make a book out of it for them! The offer came without any friendly interposition, as far as I could discover, and without any solicitation on my part; and I regretted that it was not the sort of thing I could do. Still, the mere fact that the Foundation, with the whole academic world more or less at its disposal, should step out of its way to ask an outsider, is rather interesting. All the Foundations have become a little touchy and self-conscious, as a result of the excellent satiric portrait of the Rockefeller Institute in Sinclair Lewis's Dr. Arrowsmith; and every time they spend a million, they become almost apologetic.

A few more bits of news. One of them is that Mr. J.E. Spingarn, the scholar who has made Croce's works known in America, has asked me to help him build up a little intellectual center in the village adjacent to his estate, 83 miles north of New York. He has nine cottages; and is willing to put them at the disposal of such writers, scholars, artists, or what not, as we can agree upon; the cottages to be rented for a dollar a year. He has hopes of eventually making the little village a school, in the sense that Concord was one. It is a jolly opportunity: and within a year or two we may be able to add the "Troutbeck School" to the list of confraternal associations, and perhaps even arrange an exchange of students and workers! Spingarn broached this to me last week only; so it is all in a nebulous state. The news of your various colleges, American and Hindu is good. Perhaps one of the ways of overcoming the economic difficulties in such a student-group as you are projecting is to rely partly upon the general fellowships that certain American foundations are offering. The Guggenheim fund, whose application blank I am enclosing, and the Rockefeller fund are now supporting students who have a wide latitude in choosing the place and kind of work they wish to pursue: they are real fellowships for study, and not specifically for university study. The offering of such facilities as you have in Montpellier, or as we may have in Troutbeck, may appeal to the directors of these foundations: who, being college men, know some of the limitations of the conventional routine.

Your letter to Paul Kellogg was quite justified; and I trust it will bring a favorable answer. If not, I'll gather the articles together and see whether they will break through the publisher's prudences and cautions. Stein's Housing and Regional Planning Commission has temporarily finished its work: the Governor is endeavoring to put a bill through the legislature which will provide for a Housing Loan Bank, which will provide funds for limited dividend companies, operating under state supervision, and granted the power to seize land in large blocks, under the law of eminent domain: that brings one part of our work here to a head. Stein and Wright have been planning farm villages for the American cooperative farm groups working in Russia, and introducing improved methods of tillage among prairie farmers near the Caucasus. Meanwhile, Stein himself has gone back to architecture, and is building a synagogue, a museum, and a college of technology. Both Stein and Wright will probably be in Europe this summer, to attend the International Cities congress in Vienna in September.

I know what a burden letters are to you; so I close here. We all join in affectionate greetings.

Lewis

P.S. Tell Mabel I shall ask the *Survey's* permission to use the cuts she wrote about, in super-economical fashion, in her postscript.

1. *The Golden Day: A Study in American Experience and Culture* (1926).

(*Private*)

Dear Lewis,

How goes it with you – and the Missis and the kid? I hope all flourishing? I doubt not that in any case, whether you have yet got to the country or not, you are going strong?

So have I been doing too, for I made good recovery from my illness after return to India from U.S. But now, alas, the septuagenarian trouble is fairly on – and resists treatment – beginnings of atheroma & arterio-sclerosis, & with abating working powers accordingly. So I can't but expect a fairly short period of real activity: Hard lines of course, when after all these 2 years building & planning, I am just getting out my first circular enclosed, & feeling a little more free for *Essentials of Biology* with Thomson this summer – & if possible for getting on with the corresponding *Sociology* thereafter. V.B. too is none too well – and Mrs. B. has been & is still very ill at Territet after an operation, so I am anxious about them both too!

It has been a speculation on the speculative life lasting longer than it now can, which has made me put off publication so long – until, for the most of it, it may be too late. Yet with care – & collaboration – something can be done. (My son is essentially an artist – & it is time he was on his own: we don't quite fit – and he needs to start now by himself – so I not even telling him of this, but only consulting one other old friend – not yet even V.B. who has enough to worry him at present.)

But here I come to it: is it not possible to tempt you to come to me – after your Geneva course – all three of course – next winter? That will be a pleasant change & you'll like the place with its varied gardens & beautiful heaths & woods around, & great excursions, even into Roman and Medieval past – as well as earlier & later.

In short then, I ask – will you be my collaborator first – and my executor

Collège des Ecossais
Plan de Quatre Seigneurs
Montpellier

23 March 1926

[Ms UP]

and residuary legatee – of all the stuff that suits you later? I've made a considerable clear up – for *Vivendo discimus* is a true process, & my building & planting this wild old place into many fresh kinds of order – has aided my clearing up, so my brain is at its best for every morning – though the old frame begins to fail unmistakeably at the centre. I'll pay too what I can – & I believe much of the material can be made to pay, specially with your vivid pen to aid it. Don't think then that because I went on speculating when in New York, so my manner has been in every fresh environment, that I have not yet some constructive and coordinative work in me: in fact *Studia Synthetica* & *Agenda Synergica* have both been getting on, and are so day by day. (But Dixi. Valente, pro tem: before *Valente*!)[1]

Yours,

P.G.

Since above written your letter of 7 March has come in – with its interesting news, and assurance of your coming to Geneva. (It is not such letters that are a burden.) Moreover, with these Guggenheim Traveling Fellowships, you are the very man to be welcomed for one! And even if all are allotted for '26, they'd grant one at beginning of '27. So pray do think seriously of the above proposal: you'll pull all off – & be all the better of a fresh experience in France.

I submit too, you can take your D.Litt. within the year, as work with me will, I am sure, be accepted as satisfactory & release you from more than an occasional visit & talk with this & that Departmental chief who may most interest you. Why, man, I want to see you as a University Principal (or if not see – at least foresee) & in better times than the present for *live* ones.

(What would J.H. Robinson say to a little book in his series on the Organizing of Knowledge? Classification of Sciences, etc., up to date: it can be made very reasonably interesting, as such Books go.)

Very encouraging to have such good news of kindred movements, as also of Stein & Wright at work from N.Y. to Caucasus. I hope they'll look me up here or in Edinburgh according to time of year they are over. Say here after Vienna Congress, to which I won't go. (No answer from Kellogg!) P.G.

1. I have said farewell, for now, before farewell.

---

31 March 1926[1]   *Dear Master,*

[Ts NLS]   I went around to Littell, and talked over the book; but I have still to hear from him. In the meanwhile, I send you this, as much as an earnest of Kellogg's underlying good will as of any efforts of mine. If Macmillans definitely turns it down, I will go elsewhere with the articles that now exist. I took the liberty of suggesting to Macmillans that I might write an introduction to the book – an enlargment of that account of yourself and the Tower which I wrote for the *Survey*, with what further vividness my visit to you has given me.[2]

The strain of the winter has been a little too pressing for me; and, following a cold, I broke down for a fortnight and was forced to stop working.

I have been on a strict regimen, and what with the spring sun, and the prospect of leaving this elevated ash-heap – for alas! this part of Long Island, in spite of all our skillfull housing and tree planting, is little but that beyond the immediate oasis – I am getting back to work again. The first draft of my book is within about 15000 words of being finished; and the first chapter will presently appear in the *American Mercury*. It goes over familiar ground; as you will see; but the analysis of the earlier phases of LIF, during the abstractional perod, is still a new one here; and after rewriting the material about five times, I think I have at last put it in a clear and irreproachable form.[3] I am anxious to get your opinion of it; and will send you a copy at the earliest opportunity. Alas! since I do my own typewriting, and the article is a long one, I have none on hand here, and no chance of getting it copied.

Whether I go to Geneva this summer or not is stilll undecided. The New York Secretary has just told me that he was informed by cable that a letter was on the way; so I shall find out in a couple of days, and will write you again presently.

Have you seen Radhakamal Mukerjee's new book on *Regional Sociology*? I have just glanced at it; enough to see that he has been assimilating you lustily, albeit it would seem unconsciously, since he mentions your work only at one point, and makes no acknowledgements at all. That sort of thing enrages me. It is good, however, that however anonymously your thoughts are getting into circulation; and one of these days I shall sit down and add a little bibliography to works like Mukerjee's, which will balance up accounts. He got Edward Alsworth Ross, of all people – the man who believes that sociology is merely the study of social "minds" – to write an introduction. Sickening!

Sophie and your namesake send you cheerful greetings. I trust you have managed to brave well the raw Mediterranean spring.

Lewis

1.   This letter is written on the reverse of a letter to Mumford (dated 12 March 1926) from Robert Littell of The Macmillan Company, Publishers, inquiring about the possibility of publishing a collection of Geddes' articles as a book.   2.   "Neighbors," *Survey Graphic*, 50 (1 April 1923): 44.   3.   "LIF": liberal, imperial, financial – a Geddesian formulation.

Dear Master,

By now some trickle of news about my all too numerous disappointments and ailments must have reached you: first my six weeks' illness, then the collapse of my lectureship in Geneva, through lack of American funds, and finally the cessation of work on my book, until I can get financially on my feet again. In the midst of all this, your letter of March 23, followed by one from Ratcliffe, came: and I hardly knew what to say in answer to it. Part of it was already answered, of course, by the fact that we were not going to be in Geneva this summer; but I am sill in a quandary about the collaboration that you suggest. The urgency, the opportunity, the importance, of working on the Studia Synthetica and Agenda Synergica I feel deeply; but I am overwhelmed at my own disabilities for participating in it. I do not write easily about the thing before me; it must settle into my unconscious, and get seasoned there for

4112 Gosman
Avenue
Long Island City

22 May 1926

[Ts NLS]

a long time, before I can bring it forth again in suitable literary shape; and by that time, it has become somewhat personal, and is no longer what it originally was! This is a normal process of art, and one that I can put to good account in my plays; but it spoils any active and immediate collaboration. I remember that in 1923 you pointed out to me that the various religious orders, Franciscans, Dominicans, et al., tended to specialize predominantly as chiefs, people, emotionals, or intellectuals; but that idea did not crop again in my thoughts until the other day I wrote an article on the Edinburgh School for the *Commonweal*.[1] That you have a great deal to give me I have no doubt whatever; each time I go over a writing of yours, or rehearse some phase of our alas! so limited companionship, or find some "new" thought cropping up in my writing, I realize how my debt to you renews itself and increases with each new application; but I cannot give any immediate and practical help; and to come over to Europe with the hope that I might would breed disappointment and a sense of frustration again. Nevertheless I shall try to arrange my affairs and finances so as to be able to spend a full month or two next spring with you in Europe, and together we'll go over the ground, and see what most needs doing, and what I can best appropriate, take hold of, and carry on. Please count on me to take hold of every bit that I can master: anything more than this would be a vain promise, for it would lead to a parroting of the form, instead of a true continuation in the spirit. If I can manage to come over sooner, I'll come; for my deepest disappointment over the Geneva fiasco was the realization that it would inevitably lengthen the time before I could see you again; and I know in turn what this means to you. When I come, however, it will not be to take a doctorate, or to pay attention to anything except the business in hand between us. My lack of a degree has become a valuable distinction in America. The Ph.D. is such an inevitable sign of mediocrity here that when the Carnegie Foundation for the Advancement of Art wanted some one to examine and report upon the various schools of art in America they tried to get hold of me – and this in the face of the fact that with their resources they had all the academic young men in the universities at their beck and call. I was lured by the prospect of touring all over the United States; and almost accepted for that reason; but I countered with an offer to write a critical history of the development of the arts and crafts in America *when I got around to it* – and at that stage we both left it. One of the plans that occurs to me now is to try to get the Carnegie Foundation to support me for a year while I do this work, and then to seize the necessity for tracing various origins in Europe, to run across and see you. If they pay me generously enough, that will be one way of overcoming financial difficulties. The trouble with the Guggenheim fund is that the fellowships bring only $2500 and I need a minimum of $3500 to carry on in America, with another thousand for ocean travel. Next week Sophie and I and young Geddes are going to our summer cottage. The address is: Maple Cottage, Leedsville, Amenia P.O., New York. We expect to be there until September. We join in sending you our love.

    Lewis

1. "Science and Sanctity," *Commonweal*, 6 (30 June 1926): 126–128.

Dear Lewis,

Collège des Ecossais
Plan des Quatre Seigneurs
Montpellier

8 June 1926

[Ts UP]

Yours of 22nd May. Very sorry to hear of your various troubles, but trust your country holiday is settling them all!

I suppose I may take it that Mr. Kellogg has dropped all relations with me, since he neither replies, nor returns remaining manuscripts, nor uses them; and also that your hopes of finding a publisher have failed?

You misunderstand my invitation. I want no mere ordinary assistant, but just the sort of free utilisation of my stuff as you describe, only you would see it here much more fully and clearly than from my visit in '23. Nor are the financial difficulties so serious, for I understand there are now very moderate passage fares for the American student community coming to Europe; and you would all be my guests, as I have plenty of room here, and still more at our last acquisition, the best château in the Department, which you will find inspiring with its thousand years of history, and its fine surroundings and varied activities. Thus might not even the Guggenheim fellowship be enough? – though of course I wish all success with the Carnegie one.

You know too how little importance I attach to degrees; but my reference to that was because any small book, or long paper, you might write here would be sufficient, and that you might someday find it useful to retire into some position of academic influence – even as a form of health insurance! (and meantime you could keep dark as to your D.Litt.!).

The American College scheme is beginning this summer with a small group of students (of women's colleges) Prof. Mackenzie is bringing over for July; and the Indian group are also going ahead, as per enclosed.

Frank Mears has just passed here on return from winter's work at Jerusalem University (Library and Einstein Institute); and other large developments are coming on; while I have to plan for the extensions of the new University of Bristol. Tagore, now in Italy, is on his way here; and thus the long dreamed policy of University development is increasingly taking form.

For all these *various* national undertakings, my lawyers tell me an *international committee* is necessary, and most conveniently and economically developed as an enlargement of P.G. & Colleagues, as trustees holding all these properties. To these Branford is also considering how to add his own. So may I put you down as our American partner? You and all the other partners will have no financial liabilities, beyond the small capital sum which I personally inscribe to each of you. Business matters will of course be managed by my executor among partners (probably my daughter), so the only demand upon foreign & French partners is to have their names as evidence of good faith as to international character. In fact this idea of partnership is an attempt to solve the problem of international ownership of institutions of culture, etc., somewhat like the "National Trust" in England, but now internationalised here and elsewhere if need be – e.g. V.B.'s in Paraguay and mine in Palestine. The League of Nations International Relations Committee has not solved this problem, but we think we may do so in this simple way.

With cordial regards to you all three, yours ever,

P.G.

The Maples
Leedsville
Amenia P.O.
New York

9 July 1926

[Ts NLS]

Dear Master,

Yours of the 8 of June; followed by Mabel's thesis, a very able and helpful bit of work, and finally by a letter from Branford, telling of his loss, and enclosing correspondence with you about the Coal Crisis book.[1] I was happy to learn from Branford that he was cheered by the article on you and him that I wrote for the *Commonweal*: I trust you also got a copy of it in due time. With regard to the Coal Book, the great difficulty in getting acceptance with a popular audience is that it attempts to present too much; and the effect is a little confusing. If the sociological background were taken for granted, and the conclusions and inferences were made without such elaborate preparation, they would stand out more clearly; and would, in fact, awaken curiosity about the background itself. As it is, the preliminaries keep people from reaching the heart of the matter. This doesn't call for any compromise in the matter of comprehensiveness, or of stating the moral and social issues as well as the technical and economic ones. It's a matter of tact in presentation. Branford's *Hell or Eutopia* pamphlet has this tact; and it was quite successful: the skeleton was all there, but fleshed in the matter under investigation.[2] By itself the skeleton is formidable, and, apart from involved explanations and qualifications, sometimes unconvincing. There are times when I itch to be the tyrannical editor of your writings: but the position is an ungrateful one. We have been through the same mess with our *New York State Regional Planning Report*. The data was supplied by Henry Wright: it was put into its first form by one of the editors of the *New Republic*, named Soule: this form satisfied none of us, so the director of Investigations re-phrased the whole report – and finally I revised the revision![3]

It may be that I shall not have to apply to the Carnegie or the Guggenheim funds after all. Clarence Stein, who, financially speaking, is the Regional Planning Association, has offered to give me $200 a month for working half time as editor of a series of pamphlets we propose to get out. We have outlined four so far: the first on the historical background of regionalism, the second, on the Regional City, past, present and future; the third on the probable forms of the Regional City in America; and the fourth, on the essential equipment for a new Regional City. We are attempting to discard the word, Garden City, and Regional City is our present substitute, which must carry with it the notion of a balanced relation with the region, as well as a complete environment within the city for work, study, play, and domesticity. I am going to do the first pamphlet on the Background myself and there is nothing to prevent, and everything to invite me to prosecute my investigations with you next spring. Have you any Spanish students yet? I imagine that there must be a thriving literature of regionalism in Catalunya; but I have not acquired Spanish enough to master it. Are there any studies of Spanish regionalism in French? I know most of the French books Mabel mentions in her bibliography.

As for my part in the international trust, I of course accept the honor; and am grateful. It is an excellent notion. With all your acquisitions around Montpellier, who knows but you have endowed another Oxford? Will not Montpellier attract Zionist students soon, too?

What has happend to Kellogg and to the mss. at Macmillan I have no means of knowing; but I'll try to find out the first time I run down to the city.

I am now in the midst of a final rewriting of the book. What have I done in it? Well, I have restated Comte's abstractional stage, attempted a new analysis of its origin in the ideas of space and time fostered by time-keeping and refinements of measuring – developing your idea of the timed-space, and spaced-time diagram – have shown, following Whitehead, the persistence of abstractional ideas as part of the framework of 17th, 18th, and 19th century science; and have dealt with the various intuitional efforts at synthesis, from Blake to Walt Whitman. At the same time, I have attempted to show the relation of these larger rhythms of thought to the minor ones in American community, the disastrous Civil War, the period of pragmatic acquiescence that followed; and this part of the book is a criticism of all our important American thinkers and imaginative writers between Benjamin Franklin and John Dewey (both birds of a feather!). The weaving of these two strands of thought together, showing how the first – the breakdown of medieval culture – illuminates the weaknesses of the second, the development of American life, has been a pretty difficult task; and I probably have not altogether succeeded. But this, at any rate, is the intention of the book.

Did I mention Whitehead's *Science and the Modern World* to you?[4] It's a book of first rate importance. He has an ingenious solution of the problem of mechanism versus vitalism; by showing that the categories of mechanism are useless to further modern explanations in mathematica-physics, and suggesting that even the electron is modified by the properties of its environment – so that iron in a stone is one thing, and iron in the human body is quite another, although the laboratory analysis may reveal identity – i.e. identity in the laboratory. It indicates the important modification of the old physical concepts by biology; and is quite in the line of all your own thinking – unless I misunderstand both Whitehead and yourself! Do look at it.

The youngest member of the household was a year old last Monday; he weighs over twenty-four pounds; and is very jolly and healthy and good tempered and above all things a true scientist: he examines everything, explores every place, and puts everything into his mouth. His mother sends you her very hearty greeting; his father, who is now almost as brown and healthy as young Geddes, likewise affectionately salutes you. The spell in the country has remade us all.   *Lewis*

1.   Mabel Barker, *L'Utilisation du milieu géographique pour l'éducation* (1926), her University of Montpellier thesis. Branford's "loss" was the death of his wife, Sybella, in early June 1926. Victor Branford, ed., *The Coal Crisis and the Future* (1926), comprising papers from a symposium organized by the Sociological Society with an "Introduction" by Geddes.   2.   *Whitherward? Hell or Eutopia* (1919).   3.   George Soule (1887–1970), economist and author of *A Planned Society* (1932).   4.   *Science and the Modern World* (1925).

Dear Lewis,

Yours to hand, but unfortunately my daughter took the enclosure of verse, before I had read it, and lost it – most unlucky! (Found) If it does not turn up, we beg another copy! I just saw it was great fun!

Herewith list of the *American Iona University Committee* in N.Y.C. – an amazing & generous endeavour – but needing to know more of the Highlands

Castlehill
Edinburgh

19 October 1926

[Ms UP]

before it can be assured of really taking the Highlanders with them – otherwise like Lord Leverhulme in Hebrides [...] who was like the motherly elephant who sat down on the nestlings.[1]

Is it asking you too much to join it, (as you can do by paying $5 for the coming year) and thus giving them an occasional word in season to committee & in press?

Thus they think Highlands "a small place." But as it takes mail three days to cross from here to Harris, & more to further isles, it's really as big as U.S.A. nearly, *in time*, & in complication. Indeed in stormy weather you can easily spend a fortnight in getting to outer Shetlands, etc., etc. So you *can't* centralize too much in a great building, as they'd like to do: *they must come to top-dressing the local culture*! Again, they speak of 22 professors – say then a staff of 44 – when even the little St. Andrews has 130, & so on. No sufficient consideration yet of the University spiritual, instead of the edifice, etc., structural & administrative, etc.! Get their prospectus anyway, & help them: they do mean well, & Montgomery, their president, seemed open to ideas, as money goes! Give him some!

It is a sample of a possible movement for many other regions, as you'll readily see. I'm at it in Montpellier, & at the village of (Château d') Assas, a place which will interest you, when you come.

Yours always, & with best wishes to all three,

P.G.

I suppose no more word of my poor little articles as a book? or of *Survey*? The Yale Press agent asked me lately for my *Olympus*, but not settled yet.[2] How how I *hate* writing! Thomson & I have nearly killed ourselves & each other at this one, & now we'll never try again. I can only talk – or work, but not write. Find me some one to help! 72 lately: I'm tired, & feeling old – perhaps better when I get back to sunshine shortly.

Manuscript since found; and read aloud to household, with laughter – albeit some perplexity at points. (*A stew of papers poor Macmaster's Epitaph*!)[3] Miss Defries writes me she has found a publisher for her book about me – & you have done the review already![4]

1. When William Hesketh Lever, Baron Leverhulme, was created a viscount in 1922, "of the Western Isles" was added to his title; this appears to be his only connection with the Hebrides (the islands of northwest Scotland).   2. Geddes had contracted with Yale University Press to write a book showing how the Greek gods and goddesses of Olympus corresponded with various stages of human development. He never completed the project.   3. Mumford apparently sent Geddes a copy of a poetic draft of "The Little Testament of Bernard Martin Aet. 30." This work, eventually published in prose, describes the first meeting of "Martin" (Mumford) and "McMaster," a figure modeled on Geddes. Notice that Geddes quotes "epithet" as "Epitaph."

> ... At the edge of the New Forest: Bernard beholds
> The man he has begun to call his master:
> Age has achieved the victory of a red beard
> Beneath a spreading crown of silver hair:
> Gray eyes leap to Bernard with a friendly kiss:
> A knotted hand that seems a tough old root
> Holds Bernard's hand and clasps him on the shoulder.
> The perpetual energy of McMaster's mind bulged the brain itself
> Into a forehead which became him like a crown.

Bernard longed for the slow digestion of solitude,
But was relieved to find a master looking like a Master.

McMaster says abruptly: What have your days been like:
What have you done and seen: what have you thought:
What have you got for me: what can I give you?
When Bernard puts himself and all he's seen
Before the kindness of those eyes,
He feels like children who in manhood still
Take their dolls to be the proof of their fecundity.
The days have been crowded with emptiness:
The days are the black embers of a letter
With an irretrieveable message.

A stew of paper is McMaster's epithet:
The brief diurnal flickers of daily journalism
Have neither light nor heat enough to shame a candle:
The worming through of books' experience does not season
Is scarcely worth a worm's life, still less yours;
The poor preservative of abstention is all that's kept
Your life from rotting utterly: you lived like
Clerks and academic dunces who, wound in their paper cocoons,
Prepare to metamorphose into dead butterflies:
Soldiers, though stupid, have discipline of drill:
But you have neither discipline nor the strength that can forgo it:
Brace up, my lad, you're twenty four and you
Have not yet begun to live: Now look you here . . .

A panic sobbed in Bernard's bosom: it was true.
McMaster spread a map that diminished the confusion of an impenetrable landscape.
Each contour was the shrunken reproduction of the thing itself that man could grasp.
Life active and passive: dominating circumstance
By dreams, thoughts and inventions and now
Submitting like soft wax to circumstance's mold:
Seashell and house, antheap and city, tropism and full-fledged idea
March into an organic unity; priapic beasts
And the seven gods and seven goddesses of Greece
Reveal man's biological aspiration! At every stage
The ideal is but the uttermost of Life's own reality.
[UP f. 7849]

4. Amelia Defries, *The Interpreter: Geddes, the Man, and his Gospel* (1927).

Dear Lewis,

Yours from Amenia (Summer camp?) (no date, but I suppose Sept. '26) is before me now, after recent return from Aberdeen–Edinburgh–London: so very pleased to hear you are all so well & in good activity. I'll be greatly interested to see the new book. No, the *Commonweal* article never arrived (or was lost by some resident in my absence, if it did).

I am glad you are fairly occupied with Regional & Planning schemes; and also that you look forward to coming here. (Try to give me, as early as you can, an indication of when to expect you for these houses (now two) will be fairly full, and I'll have to adjust various residents – including two family groups, like your own! I'm busy with repairs, extensions, etc., in anticipation of such growth.)

Branford has done his & my old friend George Sandeman & his

Collège des
Ecossais
Plan de Quatre
Seigneurs
Montpellier

21 November 1926

[Ms UP]

collaborator-wife – and me also – the service of bringing us together: and they are now settling in – and with determination to attack the various middens of which you have seen samples, and towards getting them into more psycho-sanitarian order > fertility > fruit to market![1] And as *both* are experienced journalists (even under Northcliffe regime!) & experienced encyclopedia editors to boot, it is a great hope! So things may look less alarming when you arrive, than at present.[2]

Try to get here by 1 March – ? I am pledged to V.B. to go to London to give June & July to *Sociological Society, Le Play House*, etc., with endeavours towards such renewal & development as may be – followed by a summer school at Education in August. If we can get the scheme of *Studia Synthetica* in order & in relation to *Agenda Synergica* to some intelligibility by that time, June, it will be a great thing. For e.g. we might then use our outlines towards some circulation to other Sociological Societies, etc. (& towards a Congress a year or so later? at which, after profiting by their criticisms, etc., we might have cleared & outlined more basis of co-operation in studies & treatment than heretofore.)

Stein's support is encouraging, but I submit there is no harm in your asking Guggenheim or Carnegie fellowship as well – since you are to be in an accredited study-atmosphere and activity. Why not put Montpellier, London, Edinburgh all in your programme for them? (And include the fortnight in Dordogne at Easter – you will find that a memorably emotive experience – to include in one rapid sweep the prehistoric & historic past.)

(I don't know to what extent you & Mrs. Mumford speak French? If not very easily, you would find it worth while to have a few conversation lessons; as these will greatly help your contacts with live people here.)

I have no Spanish students yet, nor time to look for them, which would need a journey to Barcelona & Madrid to renew two old contacts! If you wish information on Spanish regionalisms however, enclose a brief note I can send on. But won't Hispanic Society Library in New York be informed & provided? (Or too orthodox & old-fashioned?)

I have ordered Whitehead's new book; of which I've only seen a review as yet. Yes, evidently important.

Tagore writes from Belgrade saying he is ill again & ordered home by direct East route to Port Said, giving up passage at Marseilles. He'd also have been molested (or murdered?) by these Fascist fanatics on way here through Italy for a letter in *Manchester Guardian*, evidently not accepting Mussolini-Caesar's papal & practical infallibility!

What of Region-City rather than "Regional City"? (Or even *Regioncity*!)

With cordial regards to you all three. Yours,

P.G.

Yes, let me hear if you can extract those remaining chapters of Tower Talks from Kellogg, or have him publish them! Strange mortal to answer no letters – *not* that I've tried again. Perhaps when you come, your editorial experience may show me what was wrong with them?

---

1. George Sandeman, journalist, contributor to the *Sociological Review*, and author of popular

histories; author of *Social Renewal* (1913).   2.   Alfred Harmsworth, Viscount Northcliffe (1865–1922), newspaper publisher.

Dear Master,

Forgive my tardiness in replying to your letter of October, to say nothing of November's: but once we were driven out of our summer's paradise we found ourselves in the midst of something quite other than Elysium. In October young Geddes had a week of severe illness; and that was followed, in November, by Sophie's having a miscarriage; and since I had the care of the household and the baby, very largely, whilst she was in the hospital this left me very little time or energy for anything else. The result is, naturally, that we are burdened by expenses and arrears; and our plans are in a very uncertain state. I see no possibility of getting to Europe this spring. If Zimmern's plans come through and his funds increase sufficiently there is a chance of my being able to get to Europe, perhaps alone, as early as June: but this too is a dubious affair at present. Branford suggested in his last letter that I join you two at Montpellier early in 1927; but Zimmern's money is the only source I can draw on, since the Guggenheim fund demands a year's residence abroad and does not nearly cover the expenses of the family and the trip, and if I go at all, I will therefore have to bunch my Geneva engagement and my visit with you. I will write you as soon as Zimmern gives me definite news: he is now in this country, lecturing and making contacts. My book, *The Golden Day*, has just come out, and I have sent you a copy via Edinburgh.

In a little while I shall write more at length. I herewith return the American Iona Society letterhead: I alas! know none of the members personally.

Sophie joins me in sending you our hearty Christmas greetings.

Affectionately,

Lewis

4112 Gosman Avenue
Long Island City, N.Y.

14 December 1926

[Ts NLS]

Bon Noel! et Bon Annee! dit Collège des Ecossais!
Viande et Vin, bon pain et beurre, comme au Plan des Quatre Seigneurs!
Assaisonnés de bonnes idées! – à la mode de Montpellier
Même toutes poussées à outrance! Ainsi vient l'Ecosse en FRANCE![1]

Xmas Day 1926

[Ms UP]

Dear Lewis,

Your *Golden Day* was welcome, & I have read it with the greatest pleasure: it is the best sustained critique I know! – fully on the level of the French critics who relate literature to its times, & in advance of them, in having also a far clearer idea of the tendencies of the times we are living in, and thus of the significance of the critic's task, & his value in guidance towards new writers as to what the world needs accordingly.

The book has been going round our group here, and all are delighted with it. George Sandeman, my new colleague here, as adviser of studies, & collaborator too, whose name you may know as original editor of *American*

Encyclopedia (Nelsons – New York) and an old friend of Branford & mine since our early Edinburgh activities 40 years ago, has settled down here with his bright wife & promising little boys: so he is to review it for *Sociological Review*; or both of us together.

What next? If you'd do the same for English Literature now, and French too, that would be a fine trilogy – & start you as our new Brandes![2]

Or will you return to Utopias – & make one up to needs? (I believe the diagram of IX > 9 (& with 9 > IX also as activity; of Eutopists) may here be of service, in getting ideas into order, & thus in more than usually convincing form.) (Pray criticise this.)

Herewith I send you Draft Circular regarding American College here for your criticisms & suggestions – for it is well to be getting idea moving on: for at my age, though I hope not much damaged by my Indian illness, etc., after we met, one must prepare for abating activities – & I want to see this variety of projects and beginnings more of going concerns without such constant responsibility for them.

Since writing the above yours of 14 December has come in. Very sorry for your home troubles. I hope both Mrs. Mumford & young Hopeful are all right again.

I hope too you may be able to pull off your European visit somehow. After all, though the Guggenheim scheme may not be sufficient, you'd have your time for writing, etc., practically free: so it should be so far to the good in paying expenses. And if you care to steep in French & English current literature and movements, you would surely find market in U.S. magazines & press for your impressions of them: so you need not be falling out of touch with home market, but rather enlarging it in New Ways? However, know I can't advise: I only speculatively suggest – but why not make yourself known in European letters too? Yes, let me hear from you as you say.

This weary biology book not yet finished – I write with great difficulty & irregularity: still, something gets done.[3] I am engaged, by Yale Press agreement, to produce "Olympus" by May.

With cordial wishes to you each & all for 1927.

Yours

P.G.

1. Geddes combines his French greeting with the printed letterhead address: "Merry Christmas! Happy New Year! says the Collège des Ecossais! Meat and wine, good bread and butter, as at the Plan des Quatre Seigneurs! [perhaps a wordplay: 'as on the plate of the four lords'] Seasoned with good ideas! in the style of Montpellier! Even pushed to the extreme, thus comes Scotland to France!"   2. Georg Brandes (1842–1927), prolific Danish literary critic and historian; author of *Creative Spirits of the Nineteenth Century* (1923).   3. Co-authored with J.A. Thomson, *Life: Outlines of General Biology*, 2 vols. (1931).

[1926][1]

[Ms UP]

Dear Lewis,

Glad to hear from you, & that things go well with you all & with your work – & towards your next visit to Europe before long – of which I am most desirous of interesting you into a good long stay here. I forget if I ever sent you accompanying booklets & photos (postcards) of this place & of Château d'Assas – from which you'll see we've some room for you all, the more since I hope to have the College more than doubled by Easter – adding 18 or so new rooms or so to the present 15; while Dr. Advani Secretary here & of Indian College is at present in India on tour of Universities & Cities for students & friends to build.[2] Jewish House = Maison Zion = Collège des Palestins also in quiet progress there two years – & the American College I hope also promising.

I go into these practical points because you have known me only as speculator & speculative interpreter – and my visit to you in New York City was quite of that nature but here are many sorts of activities in progress, & more in preparation – & on which I shall value your fresh eye & attitude of constructive criticism – whether you see your way to cooperation or no.

I had hoped for V.B.'s co-operation here these last 10 days, in brief meeting – for sociological studies organised by the good Miss Tatton – but he is ill – harmful attack of arthritis & so sent to salt baths instead. Very unfortunate we three friends – Thomson, Branford & self – are all showing signs of the main malady – of Anno Domini. He will I hope come for a visit in November when we hope to go into possibilities of *Review*, etc., & of Le Play House, Outlook Tower, & this place – on more ambitious footing – aiming not only towards Studia Synthetica (with of course this more of Analytica & Critica) & with wider endeavour of co-operations in the IX–9 diagram, if you recall it.

I'd like to know if you have found anybody caring for graphic methods in Social Studies & to whom they appeal? In this respect, I make no progress – find no response. Yet see how the essential [principle] of the preliminary sciences is so essentially old verbalisms to new *imagery*. And is not the like manifest also in letters? Your own *titles* seem to me so many indications of this.

Let us know your plans as soon as may be – & consider a good long stay here, & so that Branford may be able to come & stay as long as he can. Note too that for us July & August are too warm here – & that we need cooler quarters – probably in own island, as in past years.

1. Undated letter; Mumford's note: "Probably 1926."   2. Gopal Advani had been Geddes' student in India and came to assist him at Montpellier.

Lewis Mumford AND Patrick Geddes

THE CORRESPONDENCE

1927

Dear Lewis,

Collège des
Ecossais
Plan des Quatre
Seigneurs
Montpellier

12 February 1927

[Ms UP]

In the midst of many cares I can't recall whether I have really answered yours of 14 December, though I think so; yet I may only have thought the letter, without writing it, as sometimes happens. Surely I acknowledged the *American Caravan*, and found it brightly laden – though some of the stories did not greatly appeal – and I read few – being more & more absorbed in my social & other puzzles, & in large building extension, for this place, & the Indian College being worked up in India by Dr. Advani – while for the possible American one, I have also made a fresh design – followed by acquisition of 7½ more acres of the heath to west of this, for future extensions; also large & costly repairs, etc., at Château, which all cost time as well.[1] *Olympus* still lags – no time to dictate even, if I could do that as I'd wish – & the book with J.A. Thomson is delayed by his stress & health: I fear none of the best. V.B. too has been very ill – but I hope now recreated by a stay in Switzerland – at his favourite cure. But what a severe winter this is in Europe, snow even here today again!

Very sorry too for your cares! I hope you may be able still to come over here – & make some stay, as well as join Zimmern in Geneva? I have to be in Scotland by end of June – & away all August at least, but return some time in September. Here you'd have peace; or at Assas alike; since few, if very few, others. But of course I'd like to be here with you & as long as possible.

I don't think Spengler was sent to *Sociological Review* for review, but I'm asking Branford.[2] Certainly Dawson would do a thoughful critique – but somehow he seems to be dropping out of *Review*, whether too busy with his History Chair at Exeter (a new start), or having altered his views & diverged from us, as a good Catholic may not unnaturally do, though we are not so unsympathetic as may appear. Great thing the Pope's settlement with Italy – as he can't now be kept out of League of Nations for instance; & other arousals will follow.[3] Curious to see what he'll do with his £20 milllion – he can't make such bad use of it as any secular government would do! Even if he spent it all on plaster saints!

I work away at my IX–9 & other graphs before breakfast but get 0 written! Still, I cherish the dream of getting something done some day!

P. Boardman prepares to come over here – but to work in garden, etc. – as his eyes have given way – & he has to leave off books![4]

I hope your "Melville" book has got finished? I have read nothing of his as yet. What next?

Interesting visit yesterday from Waldemar Kaempffert – formerly efficiency engineer – now preparing Techno-Social sort of museum, etc., with three millions from Rosenwald, a public-spirited Chicago millionaire.[5] Very open-minded & progressive is Kaempffert, also been studying (in the too rapid American way!) the like attempts in Europe – of scientific & technical character, as yet deficient on the social side, as he would fain *not* be. At any rate he'll make a good beginning.

I've also a bright fellow – John Fuqua – from Chicago, formerly at School of Economics in London, who came here for Christmas holiday, but won't go back! But as yet small sign of the live students I want! Too simply absorbed in their respective more or less conventional round, & difficult to talk out of it!

Write soon, & let me hear how you are all getting on. Young Hopeful seems indeed worthy of that name – so must be a great joy to you both! Give him my love.

Ever yours,

P.G.

1. *The American Caravan: A Yearbook of American Literature*, ed. Van Wyck Brooks, Alfred Kreymborg, Lewis Mumford, and Paul Rosenfeld (1927).   2. Mumford reviewed Oswald Spengler, *The Decline of the West*, Vol. I, in *New Republic*, 156 (12 May 1926): 367–369.   3. Although a professed unbeliever, Mussolini made numerous concessions to the Catholic Church in return for its support beginning in 1922. Under the terms of the Lateran Treaty adopted in 1929, the Church recognized the Italian nation and its occupation of Rome; in return, the Church was granted the Vatican City as a sovereign territory, was recognized as the official state religion, and received a substantial indemnity.   4. Philip Boardman, who was then a young American student, first met Geddes in 1925 and subsequently studied with Geddes at the Collège des Ecossais, 1929–1930. He later published two biographies of Geddes.   5. Julius Rosenwald (1862–1932), President of Sears, Roebuck & Co. (1910–1925) and philanthropist who established the Chicago Museum of Science and Industry (1929).

---

23 June 1927

Chilmark
Martha's Vineyard
Massachusetts

[Ts NLS]

Dear Branford,

I am addressing this letter to you; and trust that you will pass it on to P.G. since these accounts of my doings are of common interest to both of you, if they have any interest at all. When Zimmern left America I had given up hopes of going to Europe this summer; a month later, however, he bade me come over; but it turned out that the funds were insufficient, and that the work would be incessant, not only lecturing but tutoring the whole summer; and at the pace set in Geneva that is an exhausting business, not merely intellectually but because of the drain of social intercourse. I was in no shape to do it, particularly since I could find no way of decently disposing of Sophie and Geddes for the summer without taking them with me; and I reluctantly said No. Had I gone, I should not have been able to snatch more than a few days to see you and P.G. – and that was one of my main reasons for wanting to go across. It was a hard decision to make, however; but as things have turned it out, it was justified, since the last two months have been an incessant strain, what with lecturing at the Robert Brookings School in Washington, and at the University of Michigan, and compiling the bibliography, and, on top of all this, moving our household to new quarters.[1] My permanent winter address is now 4002 Locust Street Long Island City, a little five room house in a quiet cul-de-sac; but the above address should serve till the middle of September – a little bungalow, in the midst of rolling moorland, within sight of the sea. I have much work planned for the summer; and no financial cares whatever in sight until October, a new and surprising condition in my life, due partly to the bibliography, and partly to the fact that *The Golden Day* has already sold 4000 copies. The response to it, by the way, has been very good from the youngsters of from 16 to 22 I have met in the Universities; this, you see, is the post-war generation, ready again for active adventure and enterprise, without the disillusion and cynicism and fatigue of those who are ten or fifteen years older. This is the soil for our biosophy! One lad I have met – he is sixteen or seventeen and came all the way from Philadelphia to see me! – is exceptionally bright and able; I had advised him to have a Wanderjahr with P.G. in Montpellier; but this summer he is one of a body of students who are going to Denmark, under the auspices, I think, of the University of Copenhagen, or at least of one of their professors. This lad is twice as intelligent as I was at his age; and much better equipped for active life and thought; he inclines towards architecture and city planning; and for all I know may be fit for many other things. Meeting these boys and girls has been a great encouragement; they are such a welcome contrast from most of the tired wistful souls of my own generation. Lindbergh is a true sample of that generation at its simple best: tremendously interested in their *job*, with a sense of vocation not easily corrupted by the dominant pecuniary values: an attitude that comes like a cool west wind in a land befogged by salesmanship and advertising and organized Babbittry. The very "wildness" of this generation is a a sign of its life! The girls settle down to motherhood easily; and as far as genuine sexual morality there has been a raising of the common level, even though this has meant the sacrifice of the *technical* chastity of the girls. Unless I read signs wrong, there is a far healthier inner life, and ideals and practices are much more closely harmonized than they used to be. This of course is a general picture; you must allow for exceptions and divergences and lapses; but the effect is of life resurgent, and not, as some of the more acrid and unsympathetic elders have pictured, of life decadent.

I am looking forward to your criticism of my general scheme of the bibliography. On the whole it seems to work well; but I am playing with several modifications, and I will submit them to you when I send you the list of the first two thousand or so. In casting about for a squel to *The Golden Day*, I have almost resolved to combine the suggestions you and P.G. severally offered. He suggested that I go on to do a similar book for English literature; and I have now hit upon the notion of beginning such a study, with Shakespeare and Bacon, and concluding it with P.G. It is perhaps an audacious anticipation of posterity's judgement; but I think I can carry it through, and it will give me an opportunity to place in its proper frame the Geddesian outlook, as the climax of one episode, the Baconian demand for knowledge unlimited, and Shakespeare's expression of a life opulent; and the opening of another quest, that which you yourself have more than outlined in *Science and Sanctity*. How does this strike you? I am a little aghast at the work this thesis will involve; but I think it will be worth doing; my publishers have agreed to publish it, with twice their usual advance against royalties and, if all goes well, I expect some time next year to pay my long deferred visit to England, and if possible to Montpellier, to recover my place-sense and my feeling for English life, before I finish off the manuscript. All this of course is tentative. My arrangement with the University of Wisconsin for next winter is also tentative; it probably will not be for more than a month; and if I start work on my book next winter, it may not be possible to do it at all.

My little essay on the Background of Regional Planning, which I sent you this week, is not what I had hoped it would be: I wrote it when my energies were very low; and it is pretty poor. If you don't care to use it I shall cordially sympathize and will understand![2]

Please give me word of yourself and P.G. when you are able. Remember me to Farquharson. Sophie joins me in good wishes to you all.

Ever yours,

Lewis

1. Mumford had been engaged by the J. Walter Thompson advertising agency to compile a bibliography of three to four thousand titles for the reception room library of the company's Manhattan office.   2. Pulished as a series of three essays, "The Theory and Practice of Regionalism," in the *Sociological Review*, 19 (October 1927): 277–288; 20 (January 1928): 18–33; (April 1928): 131–141.

Collège des
Ecossais
Plan des Quatre
Seigneurs
Montpellier

12 July 1927

[Ms UP]

Dear Lewis,

I am more vexed with myself than I can say for putting off, through April May June till now, any reply with regard to your Library scheme. But I have been going through a combination of difficulties – first the *great* one of giving up smoking, with discomforts & depressions (of writing powers especially!) beyond description, & not yet quite ended – & also a series of other cares & difficulties more than usual heretofore – but on which I need not particularize!

However I hope you took my silence as of general assent – & indeed agreement & acceptance.

For minor criticisms, I doubt use of introducing the old Greek & academic terms of *Physics* & *Dialectics* (now misleading in ordinary senses). *Sciences* & *Ideologies* are surely quite enough.

Symbolic and *scientific method* do not in my thought & vocabulary quite go together – I use *notations* for mathematical, etc., and keep symbol for religious and poetic meanings. Yet I thus can hardly plead for *Graphics*, since these will be taken with Statistics & nearer Maths. The word I use is Thematimetrics – but that's my own invention I fear – but what of *Thematics*? (I think Gratry's.)[1] This is easily explained as orderly & logical development of Themes, of argument & theory – and of subjective studies so corresponding to orderly development of objective studies by statistics – the former especially dealing with quality & the latter with quantity. (Just as we no longer write "Æconomics" & "Æcology" but *Economics & Ecology*, so it is time to give up "Æsthetics," and "Mediæval," etc., for *Esthetics & Medieval*. My public don't mind so still less will yours!)

I don't like separation of

Customs/Ethnology   World

Parallel columns here thus rather confusing than helpful:

Ethics   Asia
Jurisprudence   Europe
Folklore   U.S.
Anthropology   Geography

This seems to me the point most open to criticism – nor does this parallel help, but conversely. In fact it won't do! Folk, at all levels, give you "moeurs," a "mores" or "Folkways." [...]

Then too these [morals, manners, customs, and laws] sublimate into religions, especially the moral feelings – yet all are represented. (You don't mention Religions or Comparative Religion) on this side. I sense on other. Yet this is for Comparative religion, general religion. The concrete religions need place on left. You have *Ethics* on the right; say *Morals* for it on left.

To modify your scheme (as little as possible) why not morals & religions above laws & jurisprudence, customs & folklore, anthropology. See our little *Biology* for eight subsciences of Sociology. Anthropology in ordinary sense may include Ethnography (race) but use Ethnology as almost a synonym.

I suppose Miss Defries sent you her book? I hope she got *Survey's* permission? (By the way is Kellogg never to answer & not to print remaining essays? I suppose not!) What are you doing? – What next? (I still hope you may come! I can put you all in empty rooms & space! so costing me 0; & you can pay your own food bills, so feel independent – yet spending little.)

Yours ever,

P.G.

Herewith Circular. Can you mention this in press & so send some live student over, male or female? I am mentioning your library in my report on Bibliography-classification of sciences & arts-Universities, etc., etc., to Toronto Conference of Educational associations.

1. Auguste Gratry (1805–1872), French philosopher; author of *La Morale et la loi d'histoire* (1874).

Dear Master,

It was a pleasure to get your letter of the twelfth. In the meanwhile I had written a long letter to Branford, with the request that he pass it on to you; but I'm not sure that there was time for you to get it, before your writing me. The spring was an unkind one to us: for a time we were in financial straits and Sophie was forced to take on some proof reading, and though by a sudden leap we got on our feet again, with a little something to spare, it was only by exhausting ourselves with numerous engagements to lecture and write. By the time Zimmern came through with an offer which made it possible to go to Geneva, I hadn't the courage or the will to do it: I was quite fagged out. We moved to somewhat larger quarters in Long Island City – a five-room house – early in June, and shortly after came up here. The last month has done much for all of us – salt water bathing, moorland walks, plenty of fish to eat, and long untroubled nights of sleep! I am back again at my best level: and am working regularly, tidying up many pieces of work which I had begun during the last two years, including a play, and the long olla-podrida, half poem, half prose narrative, of which I sent you a fragment last October.[1] *The Golden Day* has gone very well: far beyond my expectations: it has already sold 4000 copies; and the Oxford Press is getting out an English edition. Following up your suggestion, I have arranged with my publishers to do a similar book upon English culture; and I am almost sure now that I shall be able to get across next summer, as indeed I will have to, in order to recover my place-sense and folk-feeling, without which any

Chilmark, Mass. (till 15 September after that, note new address: 4002 Locust Street, Long Island City)

26 July 1927

[Ts NLS]

such survey would be a pretty barren one.[2] That is the earliest it will be possible for me to go; I must remain on this side during the winter to take advantage of various profitable opportunities for work.

I have digested your criticism of the first hasty outline of the library; and I'm happy to say that in most cases I've anticipated your dissents. The parallel columns indicated for Geography and Anthropology did not have any schematic significance in my mind: I was thinking merely of the library shelves! My greatest difficulty so far has been to break up literature into its functional divisions and to see where they belong within this grouping. If music, painting, and sculpture are together, should not poetry be there, too? But if so, where do novels go? The Greeks did not take the Milesian Thales seriously; so they didn't inform us which of the Sacred Nine wrote them; and if they don't belong to the Sacred Nine, they nevertheless do most bulkily exist.[3] Where should one put them? One might of course break the novel up into the biographical novel, the sociological novel, the poetic novel, the historical novel, and so put them, in an adjacent compartment to these main categories. I am thinking aloud. Please give me the benefit of your own thoughts.

Going back to the sciences. As arranged in ascending order from mathematics to sociology, they recall Lloyd Morgan's doctrine of Emergent Evolution: so far a good correlation with Comte and with your own theory. At each stage, that is, a new quality emerges in experience; although for Morgan the final emergent is God. It seems to me that this fact comes out a little more strongly if one arranges the sciences in this fashion:

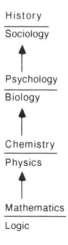

History

Sociology

Psychology

Biology

Chemistry

Physics

Mathematics

Logic

Then the upper science in one category is the connecting link with the lower science in the next category. If this is so far true up to Sociology, is it true that History is the Emergent of Sociology? It is not so, it seems to me, when History means merely chronicles and events; but when history signifies the accumulation of the social heritage, it *is* emergent: and is this not what Croce means when he says that Philosophy is History – a dictum which is otherwise a meaningless paradox. This gives the rationale, does it not, for the box "Philosophy of History" which if I remember right belongs to the Eight sub-sciences of Sociology as worked out in your book-case diagram.

Thematics is a good word: it was stupid of me to forget it: and I shall use this in my final diagram. It may be that I will publish a short explanation of the scheme, when it is all finally arranged. In another week I hope to have in hand the complete list of books: I'll send it to you for criticism in detail: and particularly for the addition of European books I may have neglected. The chief list I've worked from has been Sonnenschein's 100,000 Best Books.[4]

Your little namesake thrives well up here. Sophie joins me in very affectionate greetings.

Lewis

P.S. I did not receive a copy of Amelia Defries's book yet. She persists in writing to an ancient and obsolete address; and this may be the reason for it.

P.P.S. I have seen nothing of the *Survey* people for a whole year. Kellogg is impossible!

N.B. I'll do my best to write a piece about the College.

1. "The Little Testament of Bernard Martin Aet. 30," first published in *The Second American Caravan: A Yearbook of American Literature* (New York: Macaulay, 1928), pp. 123–169; reprinted in both *Findings and Keepings* and *My Works and Days*.   2. Mumford never pursued this project; his annotation of August 1972 states: "I had quite forgotten this!"   3. Thales of Miletus, sixth-century B.C. Greek philosopher who sought truth in science; he held the view that the world originated from and will return to water, and perhaps this explains the allusion: that is, the Greeks did not like to reduce things to one, general category. Sacred Nine: the Muses.   4. William Swan Sonnenschein, *The Best Books: A Reader's Guide to the Choice of the Best Available Books (About 100,000) in Every Department of Science, Art, and Literature* (1910–1935).

Dear Lewis,

I've been away, & rushed, since return: & now I see I must post packet without further delay: hence I can do little – & the more since I've read so little for long!

I agree with V.B.'s criticisms & suggestions. (I hope you'll correct titles all through with *capitals* to essential words.)

I'd add Frazer's *Golden Bough* & his *Anthropology in Bible* (or some such title). In *Biography*, I don't notice Lytton Strachey – nor Boswell's *Johnson* nor Lockhart's *Scott*, each a model in its way. & Carlyle's *Past & Present*. (*Abbot Samson* surely classic!) I'd separate sub-sections in many cases – e.g. Religion – e.g. Historical, Mystical, etc., etc., or again art, as Architecture, Painting & Sculpture & Town Planning apart (may I recall here my *Cities in Evolution*?) & suggest *Dramatisations of History* and a *History*, rather than Education – since *interpretative* also. Add Stanley Hall's *Youth* (as shorter than *Adolescence*), Travel, etc., Doughty's *Arabia*. Literature – I'd have liked more of French literature – Victor Hugo for popular awakening (Say *Légende des siècles*) also Hérédia's Sonnets, etc., Verhaeren & something of Henri de Regnier – e.g. *Médailles d'Argile*. Chesterton's *White Horse* & *Poems* are his best work I think. In psychology I think Smuts' *Holism* likely to be worth insertion, though I've only seen review.[1]

In every subject why not mention one or two leading reviews & journals

Collège des Ecossais
Plan des Quatre Seigneurs
Montpellier

22 September 1927

[Ms UP]

– e.g. Why *Sociological Papers* & not *Sociological Review?* & so on throughout. So for Education, etc., etc.

Sorry to be so little use – yet – Yours ever,

P.G.

Literature: Add The *Golden Day*!

1.   James G. Frazer, *The Golden Bough* (12 volumes, 1890–1915) and *Folk-lore in the Old Testament* (1918); Lytton Strachey, *Eminent Victorians* (1918); James Boswell, *Life of Samuel Johnson* (1791); John Lockart, *Life of Scott* (1838); Thomas Carlyle, *Past and Present* (1843), containing a chapter "Monk Samson"; Charles M. Doughty, *Travels in Arabia Deserta* (1888); Victor Hugo, *La Légende des siècles* (1859, 1877, 1883); Jose Maria de Heredia, *Les trophées* (1893); Emile Verhaeren (1855–1916), Belgian poet, author of *La Multiple Splendeur* (1902); Henri de Regnier, *Medailles d'argile: poèmes* (1900); Gilbert K. Chesterton, *The Ballad of the White Horse* (1911), *Poems* (1915); Jan Christian Smuts, *Holism and Evolution* (1926).

Collège des
Ecossais
Plan des Quatre
Seigneurs
Montpellier

20 November 1927

[Ms UP]

Dear Lewis,

(1) Mr. Joseph Ishill – The Oriole Press – Berkeley Heights N.J. – has sent me his wonderfully fine books – respectively entitled (1) *Kropotkine* & (2) *Elisée & Elie Reclus* – which pray certainly add to your list of *Biographies*! (soon to be out of print).

22/11 – Since above written, Victor Branford has been here – as well as Professor Ellwood of Missouri – and the latter gave a lecture to the University on recent Sociology in U.S.A. The great point however was that I persuaded him to plead that University here should have not only a chair of Sociology (at present there are only sporadic courses, by one or other legal, economic or philosophy professors) but a regular Department – & this as *Institut Internationale de Sociologie* – & memorial of the founder. We had some talk with professors a few days later, & the idea is moving – but may be long enough of coming towards action. Still, I'll keep at it as part of the scheme of *Cité Universitaire Méditerranéenne* – of which this place, with its associated Indian, Palestinian & I hope by & by American beginnings, should become nucleus. (We have a live Mayor, though torpid principal.)

They say there are getting over 5000 American students in Paris! – of course counting those at Art, etc. – but surely an exaggeration! Still a great many – too many for any one University to absorb – and thus likely to develop disappointments! Our plans of 2 years ago, for 50 residents, here, are thus too small, if (as should be easy, when we get some live Americans interested) we get some of at once the overflow & the pick of these great numbers. Since above written letter from lady Professor of English & President of "Deans of Women" writing that she is coming over in summer & to settle in to thesis, etc.

More & more this place is hopeful, if I can escape ruin by its initial outlays still in progress – & now about $40,000, out of which I owe $6 or 7000 to friends (Victor Branford, etc.) & more (say 9,000) to the Bank, as security of my life-policies. I have thus reached my limit – and my 8 or 9 residents are not enough to pay, so that it's a very difficult thing now to make up the difference of loss – though not so much real loss as capital outlays continued

– as on gardens, Château, etc. – & now lately a *school* (of which more might be said – but not urgent).

There are other reasons for trying to get a move or two on as soon as may be – e.g. American College & Indian College – which would each relieve me of a proportion of my heavy outlays for sites, gardens, etc. – and also give more importance and attractiveness to the whole scheme, & to British students, still so few out of even my small numbers!

But beyond these business cares of course, the ideas of the place keep me going. Thus beside our six theses writers for doctorate in science or letters, there is a bright boy, & a musical girl (each 18), each fresh from their famous & bad schools, and having a breathing space between school & college. This works encouragingly, for the elders educate these younger ones, who tend to put away childish things more rapidly than when they go on from 6th form at school to 7th, 8th & 9th at College!

Besides the scheme of rationally attractive & ambitious thesis-writing, by able young people, now satisfactorily in progress (as indeed from the first), I am trying to evolve and offer opportunities of *tasks* – with their *practical* education values. Thus our botanist woman is helping to plan out our type botanic garden (on my evolutionary theories largely); & the (latest cubist!) artist is at work decorating our new village school nearby – while above botanist, & another one, are also at *Jardin Scolaire*, which will be attractively floral – & edible too! A party are coming from Le Play House, etc., for Surveys at Xmas & Easter – & this may lead to useful impulse impulse to village & neighboring ones.

(I mention these things because have no doubt much going on in such ways around you, of which I should like to hear more – or should value newspaper scraps & cuttings.)

Glad to learn per V.B. that you liked Miss Defries's book, & are even good enough to propose to review it, & give a publisher a push to take over sheets! Poor Miss Defries tells me that of the 1000 printed, 600 remain unsold! – Indeed only 240 were sold, the rest being review copies, etc. Reviews good, she says (practically only one exception) but no interest in public!

Get Mrs. Taylor's *Leonardo the Florentine* (Richards – London '27) the most elaborate piece of word-painting I know of! Too long, too rich, etc. – but all the same most extraordinary! I suppose you got her *Aspects of the Italian Renaissance* 1923 (& I trust did not disapprove my review in *Sociological Review*? 1924 I think). She is now to tackle the *English Renaissance* – I asked her to write *Provence* (which just borders here) & that may come in time – if time lasts for us! Expensive – 30/ = $7.50 – but to my mind extraordinarily well worth it.

Regards to Mrs. Mumford & Geddes.

Yours,

P.G.

Advise me on this point. You'll see in printed circular herewith I say a little about Sociology & practical application. Why & how is it that as yet, in all of my contacts widespread, there is no case as yet of response to that?

LIBRARY

Your library scheme has been left me by Victor Branford on his way to rest at Territet for a month. I was sorry to be so brief in my return of it but I understood that it was urgent – & I was (& still am) much too over pressed to do it justice. I can't afford a Secretary unfortunately, & correspondence is heavy, while neither the *Biology* with Thomson, nor the *Olympus* for Yale, has got anything added for many months! It looks as if I'll have to disappear, with all my essential say still unsaid!

Still, I welcomed your treatment – very good! It involves long talks however to coordinate your & my outlines more fully. Pray look at my & Thomson's *Biology* – for beyond outline of main sciences, their subdivisions are rationally outlined – also the historic concept of the literature as *Heritage* – with Precursors before the (Editor-)Initiator – & continuations, etc. – and all in *deepening* shelves of analysis – in every field – structural, static, formal, on one side to functional, kinetic, & vital on the other, & similarly in ascending shelves of Synthesis.

Of course I admit that this can't so fully be done with literature as with science – & that many scientific documents even would have to appear in several references – so this scheme is not fully practicable for your purposes – yet at times such may be helpful – and suggestive to the reader when it can be used here & there.

My more complete classification is as per accompanying sheet. (Reference to diagram in Biology will show treatment of "*materialisms*" & "*transcendentalisms*" of Comte – as "*legitimate*" & "*illegitimate*" – which can be next filled up in each.)

The *eight-fold subdivision of each science* & the placing of the bibliography upon its analytic & synthetic shelves is a scheme I'd fain work out with cards, & as model. Then too there has to be the essential & complemental rearrangement of all this arrangement still too static – since as *Past* first, then *Present* & only finally *Future* – & needing to be mirror reversed – or turned right over – to *Becoming, Being, & Having Been.*

In the garden here I am trying to work into my apparently purely practical & constructive terracing & purely esthetic planting, a scheme of graphs expressive of as many general views as may be (each a philosophenweg) thus literally for peripatetic teaching & though stimulus – (a) from primitive cavern folk to early Greeks – next to Olympians & Parnassians – (b) something of medieval thought & of Renaissance – (c) later philosophies, e.g. Hume, Kant & Hegel – to Comte, etc. – and Platonic as complemental not exclusive (as pencilled on diagram). Sciences too as leading to arts, yet also developed from their experience & aid – theoretic & pragmatic thus combined – and so on.

On other terrace-floors are coming *Life Theory, Social Theory* & so on – and on to IX–9 as per my diagram of Social Transition (made first clearly for a Cornell lecture). Thus in short, an endeavour to cordinate many graphs intelligently.

But alas, I can't express all this! My dear fellow – you *must* come! – and save what you can – essentials – of this old man's thought – at this rate *never* to be adequately written! You only saw me relaxed, absorbent, dispersive – but – with your stimulus especially, I can still make my last rally here. Otherwise

I fear not – nice young folks of promise here but all too immature, & leaving when thesis completed.

I stick to my forecast – that *you are needed for a University headship by & by*, & of not mere administrative character – but – frankly! – what I should have been offered – somewhere, congenial! And I want you to be especially my heir – the leading legatee. There are others, but none so well prepared on the whole, & others needing your leadership in the whole too. For overcoming objections from one side, but above all as fuller discipline on the other, there's no place so good as France – & this is less crowded & distracting than Paris. You see I keep thinking what I've already repeatedly said – & always more clearly. It is not a merely egoistic suggestion – though of course I'll value such cooperation & stimulus as you can best give. Circumstances – and even residents here – make feasible more than I can explain in letter limits. Yours P.G.

Dismiss all idea that in suggesting a doctorate here, there is insistence on that – though you for your part should understand that as involving only a translation of some bit of your work – useful over here.

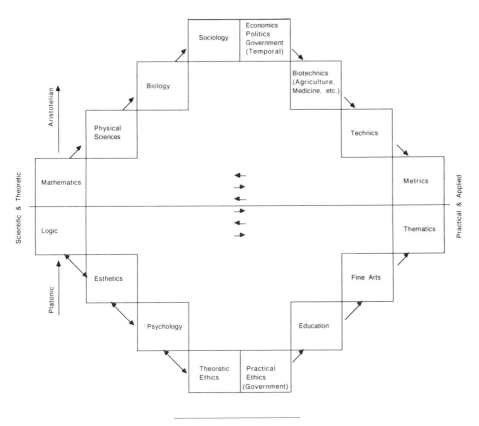

Dear Master,

I have owed you a letter these many weeks, if not months; and meanwhile life has been flowing by all too rapidly for me. We are now well settled in a little five room house in Sunnyside: We front on a terrace and our rear faces the interior of the block, which is common for all the dwellers around it, under the

4002 Locust Street
Long Island City, N.Y.

13 December 1927

[Ts NLS]

plan that Stein and Wright have worked out for Sunnyside. Little Geddes goes to a nursery school which the parents have founded and organized; and our lot is a far happier one than it has been for many a year. We are within twenty minutes of the heart of Manhattan; there is no wear and tear in going back and forth; there is plenty of sunshine and air; all the open space one needs; and in our little handkerchief of a garden, front and rear, we already have 150 bulbs, tulips, crocuses, and daffodils started against next spring; while in the cellar is a box of earth, ready for the earliest spring planting! I have carpentered at my leisure, having at last a workshop in the cellar; and have now plenty of space for my books, besides getting out of carpentry just the manual stimulus one needs to avoid transformation into a book-worm. Our common projects go on pretty well here. The company that built Sunnyside have now acquired five hundred acres in New Jersey, which they can plan from the beginning, and thus effect economies which were impossible on the gridiron plan used in the city proper.[1] Henry Wright has worked out a plan which will turn all the houses onto the garden space (which used to be the rear) making the actual rear mere traffic lanes for automobiles, and forcing the pedestrian to walk along the garden paths. By using block plans at least twice as big as the present New York ones he achieves spacious parks within, cutting down the cost for highways, asphalt, extravagant utilities, and can arrange the residence areas in such a fashion that children under 12 years will not have to cross a single traffic lane. What this means you can judge from the fact that the decrease in death rate from measles in New York is more than counterbalanced among children by the growth of deaths from autos. In fact, as Wright at present plans it, it will be possible to take a half-hour's walk through the community without once leaving the gardens to cross a major traffic thoroughfare. Wright's plans have astonished the park and playground experts who have been worrying how to get the necessary funds for their areas: for on cheap land he has no difficulty in more than meeting their expectations merely by cutting out useless and harmful streets, with all their attendant utilities. Benton Mackaye, for his part, has been working hand in hand with the Massachusetts state forester, preparing plans for permanent flood relief, combined with afforestation and the development of electric power; and Clarence Stein has been designing the new school of technology in Pasadena, California, as well as a museum of Art, and Felix Adler's new vocational high school in New York. As for myself, I have, besides preparing the little biography of Melville I am now about to write, been revising the bibliography I sent you and getting together an essay on the arts for a symposium that Charles Beard, the historian, is editing on the present prospects of Western Civilization in relation to science and technology.[2] The last has been the hardest task of all; for there has been very little critical work, since Taine, on the relation of the arts to the milieu – with the exception perhaps of Bücher's book on *Labor and Rhythm*, Jane Harrison's *Art and Ritual*, and the introduction of a countryman of yours, D.S. MacColl, to *Nineteenth Century Art*.[3] Incidentally, as a result of my writings on the arts, one of our biggest industrial corporations, the Dupont Nemours Company, which manufactures about seventy byproducts of gunpowder, asked me to become a sort of consulting art director, to criticize their products. I was greatly tempted, but it would have taken too much time to have mastered the job to my own satisfaction, and I should have had, in the process, to neglect things which interest me more keenly, and which seem, on the whole, more important. It's an interesting sign of the times, however; not altogether unlike the commission to do the bibliography and all the more significant because, far from making any effort to ingratiate myself with the ruling divinities, I have never missed an opportunity to shy a brick at them.

As things shape now, I think I shall be able to go to Europe next spring or summer; but when, and for how long, depends upon whether or not Sophie has conceived. The miscarriage a year ago seems to have induced a temporary sterility; at least, we trust that it is temporary. I have tried to induce some students I've met to visit you in Montpellier: one of them was a Dutch architect named Siebers who will travel in Europe in the spring, and another was an American student, now studying law in Oxford, but very much alive in spite of that. But perhaps I shall reach you before they do. Miss Defries's book about you is much better than when I saw it in manuscript: is that an optical illusion or has she actually worked over it? I have not given up my hopes of writing a connected "Life" which will supplement and round out but not in any way replace her volume. Please drop me a note about your plans and doings; in the meanwhile Sophie and I join in sending you our happy greetings for Christmas, and our affectionate hopes for your continued health and activity during the new year.

Lewis

1. Radburn, the "Highwayless Town" near Fair Lawn, New Jersey, was planned by members of the Regional Planning Association.   2. *Herman Melville* (1929); "The Arts," in *Whither Mankind: A Panorama of Modern Civilization*, ed. Charles A. Beard (New York: Longmans, Green, 1928), 287–312.   3. Karl Bücher, *Arbeit und Rhythmus* (1924); Jane Harrison, *Ancient Art and Ritual* (1913); D.S. MacColl, *Nineteenth Century Art* (1902).

Lewis Mumford AND *Patrick Geddes*

THE CORRESPONDENCE

1928

Dear Lewis,

Le Play House
65 Belgrave Road
Westminster, S.W.1

20 March 1928

[Ms UP]

I've just come here – for ten days or so – partly on business with my wife! Lately Miss Lilian Brown (whom you may have met?) but especially for a congress of student deputations of Universities at Oxford – organized by a co-operative spirit (T. Macadam) whom you must come to know, & intended to raise all the questions we have discussed – as "*quo vadis?*" to the student.[1]

Here I learn from V.B. that you are disappointed at not hearing from Zimmern as to Geneva. As it happens I had just called on him at Intellectual Relations Bureau (Paris) & almost his first question was to ask if I knew you (we had not met for years & years) & then to speak of you with utmost wamth, & as best man he had struck, alike for Geneva, & for criticism, etc., & future as well! I said – "Why not at Geneva again" – & he said "No money! Can't raise it! I've been trying all my American & other connections as yet in vain, so I've delayed writing him. Can you help me?" Said I – "What's needed? Isn't it necessary (1) to pay his needful fee as before?" Said he "Now he needs more, & I can't yet get even the former one." Said I – "What of Guggenheim or other fellowship?" (This he did not seem to know much about.) "For my part" said I "I want him to come to Monpellier & can welcome him & Mrs. Mumford & kid as guest for as long as the can stay."

Well now – pray *think of it* – & *try to come* – Look you I'm old (despite rejuvenescence powers!) so is V.B. We need your co-operation (and you may still utilise ours.) Come along – somehow!

Yours cordially *P.G.*

Excuse haste & Awful writing materials! I've been making progress at & for University developments but of those enough above till more developed.

1. Geddes married Lilian Brown in February 1928 (his beloved first wife, Anna, had died in India in 1917).

Dear Master,

4002 Locust Street
Long Island City

8 April 1928

[Ms NLS]

I must apologize for this all too tardy answer to your last two letters; but before the apology, let me give you our heartiest congratulations on your marriage. I had the pleasure of meeting Miss Brown in Edinburgh, and I look forward to meeting Mrs. Geddes in Montpellier – if that is where you will be when I finally reach Europe. It is impossible for me to get away to Europe early in the summer; that is something that would only have been possible had Zimmern obtained his hoped for support or if I had obtained a Guggenheim fellowship. I did not apply for the fellowship last autumn because Sophie and I had both planned on her having a baby this summer but like the best laid plans of mice and men, the baby is still unconceived, and all the arrangements we had made to conform to its coming have fallen through so far. I now think it will be possible for me to go over for a brief visit during September and part of October; this is not definite yet, for the financial arrangements are always in my case a little fluctuating and uncertain, and it depends largely upon what new arrangements I may make for my publishers on my next book; but the choice now seems to be between a short trip next autumn and a longer one the following spring, and if the autumn trip is possible I will make it – leaving the spring to take care of itself.

I am very glad that you had a talk with Zimmern; it reassures me; for though he has always been chary of writing letters, his long silence had made me fear some sort of alienation – something that my vacillations and my difficulty in coming to a decision last year might well have given a little color to. Zimmern is an admirable man to work with and my feelings towards him are very warm indeed.

Your criticisms of my bibliography have been very helpful: the deeper I go into the job the more clearly I see that no decent selection or arrangement can be made without much more prolonged and exhaustive study than I have yet given it; and I shall therefore keep the whole scheme under my hat for the next few years, hoping that in the meanwhile we can have a little time to discuss the problems in conversation. A fortnight ago I gave a lecture before the New York University Philosophical Society in which I used and explained the diagram that I showed you: I put the whole thing forward as a game on a philosophic chessboard, told them about your diagrams and "games," and was rather surprised to find how eagerly both the instructors and the students caught on. It was the first time that they had ever dealt with philosophic problems synoptically, I think; but two of the instructors wrote me afterwards that the students were continuing the discussion among themselves and making diagrams of their own. I recommended your Charting of Life to them, and for all I know may have created a few Geddesian disciples!

The other day in making an esthetic analysis of architecture I found that my ideas had automatically pushed themselves into your sense–experience–feeling diagram and that these in turn greatly aided the analysis. In this way:
Place sense – Materials (Color, Texture)/Site
Work–experience – Form (Structural organization in accordance with techno-
      logical skill; post-and-lintel; ogive; steel frame, etc.)
Folk-Feeling – Ornament (Painting, sculpture, symbolism)

Thus the Ruskinian criticism of Renaissance architecture as poor in form, and lacking in the structural skill of the medieval builders was not correct: a round arch may be quite as good engineering as an ogive: but the criticism of it as *feeling* – its pride, its coldness, its sterility – was altogether sound; and the further criticism that the Renaissance architect often ignored site-values and despised excellent local materiale, i.e. that he had no place-sense, was also fundamentally true. Similarly, modern architecture is weak chiefly because only a small group of architects have recognized and embodied all the technological and therefore for-mal changes that have come about through the use of steel, concrete, and glass: these materials are used in forms appropriate to quite different materials and methods of work. This only for illustration: I will expand the point some day, and make an original contribution to esthetics!

You have doubtless heard from Branford about the various contributions our Regional Planning Association is going to make to the new *Britannica*, on Housing, Town Planning, and Regional Planning: I am going also to see if an article on Regionalism cannot be included. Their methods of organization on the American side have been very sloppy; they have required articles under the first letters of the alphabet to be finished in a month's time; but I suppose there it is bad grace on my part to criticize, since such hit or miss ways are partly responsible for the Regional Planning Association's share in the work.

I trust this letter finds you in good health. Please remember me warmly to Mrs. Geddes.

Ever yours,

Lewis

4002 Locust Street
Long Island City, N.Y.

11 December 1928

[Ts NLS]

Dear Master,

I have been meaning to write you these many months: but in the meanwhile all sorts of domestic misfortunes overtook us and laid us low, and we are just, slowly, beginning to recover.

The summer was for the most part a good one: I worked hard at my biography of Herman Melville and tussled with all sorts of capital problems that arose in connection with his life and work: my health was perfect, thanks to long walks and swims when the day's work was over. But just as I was finishing the book Sophie was threatened with a miscarriage which finally took place; and it was necessary to pack up our household hastily and come down to New York: since when, with one complication and another, we have been living in a temporary nightmare of illness; and both Sophie and I are pretty well done up; and though we go to bed at nine, and do little enough during the day, that little exhausted us: so it will probably be another six weeks before we are on our feet again. I tell you all this, not to make our woes worse by marshalling them up again, but in order to account for my otherwise inexcusable silence. Sophie still wants another child; but this is her second miscarriage; and since she took the most scrupulous care with herself, it is all a little mystifying and discouraging; and one of the best obstetricians in New York is just as puzzled as we are about it.

I have heard various happy reports of Montpellier from the Whitakers and the Spingarns and the Steins; and I thank you for your hospitality towards them, and only wish I could send you some students. I am about to renew my contacts with the universities by going out to Wisconsin for a week or two in the spring, to talk to the students of Meiklejohn's experimental college: so perhaps I will hit upon some good youths, ready and able to take advantage of Montpellier.[1] During the summer another young American disciple of yours, Philip Boardman, came to see me; and I have been in correspondence with him since; but he has a long way to go. I have just had a call from one of Zimmern's secretaries, asking me what the prospects of my going over to Europe next summer were; but I have made so many plans, only to see them collapse, that I am a little chary of making any more. Both of us want very much to go, however, and if there is any possibility of our doing it, we will.

I ordered a copy of the *American Caravan* to be sent to you; and if it has not reached you, I trust you will tell me, and I will wrap up the next one myself.[2] The book is a sort of cross section of American life at present; and it does not pretend to represent the wishes or interests of the editors.

The plans for the first garden city, Radburn, in New Jersey, about which Stein must have told you something, are now going through; and the first section will be opened in the spring. The fundamental plan, that of the big block broken up into residential cul-de-sacs, was a common method of subdivision around Boston a century ago; and it is a great joke on John Nolen, that he should have lived in Cambridge all his life without realizing the economy and beauty of this pattern, for it is repeated again and again in Cambridge, in all the better residential quarters.

I have been reading the second volume of Spengler lately, and I have been struck with the many profound similarities – and similar profundities – between his thought and yours. Despite his arrogance and Germanic esotericism, he is a very pregnant thinker; and I feel he is worth much more attention than the Sociological Society has given him.[3] Christopher Dawson ought to do an article on him; better still, you yourself of course or Branford. Rachel Taylor's *Leonardo* has become a best-seller here, and but for my illness, I should have reviewed it this autumn.[4] My *Sticks and Stones* has, curiously, been having an influence on some of our American historians, and in a new series of studies of American life, the authors give as much attention to monuments and cities as to documents – although plainly they are not yet as thoroughly trained to study the first as the second. I regret that Branford's ideas and yours have to filter in, with alas! dilution, through my own work; but better that than no influence at all. As for my next work, it has not yet shaped itself in my mind: willy-nilly, I must lie fallow for a while. Did you get your book on Olympus and Parnassus finished?

For all our upsets, little Geddes flourishes: he is sturdy and intelligent, very independent of other children but quite able to work with them, with a riotous imagination, a long memory, and excellent constructive ability with his blocks. Sophie sings to him and I recite verse to him when I put him to bed at night; so between us, he knows many ballads and pieces of poetry. Chevy Chase is his favorite, although I have never got beyond the first ten verses with him; and just last night he suddenly announced that he wanted to go to Scotland where the doughty Douglas lived! So I showed him your picture, and said that you lived there, too, and perhaps next summer he would see you!

Sophie joins me in warm Christmas greetings to you and Mrs. Geddes.

Ever affectionately yours,

Lewis

1. Alexander Meiklejohn (1872–1965), controversial president of Amherst College (1912–1923); invited to Wisconsin by the President Glenn Frank, where he founded a short-lived experimental college (1926–1932); author of *The Liberal College* (1920) and *The Experimental College* (1932); see Mumford's letter of 1 June 1929.   2. *The Second American Caravan: A Yearbook of American Literature* (1928).   3. Mumford's review of *The Decline of the West*, Vol. 2, appeared in *New Republic*, 58 (20 March 1929): 140–141.   4. Rachael Annand Taylor, *Leonardo the Florentine* (1928).

Collège des
Ecossais
Plan des Quatre
Seigneurs
Montpellier

30 December 1928

[Ms UP]

Dear Lewis,

I am indeed sad & sorry to hear that you have each been shaken in health – but youth is on your side, & you'll alike soon return to it. Best wishes for a better 1929 – & so on increasingly.

All the more I prescribe for you "both all three" our choice of sanatoria here – & where our famous doctors (with Sun on their side!) go on accomplishing marvels! – literally taking world records on various directions – including those concerning you both.

And if ever you mean to come, soon's your time – for I can't but feel ageing a bit faster than I like.

In a recent letter from Zimmern, he spoke very warmly of you; and I am sure if he can manage his summer school in Geneva he will ask you again: so between these two places, you should have a thorough chance.

I should have acknowleged long ago the mighty volume you have so generously sent – a caravan indeed! Lots of varied reading! My wife was particularly amused by MacMaster![1] But now you owe me some help to put some method in my madness!

Victor Branford is at Territet again after a very severe illness – arthritis, lamed on sticks! Poor dear old chap; sadly over burdened! What can be done to save Le Play House, his white elephant as much as ever – with Sociological Society, review, etc. Can't we recast that, so as to get some circulation to pay its way without his doing so?

Glad to hear Mrs. Taylor's *Leonardo* is selling. She has worked long & hard, & is now at English Renaissance.

I have given your name to Dr. J.H. Cousins, formerly of the Irish poets – now a theosophist, but the sanest & least uncritical of any I ever met.[2] He is to be in N.Y. for some time – & is worth talking with: he gives me the impression of a live educator, bringing round his doctrinaires to common sense, yet without loss of idealism – indeed towards better direction of it.

Here we are building on 20 new rooms, plus dining hall, library, etc. – a daft adventure – with only one Scots student in the place! But for others we should have had to shut up long ago ("no man a prophet," etc.) where I'm forgotten after so many years away.

All this practical work – & much correspondence – still keeps me from writing. But I have the early hours to think – &, for instance, that story of IX–9 is always coming more clearly.

But to keep going, I'll have to try to sell my Town Planning Exhibition. Any ideas as to how to do that would be welcomed!

I hope your *Herman Melville* has got finished & published. I have only heard of his work as distinguished, but never seen it – should not your publisher bring out a new edition of him – or at least of his best – so that each should help the other?

I think your coming to France – with a good plunge into contemporary French literature would have a good critical result – distinguishing between the sickly – & more or less perverted *lycées*, never quite growing up to escape from refined analyses of their own sensations – & the vital seeing & thirsting, feeling & interpretative, fore-sighted and fore-feeling minority. I think of Proust,

André Gide, etc. as of the first, & Duhamel, Romains, Rolland, Hamp, etc., as of the second.[3]

French literature has peculiarly responded to current philosophy & science, & it seems to me that if we could express these in more unified & advanced forms, this would react anew, aiding the second group, & even also the first in their younger & emerging writers of course.

Does French literature now much affect yours? I do not notice this in *Caravan*. But it certainly leads most other literatures, does it not? Even more than I feel it at present deserves? But "in the Kingdom of the blind, the one-eyed is King!"

With our cordial greetings to you both all three, always yours,

P.G.

1. "James McMaster" is a fictitious version of Geddes appearing in "The Little Testament of Bernard Martin Aet. 30." See Geddes' letter of 19 October 1926.   2. James Henry Cousins (1873–1956), Irish poet and writer on India.   3. Geddes was obviously well-read in contemporary French literature: Marcel Proust (1871–1922), author of *A la recherche du temps perdu*, 7 vols. (1922–1931); André Gide (1869–1951), author of *L'Immoraliste* (1902); Georges Duhamel (1884–1966), author of *Vie et aventures de Salavin* (1920–1932); Jules Romains, pseudonym of Louis Farigoule (1885–1972), author of *Les Copains* (1913) and *Les Hommes de bonne volonté*, 28 vols. (1932–1956); Romain Rolland (1866–1944), author of *Jean-Christophe*, 10 vols. (1904–1912); Pierre Hamp, pseudonym of Pierre Bourillon (1876–1967), author of *Vin de Champagne* (1909) and *Le Rail: vieille histoire* (1912).

Lewis Mumford AND *Patrick Geddes*

THE CORRESPONDENCE

════════════

1929

I very sincerely hope both of you are in good health now – & I trust wee boy needs no asking for! What are you writing now? R.S.V.P. soon.

Yours always,

P.G.

I had a short day's visit from Waldemar Kaempffert, re. *Techno-Social Museum* (as I may call it) for which he has large sum to spend (per Rosenwald, amiable millionaire of Chicago). He has good ideas, but they stand more putting in order. (Do you know him? Draw them out . . .)[2]

I've also had a visit from Mr. Whitaker – whom I missed last year. A very bright spirit – delighted to make his acquaintance! I'm going down to return his call, & make his wife's acquaintance too.

P.G.

1.   "Ways of Transition – Toward Constructive Peace," *Sociological Review*, 22 (January 1930): 1–31; (April 1930): 136–141.   2.   The lower part of the manuscript page has been cut.

4002 Locust Street
Long Island City, N.Y.

31 March 1929

[Ts NLS]

Dear Master,

I fancy that by now Philip Boardman has arrived, and has explained by word of mouth my tardiness in answering your last two letters. Whilst Europe froze, America became temporarily a hospital, not merely through influenza, which proved to be mild, but through an unusual virulence in the germs that attacked the ears of children, Little Geddes being only one victim among a great many this winter in New York. He had a very close shave of it; the mastoiditis had eaten almost to the brain, when they operated; and, partly through the bungling and professional stubborness of our specialist, who was annoyed at the fact that his own diagnosis was persistently behind that of our family physician, he had a very slow recovery – taking more than four weeks completely to pass out of danger; and even now there are daily trips into the city for dressings. Coming as this did on top of a whole winter of illness and prostration, it gave a coup de grace all round, or at least for a while seemed to; but fortunately, as William James pointed out, one has energies in reserve for such moments, and both Sophie and I began to improve in health when we could no longer *afford* to be ill in the midst of Geddes's crisis. The picture I enclose shows him last summer; I forget whether or not I sent you a copy earlier in the winter. He has gained weight, in spite of all he has been through, and is in better shape than we had dared to hope; but our plans to spend two months in Europe, leaving him behind in America, are now out of the question; and it will be May before I can decide whether I can go to Europe by myself without leaving too great a burden on Sophie. It is ironic that this impediment should have come just at the moment when we are at last financially free to go! My *Melville* was taken over by one of the big book clubs, and as a result, I shall certainly make seven thousand, and possibly ten thousand dollars from its sales. I trust that you received the copy I sent you. Zimmern has asked me to give a week's seminar in sociology in Geneva in the middle of August: and I want very much to spend some time with you, so if it be at all possible, I shall come over, if for no more than a month or six weeks. I am still mulling over the theme of my next book; and have not yet decided upon it. I shall spend a week with Meiklejohn in Wisconsin at the end of April and trust I will have something interesting to report about that experiment. My thinking, as you can imagine, has been broken up by the series of worries and harassments this winter; and I am impatient to be at work again. I will write you at greater length in May. Sophie joins me in very warm greetings to you.

Ever affectionately,

Lewis

P.S. The enclosed review is from the pen of the best of our literary critics: a man who has done more than anyone else to pave the way for a maturer literature in America.[1]

1. Perhaps the review of *Herman Melville* by Mumford's friend Thomas Beer in the *New York Herald Tribune Books* (10 March 1929): 1.

Dear Lewis,

Montpellier

10 April 1929

[Ms UP]

I've been meaning to write again & again to express my warm appreciation of your *Melville* – a fine contribution to the new movement of biography! – and of use, action, impulse beyond it – & in literary history too: above all I trust an "awakening" or a widening to many readers! But you have I doubt not so many appreciative reviews that you are tired of them! – And already at the stage of *What next*? or even of *This now*! So what is it you're up to? – at what new stage are you breaking out! A recent London book (Raymond's *Through Literature to Life*) appreciates Melville highly, but only briefly.[1]

Are things right with you all three in health, etc., now? Are you all coming over to Geneva together? And if so, *when here*? (You can either stay with us, as we hope – or in a separate cottage as you may choose – now on arrival, or after trying the associated life.) But when – that's what I need to know.

I have to be in London by *26 June* & in Perth on 28th then several weeks at least in Scotland & perhaps in London several more – that takes August. From Sept. 1 to 11 I'm engaged at Abbaye de Pontingny in Burgundy – Paul Desjardins' *Decade* – relation of Science to modern thought & life, & its possibilities towards sociological & psychological progress. *Then here* for autumn & winter. Can you come to us then, i.e. from mid-September onwards? – how long? Why not try this Mediterranean region as stimulant to your own development – apart from my world, & for your own? After, all how preponderatingly our civilization is Mediterranean, more than North & West realize nowadays! But its action on *all sorts* of minds – Goethe or Browning, Ruskin or Mill – suggests that it might have no less on yours! And both here & at Assas you can have peace (even solitude when you desire at latter!). But returning to my own desire for your visit – in which you'll find me not so much the inactive observer I was in New York. I beg you'll note I'll be 75 this autumn

– & also that Victor Branford has been alarmingly ill – & I am very anxious about him – though he's better for time being. Also J.A. Thomson in similar case! So *if ever you mean* ... (etc.) *now's the time*!

And I believe V.B. would try to come here for a time, if you do; and we should hatch out something together. Neither of us has any one else of your calibre to look to for collaboration – & we are seeking for heirs & executors – as yet without success! Of course we do not thereby suggest your sacrificing too much time or trouble to us – but if you could give us a hand, it would also lead others to do so, whom we don't know yet, nor find for ourselves! – whether from Edinburgh or London or here! It would encourage V.B. to a new start (& he badly needs it – in some ways more than I, alike on grounds of health, worries, & over pressures, though he gets along wonderfully in spite of all).

See my IX–9 paper in forthcoming *Socological Review*, & say what you think of it, and as basis for book? with V.B.? I believe you could start me on writing more! And by that time the present interruptions from new buildings, gardenings & plantings will be over, & students don't come in any number till 1 November.

I am trying to finish *Olympus* for Yale Press. Thomson & I have both lagged with our big Biology – & not done all we should! Could not we do a book together? I'm sure it would help me very much – & not disappoint you so much as I may have done in my unproductive mood at New York. Synthesis *is* advanced a bit – & towards action too. Now that I've a fresh fulcrum here, I want to use it!

Again, I've been struck lately by how much both V.B. & I have been working on parallel lines to old-word occultists, Rosicrucians, etc., etc., with their "*Grand Arcanum*" & "*Grand Oeuvre*," etc. – & this even to graphic details! (There would be an interesting paper in this – old thought on new spirals – new thought on old spirals too.) Did you come across much of that when at Utopias?

Yours ever,

P. Geddes

R.S.V.P. soon

1.   Ernest Raymond, *Through Literature to Life: An Enthusiasm and an Anthology* (1926).

4002 Locust Street
Long Island City, N.Y.

1 June 1929

[Ts NLS]

Dear Branford and Geddes,

I am writing to you jointly, because this letter concerns you both. It has been impossible, up to now, to make definite plans for the summer, because of little Geddes's slow recovery, and the uncertainty of Sophie's own health under the stress and strain. I now have passage engaged on 19 July and should reach Southampton on 25 July. This would make possible six days in London before the Geneva lectures: but if it be equally convenient to both of you I should prefer to meet you during the last fortnight in August, for all or part of that time. I must return early in September in order to prepare the Moore

Foundation lectures which I am to give this fall in Dartmouth; and I shall have no opportunity to do this before I leave. If you prefer the week in London at the end of July, please cable me "Mumford Amenia N.Y."; otherwise I shall make plans to see you in August.

My ten days of lecturing in Meiklejohn's Experimental College at Wisconsin and in Chicago were very fruitful; particularly the first. He has about a hundred sixty students, of average grade and intelligence, purposely *not* picked for scholarship. During the first year they are introduced to Greek civilization; in the second, to American Civilization. There are no set classes. Each instructor has ten or twelve students under him; and they meet for weekly conferences. They hand in small papers every week, and two main papers each term. At first the students were a little disrupted by the freedom and the absence of formal regulations; but by the end of the first year the best of them had found their pace, and some of these heretofore considered (in high school) as dull and useless became among the very best. The program has been experimentally worked out, and it will take another two or three years before it has finally shaken down into form; but already it has developed independence and firsthand judgement, and thoroughness and consecutiveness will, I think, follow. They are an eager inquisitive group, and I had many a good tussle with them after the lectures. Their second year's work was a little at sea until one of the best instructors, a man named John Gaus, suggested that they give each student the project of a regional survey of a city or neighborhood familiar to him.[1] The results have been excellent. After an initial indifference, under the notion that they were merely collecting statistics, they settled down to the task and many of the results were very creditable. This has not merely promoted a regional consciousness among the students, who come from various parts of the country: in certain cases it has even roused an interest on the part of the elders, who have participated in the hunt for local data! Gaus, by the way, came across Geddes and the regional survey through Wood, the head of one of the Boston social settlements; and I naturally told him about the other things you have been doing at Le Play house. You have an active outpost in Wisconsin; and Gaus, who came from the University of Minnesota, tells me that similar ideas have been in ferment there. I wonder, by the by, if either of you have run across Phillips Bradley, an exceedingly able and fine Amherst professor who is now in Europe.[2] He has gotten much from both of you; but may have found some diffficulty in getting in touch with you.

It was a great delight to read the completed IX to 9, which I saw first in the making; and I am passing it on to Gaus and to Benton MacKaye; likewise to my "rival," Dr. Murray, a Harvard psychiatrist who has long been at work on a life of Melville.[3] Murray, it happens, is an even better pupil of Geddes than I am – although an unconscious one! He began as a physician, specialized two years as a surgeon, went over to bacteriology and worked in the Rockefeller foundation, where, by accident, he heard about psychology, and finally, after a walking trip with Jung, went in for psychiatry. He is, I understand, a very capable practitioner; and in vision and outlook is a natural Geddesian. You should see the consulting chambers he is equipping for himself in his country home – doing a great part of the painting and carving himself. One enters a dark, seventeenth century room, with chains, handcuffs, and instruments of torture hanging on the wall: the room of repressions, the burdens of past evils.

Under treatment, the patient rebels against this past and climbs a flight of stairs to leave it behind him. His next consultations take place in a beautiful flame colored room, black, red, orange, yellow – the stage of his individual *Götterdämmerung*,[4] where all his old idols are consumed. He is now ready to work out his problem more concretely, and he passes over to a great atelier and workshop, where he may paint, carve, carpenter, sculpt, working out his fantasies in maturer form. That over, he goes down stairs to a modern room, from which he either goes out into the world, to the same society he left, or works out his salvation in more withdrawn fashion, by passing through one further tunnel of trials, with a skeleton barring his way at the end of it, until he climbs a tower stairs, and emerges with a new view of the world. A real Geddesian, although it is only now at my instance that he is reading Geddes!

I trust this letter finds you both in good health.

With warm greetings from Sophie and little Geddes –

Lewis

1.   John M. Gaus (1894–1969), professor of political science at Wisconsin; author of *The Frontiers of Public Administration* (1936).   2.   Phillips Bradley, political scientist at Amherst and later at Queens College.   3.   Henry A. Murray (1893–1988), distinguished American psychologist and Melville scholar; served as director of the Harvard Psychological Clinic; Murray helped Mumford locate material for his Melville biography, and they became close friends.   4.   "Twilight of the Gods"; title of the last opera in Richard Wagner's Ring cycle (1870).

Amenia, New York

26 August 1929

[Ts NLS]

Dear Master,

I had hoped that my European trip, however brief, would at least break the long round of illnesses and disappointments which we had been through during the previous ten months: but my tonsils, which had troubled me during the winter, erupted again on the steamer on my way over, and brought about such an acute depression that I needed every bit of reserve force to keep my wits about me. The attack lasted about four days and with a little care I got through my lectures in Geneva without any difficulty: but it was impossible for me to go on, for, had I stayed my allotted time, my operation would have had to be put off till too far in the autumn; and I was loathe to do this, since little Geddes must have his tonsils removed, too, before the bad weather comes on. I am not sorry I returned: but I feel chagrined that the main object of my trip, that of seeing you, remained unaccomplished. I undertook the trip with misgivings; and it is likely enough that my breakdown on the way had a psychological as well as physical origin: what we all needed, after the devilish winter, was a complete rest; and it wasn't till after we came up to the country in June that I began to discover how great the tension had been and how we had been undermined by it. Hence my silence: for every day I was debating with myself as to whether to go over or not – all my instincts were against it. The instincts were right. I wasn't fit for travel. If we have a quiet and salubrious winter, I shall go over for a six weeks visit in the spring: and if not – well, then we must do what we can to bridge the distance by letter. The lectures in Geneva were a real stimulus; and Zimmern's school is now in excellent working order: my talks were an exposition of the difference between pre-sociological and sociological thinking, in relation to the problems of international society, most of the students, particularly the Czecho-Slovaks, were headed pretty definitely in the right direction. I had an opportunity to talk a little with Philip Boardman who told me about the thesis he intended to do on your educational philosophy: he is in a rather serious state, I fear, due probably to the psycho-physical tangles of adolescence, and I advised him to devote himself to an almost wholly vegetative life, until he had stored up a little energy to go on with. His solemnity is appalling; and he needs badly to fall in love with some hoydenish young lady, who would dance it out of him. The Zimmerns had been nice enough to put him under the care of a sage old French doctor they know; and I discovered that his advice had, point for point, anticipated mine. What have you been doing this summer? Most of my reading this summer has been an attempt to catch up with recent science: Eddington, J.S. Haldane, and L.J. Henderson, whose little book on the *Fitness of the Environment* seems to me a quite revolutionary work, in that it uses the data of bio-chemistry to establish the proposition, abandoned to the theologians since Darwin's time, that the universe in its chemical and physical constituents, is favorably disposed towards life.[1] Do you know it? They told me at Harvard that Henderson, and Wheeler the entomologist, are the brightest stars in their galaxy now. . . . Sophie and Little Geddes had a fortnight at the seashore; and he is now in excellent shape. Did I tell you that, by a miracle, his hearing is quite unimpaired and his ear drum completely healed? Despite all Sturm und Drang,[2] we have not a little to be thankful for.

My trip convinced me that it would be unwise at the present time to begin on another book which would require the arduous work that my *Melville* did: so I will spend the winter putting in shape the lectures I shall give at Dartmouth after Thanksgiving on The Arts in America, 1870–1930.

Sophie and Geddes ask warmly to be remembered to you. He was playing that he was a horse this morning and the horse was looking after the two little baby birds (Sophie and me). I said that I had never heard of a horse taking care of birds. But he answered: yes: this is a horse with wings: it's Pegasus. . . . He has an enormous interest in living creatures, mice, moles, fish, insects, as well as the barnyard creatures, and will spend half an hour at a time following an ant about: in short, he has a distinct bent toward following his namesake's footsteps!

Affectionately,

Lewis

1. A.S. Eddington, *The Nature of the Physical World* (1929); J.S. Haldane, *Mechanism, Life, and Personality: An Examination of the Mechanistic Theory of Life and Mind* (1921); Lawrence J. Henderson, *The Fitness of the Environment* (1924).   2. "Storm and stress," name given to period of literary ferment in Germany in the late eighteenth century.

Collège des
Ecossais
Plan des Quatre
Seigneurs
Montpellier

14 September 1929

[Ms UP]

Dear Lewis,

(Yours received on return here today.)

Very much relieved to hear it is only your tonsils that have troubled you, as these are easily dealt with (in fact you might have had your operation here, at trifling cost – & recruited with us!).

Very glad indeed that Geddes is now safe from his trouble; & turning to naturalist also! And that Mrs. Mumford too is in better health. I'm sure they'll soon put you right! Let me hear again how you get through. A period of illness, or rather its *convalescence*, has many values.

From more than Boardman, I heard at Paul Desjardins' *Decade* at Pontiguy, how your lectures were appreciated. Branford may come here for winter; & if his health strengthens, he may do something. Desjardins' (Decades) symposia in August & first half September may fit into my *Grinzaines* in second half of September & first half October. So with mutual advantage – for their refined & elaborated culture – mathematical-physical on one side & humanistic on the other, needs simple *Life*, & life *Theory* between – both social & biopsychological – without which we have mechanistic science on one side, & too verbalistic philosophy, & literature, etc., on the other. He & I were wondering whether Zimmern's school would possibly also so far co-operate? – as we'd thus have between us a better Summer School than others – in fact a veritable vacation term of the University in Evolution, and Militant. What say you? Pray Advise!

That term reminds me that C. Ferguson, of *University Militant*, of whom we've lost sight for many years, wrote Branford (& partly also for me lately) especially about his financial schemes, of which he thinks Major Douglas's a misunderstanding. Is he accomplishing anything? I don't attempt Finance![1]

Do you see anything in my recent papers in *Socological Review* – all that stuff I talked to you – or rather part of it! & more as well. Suggestions towards reprinting as book would be welcomed.

I suppose Kellogg won't return or use those old *Survey* papers – discouraging! Do you see any outlet for any of our stuff in America? I promised Yale Press my *Olympus*, but it is not yet ready, owing to *Biology*. Book of endless toil with Arthur Thomson.

This place is growing – but more in buildings & gardens than numbers! Send us a live student when you can, & people on sabbatical year! Must send *bulletin* soon.

Always yours,

P. Geddes

1. Major Douglas, author of "The Mechanism of Consumer-control," *Sociological Review*, 8 (January 1921): 33–35.

Dear Lewis,

Collège des
Ecossais
Plan des Quatre
Seigneurs
Montpellier

20 October 1929

[Ms copy Oslo; Ts NLS]

Very kind of you to write for my birthday, as you have done! Encouraging too, towards my last pull (though never *through*!).

But why – why – say nothing of yourself and your recovery from operation? I trust so complete that you had forgotten it? – and so are now a monster of health, a Super-Hygienic-Man? Still, I'd be glad to have that hypothesis confirmed by your experience – and my own observation too.

And what are you writing, now that that *The Golden Day* is on its way, and *Herman Melville* too? Again congratulations on them both, as on the *Utopias* of previous promise.

I am still toiling with Thomson, as lately for a too short time in Aberdeen, at that Biology book. Very hard to coordinate 1,200 miles apart! – but I hope there will be some life in it – though not all we'd wish. That *Olympus* for Yale Press has to be next job – also shockingly delayed – and then I hope that diagram of IX–9, which you may remember I was getting clear when with you and in Cornell Lecture, but which has grown to many more sheets than I have time to look at, and yet which are worth deciphering. Tell me what you make of its outlines in recent papers in *Sociological Review* and think that dual diagram over! For either I am quite mad, or have got main Keys towards opening the 9 (or 18) doors from the present Industrial–Political–Militant age and towards *Revivances*. What say you? *Critically*!

But more and more I see also that the entire inability of V.B. and me alike to make any appreciable impression on our public – an occasional brother ("crank"!) like you, excepted! – is not so entirely due to our deficiences of style and presentment as we are so often told by those more or less sympathising, but to the solid and well-integrated character of the whole IX world system, and its educated and popular fixation in that accordingly. More and more we have to realise how schools and universities – press, etc., of course too – are definitely arresting and congealing their victims into the moulds of IX, and that to change them is often as impossible as to dissolve cast-metal in our social tea-cup. Very seldom do I get a student who can thaw, though here and there (sometimes I fear to congeal anew, more or less, even here – & I fear still more fully on return!). Still what can I do? I get on with such building and gardening, and planning for the future as I can? – and with such note making too? (For this moving already, as many as usual, and some better ones.) Yet is it not a pity that no one who can really help us with the Transition – IX–9 – has ever turned up, since yourself long ago!

You'll be sorry for Branford – worse burdened than me!

Tagore has sent us a poem for its found-stone fête. I think I sent you sketch elevation of my plan? and one or two other in project. But buildings are no more than hotels or barracks, while without the thought of making the University anew, amid Pallas's olives. But our British and American visitors often don't know an olive from an oak, save in spelling! *There's* the verbalistic "education," which would shudder at *olliv* and *oke* – but is indifferent to them alike, and still less sees Pallas or Jove in them, though there they are!

So how to get a move on, in the matter of ideas? What is going on with and around you. Have e.g. Meiklejohn or Glenn Frank got ahead? – and if so, how? What of "Ely's Town Hall"? What others are out for living change? Who are coming on, after Dewey, Robinson, etc., and Veblen, Stanley Hall, etc. Why not write a brief article to *Sociological Review* and tell us of them. Our previous papers in American Sociology have been retrospective. You could tell us of the younger men & later thought.

I can't make much of American Sociology of late years or just now, nay more than of French – to me all to abstract (when not too simply concrete!). That book – whose was it? Comparing Rustic and Urban, Country and Town, was too townsmanlike. "Whither Mankind?" mostly very naive in acceptance of IX, with mild improvements. Tell us of something better!

With best wishes to you all three,

Always yours,

P.G.

N.B. Now that our Indian College is building (£2,300 in one subscription paid from Bombay today) it is time to prepare for beginning the next two – *American*, and *Nordic* probably – but who will make it known? If you can, say a word to Professors on "Sabbatical year," whom this place suits particularly well, for natural sciences, Romance literatures and languages, for history, law, medicine and education – in fact most subjects, but engineering. Even a Faculty de Théologie Protestante, for which we have had a Scot Minister already, and excellent Conservatoire de Musique, opera, etc.

Lewis Mumford AND Patrick Geddes

THE CORRESPONDENCE

1930

Dear Master,

4002 Locust Street
Long Island City, N.Y.

27 January 1930

[Ts NLS]

Our last letters crossed, as far back as October, and since then I have been busy preparing for my Dartmouth lectures, giving them, nursing Sophie for six weeks through a bad infection she had at the back of her neck, and, in the meanwhile, mulling a little more carefully over the book that will perhaps result from the lectures. Except for Sophie's illness, the winter has so far been kind to us; and I have rebounded from the depths of physical misery and mental despair that marked the middle of the summer to a high and consistent state of health, marked by an outburst of verse writing. I remember the verses you wrote during your convalescence from fever in India; and wonder how far one ought to court a little illness occasionally, just in order to reap its compensations! Have you seen, by the way, Egon Friedell's post-Spenglerian *A Cultural History of the Modern Age*.[1] He emphasizes the optimism of pathology, holds that the great social advances have resulted from the periods of illness and maladjustment, and even turns Darwin's thesis upside down by erecting the paradoxical notion of the survival of the unfit! Instead of presenting the economic interpretation of ideals, like the Marxians, he presents the ideal interpretation of economics; and though there are numerous crotchets and mystifications in the work, it is worth while skimming through; if only because it is a symptom of recovery, far though it may be from the bi-polar interpretation of life and society to which we hold. J.A. Thomson was kind enough to invite Sophie and me to meet him and and his wife when he passed through New York; and we had a good talk. My week at Dartmouth gave me great hope; for Hopkins, the president, has slowly transformed, and is still transforming, the institution from a country club for young barbarians into a serious place of work, and in doing this he is breaking down the barriers between the various departments, and re-integrating the whole curriculum.[2] They have asked me to spend a whole semester with them; and though I cannot do it this spring, I shall try to set aside time next year to do so. They are limited in numbers to 1600 students; and the president has been cutting away the red tape as fast as those in authority under him will permit him. He is capable of doing unusual strokes; as, for example, he found a great Russian organist on his uppers in Europe and imported him, without further authorization, to Dartmouth, to head a non-existent department of music, which now has a symphony orchestra which draws members from all over the local region. Or again: a painter who had become interested in the physics of light asked him for a laboratory in which to conduct further experiments.[3] Hopkins gave him the laboratory, and the man has during the past six or eight years made very important contributions to optics, to say nothing of furthering the life of the college by his presence, although he has no official position in the university, and does not teach. Then, too, Hopkins prides himself on the number of teachers who do not have Ph.D.'s on his faculty: he seeks out his younger men before they become dis-specialized and useless. As for Meiklejohn, he defended his college brilliantly against the rest of the university – his defense taking the form of criticizing its shortcomings harder than his worst opponents would have dared – and there are now plans afoot to extend the experimental college to the rest of the university; plans which, as I told Branford, I hope will be deferred, since Meiklejohn's curriculum and method need to be shaken down a little further before they are applied on a large scale. Still, he survives, and flourishes!

You will presently be visited, in all likelihood, by a young man named Edward Twadell, whom I commend to you, although I am sorry he left America now. He has a very brilliant mind, and is nearer to us in outlook and grasp than any young student I have come in contact with – get him to tell you of the night he spent alone on a capsized boat in the Atlantic! – but he has not gone smoothly through the crises of adolescence, and is in need of medical care. His mind has a way of racing like an engine without a load; and I fear he will come to grief unless he has exceptional good luck or first rate medical treatment. Europe may do something for him. I advised him to visit Jung; and have asked Branford if perhaps Brock would be the man to look after him. If he comes, please don't hesitate to advise him about this.[4]

Little Geddes has had a good winter so far. I must tell you the latest story about him. Sophie found him one afternoon dancing up and down, his face distorted, and she asked him, a little frightened, what on earth was the matter with him. "I am a soldier," he explained, "and when soldiers can't go to war, they dance madly!"

We all join in affectionate greetings to you. Please remember us to Mrs. Geddes.

Lewis

1. Mumford reviewed Egon Friedell, *A Cultural History of the Modern Age, Vol. III: The Crisis of the European Soul from the Black Death to the World War* (1930), in *New Republic*, 73 (11 January 1933): 248–249. 2. Ernest Martin Hopkins (1877–1964), served as president of Dartmouth, 1916–1945; author of *Education and Life* (1930). 3. Adelbert Ames, Jr. (1880–1955), conducted research in physiological optics at Dartmouth. 4. Edward Twadell, a Harvard student whom Mumford had met at Chilmark, Massachusetts, in 1927; he was later affiliated with the YMCA in Montana, and their correspondence continued through 1959.

Collège des
Ecossais
Montpellier, France

29 February 1930[1]

[Ms UP]

Dear Lewis,

Yours of 12 February welcome. I have of course to sent it on to A. Farquharson – as Trustee for V.B. but of course with strong recommendation to accept your aid. It is for him to answer your questions definitely; I can't. He has all powers in his hands. (More's the pity in some ways.)

What would you say too (again I ask within above limits) to a participation in editing of *Sociological Review*? You have not answered as to that question (I think) I put you. Farquharson is getting into touch with London School of Economics – per Ginsberg, Hobhouse's successor, & with *Carr-Saunders*, Professor of Sociology at Manchester.[2] But both are more or less dull & dry – economic & statistical, more than ethical & social. It needs you to make it a success, & in U.S.A. which would make all the difference. Otherwise I'll have to drop out, & find other channels.

Alas, I had to cable Boardman – *Impossible* – & write explaining that I am pledged to Montpellier for summer term. Also – alas – that my recent illness not only drives me home – to better climate there, & less strain – but that this illness (*threatened diabetes* – though that's *between ourselves*) may probably make it impossible for me to face any American journey at all – so young America must come to me!!

I do earnestly hope therefore that you will come, & stay longer than a fortnight! It may probably be our *last* meeting – and I have got things much clearer than when I was with you. I look to you as best of intellectual heirs & executors – in fact I've no other! – and we might readily do a book together – if you'd stay *a month*. Does not Zimmern want you this summer at Geneva? I can pay something e.g. such as collaboration as may bring!

Thomson & I have at last finished our big *Life: Outlines of General Biology* – 1515 pages, 2 volumes! (63/-) now binding – published by Williams & Norgate. This develops theses of our little *Biology* in Home University Series (with *Evolution*, *Sex*, etc.). I am now ready to write on parallel lines the corresponding little *Sociology* (for Home University Series probably) & also the big sequel to Life – viz. *Social Life: Outlines of Genl. Sociology* – for which of course there is much material in past issues of *Sociol. Review*, etc., etc., & manuscripts too. (*There's* scope for collaboration!)

You have never said if you felt my IX–9 (cf. 1923 when malarking with you) as explained in papers of last two years in *Sociological Review* from V.B.'s & my paper on *Rustic & Urban*, etc., 1928 – of use & value? I get as yet no response! – yet am convinced it only needs (your) fuller & better exposition to catch on![3] All you younger American critics are more or less moving that way – but I don't think even you & they realize sufficiently that we have to move from this whole Industrial plan of Civilisation (*Mechanistic–Pecuniary Culture*) to the *Biotechnic & Evolutionary* – as different as (say) from decadent Rome to New Christianity – or any other great world change. *Pray answer me,* as to how far you agree! or differ! V.B.'s (& my) apparent "difficulty" to readers is thus mainly explained. Readers think in terms of {Military / Theological}–{Statistical, Political, Individual / Abstractional & Legal}–{Mechanistic, Pecuniary / Mathematical–Physical, Scientific} but we – quite conversely in order – & in terms still quite unrealized by them – viz. {Biotechnic/Psychological–Biological Sciences}–{Cosmo-technic/Cosmogenic evolutionary}–{Etho-Polity / Eupsychic}. Hence a new phase of Civilisation – more akin to old peasant-world which the Militant Statists, etc., well armed by the Mechanists – conquered – as from William the Conqueror in 1066 to Indian Empire to this day! And in this large view Socialists, Bolseviks & Fascists all belong to the *Mechanistic Order*, about as much as the old Industrial Futilitarians!

---

1. Mumford's annotation dated February 1971: "Important!" Geddes seems to have made a slip with the date, since 1930 was not a leap year.   2. Morris Ginsberg (1889–1970), sociologist; served as assistant to L.T. Hobhouse at the London School of Economics. L.T. Hobhouse (1864–1929), influential British sociologist; served as first professor of sociology at London University. Alexander Morris Carr-Sanders, sociologist and editor of the *Sociological Review*; author of *The Population Problem: A Study in Human Evolution* (1922).   3. "Rural and Urban Thought: A Contribution to the Theory of Progress and Decay," *Sociological Review*, 21 (January 1929): 1–19.

Collège des
Ecossais
Plan des Quatre
Seigneurs
Montpellier

1 April 1930

[Ms UP]

Dear Lewis,

My partners & colleagues (*Jerusalem University*) – F.C. Mears of Edinburgh. & B. Chaikin (both F.R.S.B.A., i.e. fully qualified professional architects, the first of genius, & patience, & the second also of sound efficiency) have been most unscrupulously "chucked" – & so I too.[1] As per enclosed cutting, you will see something of it.

After Mears had spent a good few months, with Chaikin too, in Jerusalem, & had fully redrawn & elaborated the plans of those buildings, with all precautions & adjustments, with their *definite oral acceptance*, we were startled to receive a *lawyer's* letter acknowledging *only responsibility for first sketches*, & announcing that new architects had been employed.

We do not know yet whether we remain the consulting architects or not for the *general scheme* (but it is likely that we may be ousted there also).

Who are these new architects? *Are they good ones?* – or is this the firm which planned & published in press the *extraordinary parody of our Jerusalem scheme* for a Jewish Seminary in New York, several years ago – which was rejected – happily alike for your city, & for the reputation of Israel!

To no other task in our lives have we given so much time & care, thought & plans – nor though we say it, with such effectiveness, efficiency & unity alike of plans & of perspectives & effect. Nor yet with such sacrifice of time – while it has been very disastrous to Mears' *home practice* to be so much away in Jerusalem (& with the reputation of course of being away so much more!).

If the new regime was a better one – or so good – we'd take it as philosophically as other cares – but imagine this: After we had fully designed, & I gone out to the laying of foundation stone of Einstein Institute by president of Royal Society & Lord Balfour (at his opening of University 3 years ago) – the mathematical & physical portions – which it is the supreme glory of Einstein (reflected in Zionism and University alike) to have so much more fully unified – were *separated*, the former to opposite poles of the site; and *without informing us at all*! – by the advice of a fool engineer, of Indian Public Works (a Mr. Green) – a *group* & *type* of peculiarly poor architectural efficiency as I saw throughout India – to the Principal, Dr. Magnes, who is, I am told, a learned rabbi – but whom I know from long acquaintance, of neither scientific nor esthetic content. "The fool hath said in his heart there is no Unity"[2] – but in this – as now in central features of design – he has carried it out.

Let me hear then – and soon – if you can tell me anything? (*If necessary*, you can show this, in confidence of course, to Clarence Stein.)

Let me hear from you too when you have time re. my last letter (*sociological*).

With best wishes to you each & all

Yours always,

P. Geddes

1. Benjamin Chaikin, Jewish architect taken into partnership with Geddes and F.C. Mears to design the Hebrew University in Jerusalem (Meller, 279).   2. Adapted from Psalms 14.1 ("... no God").

Dear Lewis,

Collège des
Ecossais
Plan des Quatre
Seigneurs
Montpellier (Hérault)

6 May 1930[1]

[Ms UP]

Thanks for interesting letter of 22/4. But woe's me, man, why not get your health right? At your age, & long after, I found no better or more efficient way than that of breaking off all study & writing, etc., for a fortnight – with simple regimen as follows. (a) Regular work with gardener from his start before breakfast until mid-day lunch. Then (b) tired enough for deck-chair, all afternoon (little of lightest reading beyond paper) mostly looking at garden or further landscape. (Do you know that just as "Open the window" is the essential lecture on prevention of T.B. so the psychiatric text & summary & discourse is "Let your eyes rest on the horizon" – that's largely why people go to seaside or to hill views.) Then of course early to bed. Next week able to work with gardener in afternoon too – rather hard on Monday – but all right by Friday & Saturday. Then quiet Sunday – & back to ordinary life in good form & fully refreshed. *Try it*! – eating lots of fruit, etc. Thus too I understood how the full day, hard-working man is a wonder who has an active intellectual life! I had it – but was just as inclined to laze as can be. Of course I might have come to it – so far. But two hours work or so is to my mind enough for us more or less sedentary fellows in most cases. (If you haven't a garden, do up a friend's.)

My best child education up to going to the pretty futile school (at 8 – & joining class of 11–12) was from trotting after father among his vegetables & fruit & after Mother among her flowers – as they encouraged me to believe I was helping until I really was, a little – & talked of many things besides. I suppose you do that with Geddes? (The *noticing*, & then *watching*, of growth is very important.)

But now your "Modern Synthesis." Evidently an excellent paper of yours in *Saturday Review* – never saw it before.[2] I'd be very grateful if you'd lay out a dollar gradually in posting me more of such things as you write!

I quite agree you can't be readably literary if you used my grim scientific graphs on the poor dear public! But I have just been translating that synthesis into IX–9 diagram – & found it suggestive towards filling it up & explaining it more fully. Thus the *"New" Mechanism* is, of course, the old IX of the Futilitarians, but in a later edition, historically superposed, by pragmatists – Dewey, etc., and so far well – with their mechanical and other audiences involving a progress of their metaphysics towards more definite conception of adaptation & Instrumentalism (cf. Darwin's metaphysical phrase – "Natural Selection"). Whereas in my small yet central field of Geotechnics these six years here – reclaiming heath with gardens now exuberant beyond most – & building all I can also, is of more definitely biotechnic constructive character than the Industrial mechanistic dissipations of matter & energy; & thus more educative to me at least – in evolutionary thought as well.

One grumble let me make! Pray don't always use "Science" for its mathematical & mechanistic *preliminaries*, & contrast it with Humanism & Social life & literature, etc.

It is not permissible always to talk (as the IX Philistines do & their Humanist critics also) of "Science" as merely mathematical/physical & speak as if the Biological/Social were simply humanistic, literary, philosophical, etc., and thus opposed to "*Science* too." Whereas – ever since Comte a century ago & more, they have been clearly growing to Sciences too. You know all this of

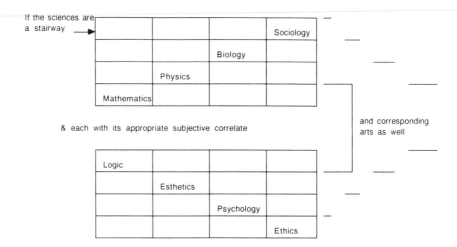

If the sciences are a stairway

& each with its appropriate subjective correlate

and corresponding arts as well

course – but you don't *here* sufficently bring on you reader! Rub it in!

Your criticism of us in *Sociological Review* is quite true so far. But only by damnable iteration has any scientific conception found its way into currency! – cf. "Natural Selection," "Germ Theory" or what you will! I plead however for your closer criticism of my (a) IX–9 paper, "Interpretation of Current events," & "Ways of Transition" – the latter offering view corresponding to Pragmatic & Instrumental in IX but more definitely bio-social, as becomes 9.

The proposition of this last paper that not simply I as experimenter, but every one, in their own way, may, can, & often does, live & act *in this Transition* (IX–9) is surely more clearly stated & exemplified in life & practice here than I ever got it before – & so far for instance as aid to understanding V.B.'s *Science & Sanctity*, which people usually don't grip & thus think of as mere counsels of perfection! Here in fact is my bio-pragmatism – *Vivendo discimus* – in advance upon *Mechanisando, Mammonando*, etc.!

Herewith I send you a booklet of Sandeman's *Practical Community* (& one or two others to pass on as you have *Sociological Review*). I objected to him that these counsels of perfection are not easily (if at all) realizable with the deteriorated verbalist egoists who are too common among our students (poor crops, these of after-war for most part!) & he has ably met my objections in new edition, for which I've written preface, & shall send you later.

Just as I began my Sociological Classification Library in Bombay with Sherlock Holmes (the scientist's observation & re-interpretation, sweeping aside the whole arguments of prosecution & of defence alike, the judge's summing up too – since they've got the wrong prisoner & mistaken witnesses, & deluded jury!) so now here is the mirror-reversal of criminal & detective in a new type of story, however coldly explained – since upon the {searching out/ charting out} of the transition from IX to the better life system of 9. As a man of letters, and a reader, you can see that better than I, more vividly & personally – you are, I am sure, you are often seeing it around you in progress! Can't you tell me of such stories & novels? I should think they must be beginning?

Again: have you realized how before my IX (itself of *many superposed pages* – e.g. Roman expansion – William the Conqueror – Reformation, etc., before Mechanical Age, & thus too with its own superposed pages.) There is the *old peasant world of Place Work Folk* (more or less continuing since Neolithic) which IX has always been conquering, exploiting, etc., as Britain in Ireland or in India, or some American expansions too – now in turn.

Again – how IX & 9 have oscillated – IX falling back from attempts at 9 – up to "the War to End War" & with a "Naval Conference" of experts keen to build more ships & deadlier ones! Is it not up to us to think out & try out the ways of the Peace–War / Helpwill & its Equipment / Hope? But *again* – how to prevent our back slidings?! (*excellent old term*) Beyond the 9 scheme & pages, comes the next – the right hand half of the ordinary Life-diagram – that of Cloister & City – & unified by the muses – of each & every "Religion" – Parnassolympian, Buddhist, Christian, etc., and this as *Re-Religion* – in action to – in Deed as well as Thought since Co-Emotionally Ethopolitated, synthetically synergised, towards better & better Imagery-achievement.

This scheme of IX–9, with its prefixed PWF – at simplest & its succeeding FWP at best has thus 4 pages in all, & not merely these two. Say 3–IX–9. Of course before 3 was the hunting paleolithic life – and before that a more simple transitional one rising from the Brute-man – as he from gentle, little early mother-animals – Mammals – who survived in their war-holes as the big Reptilian "Great Powers" tramped by (like the smaller peoples today, or often in the past – cf. Jerusalem, Athens & Rome & each at its early best – though in each case trying to be great powers killed them!).

Perhaps I am only telling you what you've already seen in these diagrams – am I? My squares are not to confine the world into my categories as some think; they are so many windowpanes for looking out into the world movement (or sometimes like spectrum analyses of complex radiations). While therefore I defend *repetition*, with *development*, I'll very much value & be helped by you putting fresh problems or forms of problems just as your paper on Synthesis does. V.B. & I & others would all be helped to *supply* if we could get any *demand*! Questions are a great help.

I think I told you in my last of my great good fortune in now – & henceforth I trust – a very good group of colleagues here. Marr[3] has already taken over both students & buildings: & to advantage. Château d'Assas has a capable organiser taking it over next week – & others are coming in due time, for 1930–31. I therefore, with their approval, after 6 years hard work here with delay of writing or further travel, etc., am to take a *sabbatical* year or at least *semester* – and Thomson has also returned from University California trimester with so glowing an account of the many live minds & interest in general ideas that he met that I enclose a copy!

As new President "Sociological Institute" (formed by definite union of Le Play House with its Surveys, Tours, etc., with Sociological Society), I am trying to prepare an address appealing to all the sociological fraternity to abate their dispersal, to circulate their various programmes & view points for mutual discussion, & thus perhaps thrash them out more clearly, in correspondence, & committees – towards a possible congress in London or Paris – possibly followed by others. Though the old Positivist Societies fossilised by "swearing by the word of the master," instead of continuing his impulse, they were a power in London & Paris such as we later sociologists are not, in forming &

leading public opinion in progressive directions (thus Beasly, etc., saved the Trades unions from legal extinction after public fury over an outrage, when I was a boy, & there were many other instances; while in Brazil, Portugal, Turkey, etc., they started the Republics). Without wishing to be quite so political, why not more social? Thus how bring together the surveys, the social workers & university departments & the academic departments (to my mind falling back too often into abstractions – of "community," "Society," etc. etc.)? And between my London & Edinburgh presidencies & a beginning of the like here, can't I play "stage army" for a little & plead for recruits!

I am too old for your cold winters & hot rooms, & for evening popular lectures, but do you think there are any openings for academic courses, of short duration, for in forenoons or afternoons I can face these. If I could do something of that kind in East or Centre in October, & in California with its climate so like this in November & December that would be encouraging, & give me the new contacts I now want – & need.

I have to confess I need to meet expenses of tour – for these buildings now represent all I have had plus overdraft, & my wife too much in like case also; for we have no pious donor! Is there any academic organisation towards such university lectures? (as distinguished from the popular lecture agencies I'm not fit for). I hope you'll accept one of those academic invitations you mention; it is a good sign that they want to be *gazogened*:[4] I wish more of ours did, but in France & Britain too much fixed as yet, though with signs of hope – for next generation!

Moreover here again consider your health, & the Missis, & the kid! We professors have a far easier & healthier life than writers & journalists – & live far longer accordingly! Condescend to the leisurely academic level, & you'll be a monster of health again – & Mrs. Mumford too – & Geddes – sturdy as well as winged Hermes!

What a long letter – poor man! – only you read fast!

Always yours,

P.G.

I wait the Blissful Organization of Kellogg you promise. The International Edinburgh Congress never printed my paper, report on Bibliography, nor even replied to my enquiries! (Paul Kellogg is not the only surly editor evidently!)

1. Mumford's annotation: "This was when I was fighting an infected tooth."   2. "A Modern Synthesis," *Saturday Review of Literature*, 6 (12 April 1930): 920–921; (10 May 1930): 1028–1029.   3. T.R. Marr was a former student of Geddes who served as business manager of Collège des Ecossais.   4. "*gazogened*": to be aerated or filled with gas.

[1930]¹

[Ms UP]

Dear Lewis,

All this you know already – but as a copy of outline of our doings & aims at Montpellier it may save you trouble – if & when you have an opportunity of doing something for us in the way of sending students – whether by some of your University contacts or in the press. For we remain too unvisited!

It is copy of a p.s. to letter to Charles Ferguson of "University Militant,"

"Great News," etc. – a friend of Branford's & mine since long ago – and an active spirit.

I shall doubtless have your reply to my last before this reaches you – so you need not acknowledge it.

P.G.

1. Undated letter in UP 1930 file.

Alas Lewis, our dear friend, Victor Branford, is gone! A wire to Montpellier on Wednesday, a week past, only reached me in Paris on the Saturday, 22/6. We came off to Folkestone at once, whence I motored to Hastings Nursing Home, & spent most of next 20 hours by his bedside. But coma had fully set in, & he died without any awakening, Sunday 23/6 – 5 p.m. – for the eyes, opening only with death-moments, showed no recognition. It was consolation anyway that he had no pain. After the arthritis (for which he was to have gone to salt-baths at Droitwich) he got kidney trouble, whence his coma, & the main consultant diagnosed something of senile tuberculosis too. He had suddenly aged, from the admirable skater in Switzerland a few months before, to look like 80! He was 66 nearly – & should have lasted far longer – had he known to take care. We buried him beside his wife in the hilltop Hastings cemetery. His pleasant "Pinders" is sold to people who will appreciate the garden.

3 Netherton Grove
S.W.10

3 July 1930

[Ms UP]

After a rest & funeral I have had two days of intense work at Pinders putting his papers in order roughly, so as to facilitate more careful study later, & utilization of his dossiers unfinished – a long job! When you come over – or even before that – you might think of taking over some congenial one? I must ask as many of his sociological friends as may be to help with this.

A. Farquharson had obituary notice in *Times* for Tuesday June 25th (which I suppose you can easily see in Library) but with too much of me in it. Herewith copy of my little notice for next week's *Nature*.

I am sure you will write such a notice, & the best of us all. Let me have a couple of copies – one for Sociological Society & one for self.

We are here until near mid-August – then at Outlook Tower Edinburgh, & next at British Association Sept. 3–10 – after which we propose to sail as soon as pssible (to N.Y. I suppose). As I think I told you, our idea is of California especially, as I have never before got West of Chicago & St. Louis, & these 30 years ago. J.A. Thomson has been there last term, & came back delighted with the bright people & students he met – so that settles us. And I need a sabbatical term, after 6 years hard work at Montpellier, & now that I have efficient colleagues there I can take it. (I *have* to do some potboiling too – after 6 years unpaid & yearly losses!) Let me have any suggestions you can.

Returning to V.B.'s work, I'll go on lecturing here at Sociological Society so as to try to stir such as I can to carry on, & perhaps get one here & there to come in. So your estimate of his work & aims will be helpful – for though the common one, that he was carrying on my ideas, has truth in it, he also went further in various ways. Thus his estimates of *Rome* & *Geneva* – his *Science & Sanctity*, etc., etc., are better than my studies of cities, & statements of ideals.

He was more truly religious – *re-religious* – than am I. Again his economic insight & financial criticism & aims of social finance were hammered out in real life in the City & in Paraguay – for of course you will have seen that he was there as applied sociologist, & learning experimentally too. Two men in one – for thought & deed.

Best wishes to you all three. I hope you have all better health now? Take care of it for *your* old age!

Yours,

P.G.

———————

[15 July 1930]¹

[Ts NLS]

I shall now write to Lovett in Chicago and to John Gaus in Wisconsin. It may be difficult to arrange for any conferences before you come over here, partly because the people who would do so are now on vacation; but I am pretty sure that the Experimental College will want you for a week, if they have any extra money in their budget. Chicago is a different matter; they have vast classes, and lecturing may be too much of a strain, both for you and the audience! But I will see what Lovett says. Meiklejohn, the head of the Experimental College, is taking my place in Geneva this year with Zimmern; you might drop him a line just to say that you shall be passing through Wisconsin and hope to see him.

As for my own state and plans. My general health is robust, indeed magnificent. The miserable experience of the last four months has been due to the fact that I have a low grade infection (staphylococcus) in a *front* tooth: it has been difficult to extirpate, but the prospect of having to take out the tooth itself is worse: so I am still under treatment. This infection was responsible for my series of carbuncles, and it has kept me from having any tonsils removed. If it once gets cleared up, I shall then have my tonsils out. I had thought this would all be over by the end of September, but I can't definitely count on it, and all my plans are in abeyance until then. Geddes is now in splendid shape; he and I are keeping house up here alone, whilst Sophie is travelling, for the sake of erasing the effect of her long series of illnesses and mishaps; and we are all now in good health and spirits – my infection aside! I am now about to do the second draft of my book on the arts; and I shall have to go to Europe this fall or the following spring, in order to gather a little fresh material from the countries I've not yet visited, and to check up and revivify my old impressions. It looks now that it won't be until the following spring; although should you decide to remain in Europe I should probably try to make it during the fall – *if* my physical troubles are over.

Have you seen the announcement of the New Era Exposition at Cologne in 1933? I think it would interest you. Dr. Ernst Jackh, a friend of Zimmern's, is the director.²

Please remember me kindly to Mrs. Geddes; likewise to Farquharson, and believe me in all sympathy and sorrow

Affectionately,

Lewis

1. Geddes' annotation: "15/7/30 visit to America?" The first page of the letter is missing.
2. Ernst Jackh (1875–1959), German political scientist.

Dear P.G.,

This letter is long overdue: but I have had a spotted and mostly unpleasant summer, interrupted by the severest cold I ever have been through; and my weekly trips to the dentist in New York only ceased at the end of August – by which time I was ready to have my tonsils out. This took place last week, and I am recovering fairly rapidly. The tonsils were badly diseased, and the continual seepage of poison into my system, even though it had not yet crippled me with rheumatism or worked any obvious large effects, had undoubtedly been a handicap these last few years.

Meanwhile, my funds have gotten low again – our domestic doctor's bills for the last year run close on to a thousand dollars, while for the year before they were over two thousand! – and it is very unlikely that I shall be able to go over to England until next spring at the earliest. You received, did you not, the six copies of the *New Republic* I sent over, containing my little appreciation of Branford.[1] It gained a little attention here, and was partly responsible for the fact that Harry Elmer Barnes wrote me the other day to ask if I wouldn't contibute a chapter to a new book on contemporary sociology he is getting out, dealing with Le Play, Demolins, yourself, Branford, and Co. It is to be 10,000 words long; and to be ready by next June; so I shall be grateful to you for any data that I might be likely to overlook which you may send on to me. I shall try to get a first draft ready by the beginning of February, and will submit it to you.

Thanks warmly for the dossier of Branford's writing you sent over. I remember discussing the possibility of publishing MacDonald's book on Franklin with him; in the meanwhile, two or three lives of Franklin have appeared, one of them, by Bernard Fay, a good one; and the chances of getting a publisher for MacDonald are almost nil – all the more so because the American publishers have suffered acutely from the current depression.[2]

Wisconsin and Dartmouth were both interested in your coming – John Gaus at Dartmouth has long known and valued your writings – and they look forward perhaps to seeing you another year. I had written fortunately to personal friends at both places; and your cancellation of your trip was no inconvenience whatever to me. I am going up to Dartmouth once a month during the winter for two or three day conferences with the arts students: my work is to be quite informal, and marks a step away from the point and credit accountancy which remains such a handicap on our educational system.

What you say about Le Play house is all too true. I value Farquharson's mind highly, and our relations have always been very cordial: but I have a feeling that, if Miss Loch had remained the secretary of the Society, it would have achieved and kept a position which it can now win only with difficulty: and had it not been for the confusion of domestic and societal issues in the Sociological menage, Miss Loch could undoubtedly have remained. Did I tell you that she is the mother of a very fine healthy girl!, and apparently is now wholly absorbed in her domestic career? She lives on the Pacific Coast, so we

4002 Locust Street
Long Island City, N.Y.

26 September 1930

[Ts NLS]

Here amid this Indian Conference actively fermenting & buzzing over its vastly heterogeneous affairs, I am trying to get some interest in our Indian College at Montpellier – as by lectures in their Club or Universities generally. Not an easy matter, but there may be some result.

I had a long talk with Tagore yesterday (here in passing home from U.S. to India) as to our trying to get beyond our too simply personal endeavours, & towards co-operation – thus towards being understood as a Movement in higher education, like that (or those) so active on the preliminary levels. (Consider this. Cf. Ferguson's "University Militant." What of him, now?) We might aid this by enlisting such cooperation with others – like Glenn Frank? – Meiklejohn? Who else? – besides yourself of course! (Antioch, Columbia & Dartmouth courses of general science & history, etc.)

Get the circular or other publication of *Carleton College* in Minnesota; for scheme inspired during past 6 years by my Edinburgh friend, Ian Stoughton Holbourn, of interaction of its Schools of *visual* & *auditive* Arts with *university courses*, so that out of these three types of Education so separable hitherto there are now emerging – far more widely educated men – the scholars more creative, & the artists more cultured too.[3] Their Medieval & Renaissance types at best had all far more of each other than at present – hence Giotto, etc., & Leonardo, etc., were abreast of the culture of their ages, and conversely the scholarly types – as from William of Wykeham to Sir Thomas More, & so on, alive amid the arts.

Here Mrs. Taylor's books are to my mind so good. Have you read her *Leonardo* – her *Italian Renaissance*, etc. (new edition, Harper) others too will be forthcoming. But Holbourn is in *action. Pray let me hear what you may learn of Carleton College & its results.* Can you take it on one of your University visits? I hope these are to go on? To get a few men & women more fully concentrated in our common movement seems to me more hopeful than can be our solitary writings, scattered here & there, though of course for every reason one has to go on with these too.

I wish you could send me some such *bright* young people: as yet I have not such, either from your side or this: nor am I finding them here. The Contemporary School Evolution – as Dissolution – towards Debacle – is so manifest – the capable Reconstructives as yet so few.

You have never answered as to my (IX–9 Diagram) papers in *Sociological Review* – from V.B.'s and my *Rustic & Urban* one onwards. I'd greatly value your criticism & suggestions as to these: as notably of that of ways of Transition, as in everyday endeavours at Montpellier I am of course now trying fresh ways on these very different worlds around me here, with IX & its Decadance so dominant! But how hard to find any one of active & open thought – or purposive action – & still less both in their needed association.

Bonne Année 1931. Yours with that wish to each & all,

P. Geddes

1. Geddes edited and printed a "collection of tributes," entitled *Our Singer and Her Songs*, as a memorial to Marjory Kennedy-Fraser, collector of Scottish folk songs who had been active in the Edinburgh Summer Meetings (Boardman 417).   2. James Bryce (1838–1922), M.P., supported several of Geddes' projects; served as first President of the Sociological Society of London; author of *The Menace of Great Cities* (1913).   3. Stoughton Holbourn (1872–1935), author of *An Introduction to the Architectures of European Religions* (1909).

Lewis Mumford AND Patrick Geddes

THE CORRESPONDENCE

1931

Dear Lewis,

3 Netherton Grove
London S.W.10

1 February 1931

[Ms UP]

You have not answered yet regarding Victor Branford Memorial, etc. – now urgent. We are still here for three weeks or so, before return to Montpellier, *so pray try to let me have answer*!

A. Farquharson agrees to my asking your advice as to republication of V.B.'s works, enjoined in his will, & hindering all application of the balance of his estate until this is settled – so as Society we are in miserable & tantalising poverty: and indeed in debt: & we can't even get beyond the publication of the belated *October* number! – & for which *we have also to await your reply to this letter.* So sleep over this – & reply as soon as possible thereafter.

Just as Mrs. Annand Taylor's *Leonardo* has only sold to 1000 or less in 2 years in this country, but to 15000 in nine months in U.S.A. so (on smaller scale of course) might it not be with V.B.'s books? – *if you would tackle them*? Can you find the right publisher? Can you do some (if not all) the editing, with little introductions by yourself, or such men as you can trust – sometimes perhaps Alexander Farquharson – sometimes me? They might even be printed in U.S.A. & sheets sent over to this side – or (2) they could be printed in Britain, & sheets sent you – or (3) even in Montpellier, where cheaper, & also decently done (unless duty in U.S. be too high? Not higher than from Britain I suppose).

Farquharson hangs on to *Le Play House Press*! (very *futile* as to sales, just like P.G. & Colleagues!).

But what of *Harper*, who I hear is preparing to develop his British side, & on great scale? or some other Atlantic-bestriding Colossus? (I don't think *Macmillan*, but what do you say?) It would be a great thing if we would thus get them published here by something better than Le Play House Press, and that might be the way of doing it. Can you enquire?

You need not begin by offering the guarantee Branford's legacy allows of £11,000 *odds when realised*, but this as yet delayed by legal devilments. Publishers sell better when they have some risks to make them push!

Conversely, *you need not work for nothing.* You have often spoken of coming over – last autumn even – why not this coming spring? – to Montpellier for a time, & then in London in June – July – what say you? This matter could help your expenses

With all good wishes, yours ever,

P.G.

Can you review enclosed Booklet – & thus help a good regional cause? But V.B. first! But answer me candidly – are my two verses too bad, as some friends think!

Collège des
Ecossais
Plan des Quatre
Seigneurs
Montpellier (Hérault)

18 March 1931

[Ms UP]

Dear Lewis – (horridly *slow* correspondent!)

What can you do in the way of *lending me* what you can of your old notes & gatherings for *Utopias* Book? (They'll be taken care of, and duly returned.) What was the name of that other book on *Utopias*, which appeared more or less at same time as yours? Pray *order it for me* too *with a fresh copy of yours* from your bookseller, & I'll duly remit to him.

As President of three Sociological Institutes – Edinburgh, Outlook Tower, Le Play House, & here – I must do what I can – as stage army! And I've been trying at the two former in these 8 months absence from here – & now am as busy here as slow yet progressive convalescence allows (from illness from overwork in London winter – for so many previous years escaped).

(This illness however has convinced me that I'm now too old to face an American winter! So nothing for it but to invite young America here! – and to get on with *Collège des Americains*, so long now in project, & even *plans*.) I'll have one good professor on sabbatical year here; & *want more*.

I was most stirred up in London by "(Adler) Society of Individual Psychology" – by far the most living group I have found – & for long! It is on scale of four or five meetings a week! – of varied students, parents, teachers, parsons even – each stirred towards bringing *Psychology* – & *that moralized, & socialized* – into *Life*. After thousands of years of good counsels, & good aspirations, here are more definitely reappearing and for the levels of everyday secular life, the needed scientific & practical *Rules* – of Life – of old heritage developing to best of monastery & convent of old "orders." And why not of all kinds? – yet duly utilizing these – and with reciprocably influenced sociology & its applications as well, is thus already inspiring to us here: and of course I should greatly value your co-operation, criticisms, & suggestions. (If you have such a group in N.Y.C. look into it, & tell me what you find.) Reply to this, & previous letters soon.

With all good wishes, yours cordially,

P.G.

(Enclosed circular is beginning of further scheme – for summer term – & next autumn term I hope also.)

Montpellier

1 April 1931

[Ms UP]

Dear Lewis,

Again I write you – & so soon after last batch! But now that V.B. is gone, there is no one else I can talk to as to you (since Farquharson retired into his Surveys & got tangled in V.B.'s testamentary business cares).

Here I am doubly in cares & perplexities (health not improving – have had to begin *insulin* today) on one side with this as last term of "seven lean years" – with need of facing that we *can't* stand, *can't* survive, continuance of such! So personal cares as well as collegiate.

Yet with this closing period, it is something that substantial progress has been made – (1) in buildings & gardens (the latter such that the Indian Moslem philosopher who came yesterday went into rhapsody – "far better than all our Moslem Paradise"!!). Something too to have just now 2 Germans, 1 Austrian,

& 2 Czecho-Slovakians (& a Dutchman) towards reconciliation with each other & with French (here easier by far than in North), & with Jews too – which peace of botanic collaboration brings together – while besides those we have eight from Sheffield, etc., coming tomorrow for ten days ramble with them – & also with our Social Survey of Region & Cities – with their leaders collaborating. But this year not one from Scots Universities to whom we remain quite unknown. And so as for many beginnings – yet deficit, at rate of £900 a year or more, reducing only to that of £800 with this term! One good American on sabbatical year; but too limited, like most thesis-writers, to a problem too small & too indefinite, etc. Thus our very continuance is menaced.

Yet just as Biology & Sociology are now being here *better knit than ever before* (since in inter-operativity & in concrete & in generalities alike) so for other subjects & approaches, not a few. We can at length offer more than I know of elsewhere; and with more or less incorporation of the best I can learn of progress elsewhere. Pray by all means tell me of any you know of, or take part in. (Did you ever print your Library outline – which I was too busy to be able to sit down & try to help with as you asked me. Here my Essay towards Bibliography has wasted years unpublished; do you know of any publisher who would look at it? It is too far apart from Dewey's & other librarians' schemes to be at first adapted by their ordinary majorities, or their publications.)

But I need not go more into these: you can believe that these are here at length educational (*re-educational*) facilities getting fairly into order, & for varied choice.

But the *writer can't* sell his books – nor the painter his pictures! (That's how with William Sharp and Fiona MacLeod & *Evergreen*, etc., our pretty books lost me thirty years ago my main available capital, so with £2000 *overdraft* or so instead of interest on it ever since.[1] And the like for all this here now!)

Hermitage & cell, cloister, garden, nook, & wild, have been the necessary conditions for getting so far – yet for bare survival *some publicity is now needed* – and urgent! – while I am indeed convinced that it could now open for us seven more prosperous years.

Hence – & with age & health-deterioration – it is now, practically, a matter of life & death – for organization & for self as well to find this: and neither in London nor in Edinburgh, etc., could I see the least hope of *any one* able to do this: so have to look to you – *alone*!

Considering it even as a matter of business – I believe it worth offering you the utmost possible of contingent results – not as a bribe – but as assurance – could American interest be roused, & to react upon British, as it would (cf. Youmans, was it not, who just floated Herbert Spencer at his crisis).[2]

American College would thus follow – & so on. I know you are in difficulty for ready cash too. Yet try what you can, to giving us that long discussed visit! – before too late!

Yours,

P.G.

1. William Sharp, under the pseudonym Fiona Macleod, wrote several books in the neo-Romantic,

"Celtic revival" style (1895–1896); these were published by Patrick Geddes, Colleagues and Company in Edinburgh; Sharp also contributed to *The Evergreen*, another of Geddes' projects, which was a "northern seasonal" published in 1895 (Meller 98–102).   2.   Edward L. Youmans (1821–1887), American chemist and admirer of Spencer; enlisted subscribers to Spencer's books in the United States.

Collège des
Ecossais
Montpellier, France

27 April 1931

[Ms UP]

Dear Lewis,

The more I think of it, the more it seems to me that we might put our long-dreamed collaboration *on a business basis* – & thus *mutually help each others' potboiling* – so anxious a matter! In fact, just as I am head of firm of Geddes, Mears & Chaikim, F.R.S.B.A. now so long architects of University of Jerusalem, why not have a corresponding more or less (private) co-operation with you! You would thus have time & trouble in finding way & work for me, and take your share of commissions of results, just at they do.

Consider the following. This place is still sucking our capital dry (from initial investment of £230, we are now out of pocket not only my all but greater part of my wife's little fortune also: quite over £18000 in all) & losing each year, from small number of students so far. For just as a writer can't personally sell his books, nor even a painter his pictures, but *have* to depend on publisher & art dealer, so for education: we have no "selling partner," no publicity. For I have mostly outlived my contemporaries – all gone – or retired – & from 17 years absence from Britain save on *vacation* visits, I make no new contacts! You can't help me with that, but you could with the *American College* beside us, so long on plan – & on one of the fine sites of the world, and at the university which, on the whole, I have most reasons, these 40 years, for respecting & working with & for, as you will also see when you come.

Again, the start of *American Surveys* came from our Outlook Tower per Dr. Johnson, etc., about 1899, as he was sent further by Dr. Josiah Strong – etc. So what of further developments? (Kellogg lately wrote me amicably, proposing to meet on his approaching visit to this side. No one has written of our developments that I know of, since Zueblin's "World's First Sociological Laboratory," *American Journal of Sociology* about 1901.) An old student here (Philip Boardman) now of a students' organization in Philadelphia, cabled & wrote lately to ask me to come over this summer to help to start an Outlook Tower there. (Dr. Ely discussed it in 1923 for his City Hall, but 0 came of it: still the idea might be taken up, thus to crown a skyscraper in every great city. Mears has made a noble drawing for this: could you write it up, & thus find us a commission? – & share it.) For that sort of matter, I am still concessionnaire of Elisée Reclus' *Great Globe* also – & this needs association with Galeron's small Holland Globe (& why not also Zeiss's *Planetarium*?). I failed with Globe in 1900 – but times are riper now.)[1]

Again my *Cities & Town Planning Exhibition* – 4000 sheets – is worth $20,000. It would need another $5000 to bring its plans up to date, but it is full of historic, sociological & educational matter other collections do not in the same degree possess. Might not Russell Sage Foundation or some great Library buy it. I am too old to take it round cities as in former years – but a competent person (such as my son Alasdair was) could do so – & with good results for advancing progress, as in British & Indian Cities, & in Jerusalem, Paris, Ghent, etc., also.

Again, after planning (in part or in whole) for some 13 universities before this as from Edinburgh & London & Bristol to Indian Universities, Tagore, Nizam, etc. etc., & to Jerusalem, why not commissions of such kinds from and for American Universities? As from the wealthy "Duke University" to smaller ones?

And why not *Experimental Colleges* like this one – increasingly adding to its initiatives now that I have several competent colleagues, & expect more.

Again – we have carried *War & Peace Research* farthest from *Ideas at War* to my Cornell Lecture, etc. (IX–9 in *Sociological Review*), & always in progress. I have on hand a Report for the Pasteur Institute (Paris) for a *Department of Social Hygiene* – on which my old friends Duclanx & Metchnikoff (former Directors) were keen. The present head – Dr. Roux (of diptheria fame) – is more purely a physician, but is willing to consider it, the more since one or two of his professors are keen. (Here is a job of world-wide collaboration – in time!)

May there not be some help from various Trusts to objects such as above?

Again, I am preparing, in harmony with two professors of French to begin with (one in Dublin, the other American & here for his sabbatical year, & thesis), and next others, a scheme of usefully organised *theses* for inter-university collaboration. Ordinarily such theses fall dead, since mostly on more or less trivial points of scholarship, & without any great interest. But recall how a long series of British writers – (say from Arthur Young before French Revolution to Arnold Bennett, just deceased – have written what they could see & say of France – and similarly Americans – say from (I suppose) Benjamin Franklin to today. Again, how French writers say from Froissart & Ronsard – or again from Rousseau & Voltaire to André Maurois & Siegfried have written of Britain & of America too. Recall also – for Franco-German Relations – how Mme. de Staël made Germany understood in France, & how Goethe, still greatest of Europeans, understood & respected France from student days at Strasbourg to late old age in 1830.[2]

In short here is a needed line of *Pan-European* – & even *Pan-Occidental* – *good understanding*! (& the like too between East & West – as I so fully realise from Indian experience, & here too.)

Now just as you have headed into the matter of publishing for V.B. why not for such a collaboration of these contributions (say with our joint editing more or less others). We would readily get the approval of the League of Nations' International Intellectual Relations Committee, etc., & of Peace Societies & Organizations over the world – so there would thus be market for such books & for the Series. Think of it.

In fact I submit here at length one business proposition for your consideration – & which I hope we may discuss when you come over, but meantime begin in letter!

Note that we are only here till about mid-June – & then at 3 Netherton Grove S.W. 10 till August – when I go to Edinburgh – & return for British Association Centenary function in London in September – and return here in October.

If your visit could be made to coincide with any of these dates (which I *cannot alter*) I am sure we might get something done – and if you could spend some time in Montpellier (say October–November?) we could readily concoct one or two joint books.

I have written as above because I know your good work – & your difficulties also; but I think here above are ways of co-operation useful to both of us. And my time is short at present age – & now that Branford is gone (& Thomson I fear is failing) I have no colleague, executor & heir to whom I can look with any substantial hope but you.

Read this to Mrs. Mumford, & see what she thinks of all this string of possibilities.

Yours ever,

P. Geddes

1. Elisée Reclus, a French geographer and friend of Geddes, designed a huge terrestial globe, to be financed by Geddes, for the 1900 Paris World Exhibition; the project was never completed. In 1902 the French architect Paul Louis Albert Galeron designed a large celestial globe for Geddes – nor was this project ever completed.   2.   Arthur Young, *Travels in France* (1792); Arnold Bennett, *Paris Nights and Other Impressions of People and Places* (1913); Benjamin Franklin represented the United States in Paris, 1778–1785; Jean Froissart's *Chroniques* (*c.* 1400) describes his travels in England and Scotland; Pierre de Ronsard (1524–1585), French poet, served for a time in the court of James V of Scotland; Jean-Jacques Rousseau lived in England 1766–1768; Voltaire, *Letters Concerning the English Nation* (1733); André Maurois, *The Silence of Colonel Bramble* (1920), a novel of English military life; André Siegfried, *Deux mois en Amerique du Nord* (1916), *L'Angleterre d'aujourd'hui* (1924); Madame de Staël, *De l'Allemagne* (1810–1813); Goethe's *Hermann und Dorothea* (1797) is set in France.

4002 Locust Street
Long Island City

3 May 1931

[Ts NLS]

Dear Master,

The iris and the tulips are at last in bloom, and the gardens and courts here in Sunnyside are a great delight after a day in the city: but I have been working desperately hard to finish my book, and I have not had a ramble in the country since last fall.[1] I hoped to have it ready by the first of May, thinking that I might then be able to start for Europe: but neither the first nor the last has proved possible: it will be another month before my book is finished, and the constant application at the book, plus various disappointments as to essay publication, has left me temporarily without funds: so I shall spend the next fortnight at the University of Michigan and at Dartmouth, making up the deficit. This means that I shall stay here at least until the regionalism conference in Virginia in July is over. I will then see what arrangements I can make with my publishers. I am already a thousand dollars in debt to them; and the book business has been so bad this last year and a half that I am not sure that the present manuscript will suffice to work off this debt; in which case I hesitate to go much deeper into the hole. Longmans, Harpers, and Harcourts all turned down Branford's books; and my friends here advise me that it will probably be cheaper for Branford's executors to print the books inexpensively in England, taking advantage of the lower costs of setting it up and binding it: if the books are of standard size it may be possible to sell some of the sheets here – 250 or 500; but the format of *Science and Sanctity* and that of *Living Religions* are both a little off the standard one. These are little matters: but they have influence upon the prospective publisher.

I am sorry that my means of helping the Montpellier project are so limited. Boardman has written a very good account of the life there and your own point of view; and I have done all I could to help him place it; and I had thought that my article upon you two years ago in the *New Republic* might bring inquiries from young men; but except for the few I reach personally, nothing has come of it.[2] During the last year I have made the acquaintance of a very able young woman, partly trained as an architect, a Miss Catherine Bauer, who has been able to assimilate more of your doctrines and points of view than anyone I have encountered: in some ways, she is by natural aptitude a closer disciple than I am, and if she gets a Guggenheim fellowship next spring she will certainly seek you out: she purposes, incidentally, to write a history of the dwelling house in all ages.[3] She has just won a thousand dollar prize in an American commercial magazine, *Fortune*, with a very able essay upon modern architecture in the Frankfurt housing schemes: if I can lay hands on a copy, I will send it to you. But money and students in larger numbers are both problems that I am a poor hand at solving: the intellectual and financial bankruptcy of our American business system has made the first very scarce, so that Stein's garden city, Radburn, languishes, for instance, and all sorts of other endowments have been sadly curtailed, including Rosenwald's generous philanthropies. Looking back over the past, I could wish, unfruitful though wishes are, that I had been five years elder when we met in 1923: I was then too unstable and too full of unsolved problems of my own, with still too shaky intellectual foundations, to be of any real help to you then: in order to keep from being completely absorbed on the level of your feminine collaborators, I had to stand too far away; and I can sense now the loss to both of us. For I should have either aided you at the beginning of the Montpellier scheme, or have diverted you completely away from it toward something of greater immediate urgency and prospect of accomplishment – namely, the various books that you have still to write. As things have gone, I have remained on the sidelines; and such aid as I have given toward furthering your ideas has been indirect. You ask how much I have gotten from the IX to 9 developments and from the later papers in the *Sociological Review* generally? I find them enormously useful in marshalling my thoughts on any given problem, in reducing them to order, in seeing that I have not left out an essential element; but, as I have said before, they will attract a majority of people in their concrete applications rather than in their schematic outline. As concrete criticism and outlook your papers link up in sociology with Middleton Murry's illuminations in religion, with Waldo Frank's conception of a new civilization in *The Re-Discovery of America*, and with various other still unpopular but nevertheless ascendant outlooks – challenges to the present industrial and pecuniary order. Spengler, too, has prepared many people for your conception of a new outburst of culture.[4] The difficulty comes in the application; and in dealing with categories there is a tendency to lump together for convenience elements which are composed of diverse and conflicting aims. Thus, while it is true that the Bolsheviks and the Fascists belong in part to the mechanistic order, it is also true that Soviet Russia has been consistently promoting regionalism and has shown more respect for the underlying nationalities and cultures than

Czardom did, and moreover they have entirely escaped from the pecuniary culture to one of physical and energetic realities in economics. One of the reasons that your papers have met with so little response is not that they break too completely with current views, but merely that the *Sociological Review* has not a wide enough circulation to ensure their being read by the people who might react favorably to them. This brings me back to an old point of mine: the economy of writing books, as compared with publication in any periodical form. For books alone are reviewed; and books alone remain in circulation long enough to be discussed by a succession of readers at different times, and so to slowly gather their audience. C.K. Ogden has had an influence far out of proportion to Victor Branford or yourself, considering not merely the quality of his contributions but the amount of energy he has spent on disseminating his ideas: and this was because his main efforts have gone into a series of book publications, rather than into the *Cambridge Magazine* with which he originally started.[5] The original idea for the Making of the Future Series was a good one: but the series weakened as it progressed, because Branford's energies and money went to the Sociological Society, and instead of using the series to gather around new writers and to build up a school of thought, he let it finally peter out. For the sake of the ideas that both you and Branford had to give to the world, I could wish that you both had either gone into a lay monastery in 1920 or been imprisoned by the civil authorities, with nothing other than pen and ink and a library to keep you company! A garden would have kept you both in good health, and instead of communicating with a few poor disciples like myself, scattered at the ends of the earth, you would presently have found yourselves surrounded by a school. Surely, it was contrary to your own teaching, to build the buildings first and then attract the pupils. That is our own weak American method: the method that produces palatial buildings, and fills them with vacant minds. I thank heaven for Thomson's persistence in keeping you to your biological cooperation; and I wish I had had a similar early start and a similar opportunity in sociology. It is too late for practical collaboration now: our habits of work are too different, and a good secretary would be of more advantage to you than I could at my most sympathetic best: but I still hope to absorb more from you, to use as best I may, and if I come at all this summer, it will be for at least a month. I trust that your cure progresses, and I am sad over the practical problems. I told little Geddes a little while ago – he is in bed, recovering from the measles – that I was going to write to you, and he inquired minutely where you lived and sent you his love. Apart from the abscess in his nose which began the winter, he has had a good time of it for the last year, and he wakes up in the morning singing, or whistling imitations of the bob-white and the whipporwill. He is a child after your heart, and better than his father, who nevertheless remains

Devotedly yours,

Lewis

1. *The Brown Decades: A Study of the Arts in America, 1865–1895* (1931).   2. "Patrick Geddes, Insurgent," *New Republic*, 60 (30 October 1929): 295–296.   3. Catherine Bauer Wurster (1905–1964) wrote on housing and urban planning and was a member of the RPAA; Mumford was involved in an extramarital love affair with her from 1930 to 1934.   4. John Middleton Murry (1889–1957), prolific author, critic, and editor. Waldo Frank (1889–1967),

novelist, critic, and friend of Mumford; author of *The Re-Discovery of America* (1928).
5.   C.K. Ogden (1899–1957), linguist who developed "Basic English."

Dear Lewis,

Collège des
Ecossais
Plan des Quatre
Seigneurs
Montpellier (Hérault)

17 May 1931

[Ms UP]

Yours of 3rd only received today, so I may as well reply, as otherwise letters fall into files of arrears. (I am being helped by insulin, & so may hold on a while; for this disease is curious in killing young men quickly, but old ones slowly! However, we won't speak more of that!)

Thanks too for all your care & kindness over Branford's books. Sorry we can't boom them a little! Still I hope, after V.B.'s confused will has been settled, that you can be paid a reasonable fee for editing one or two of them – and it would be a great help if you can. But this won't be settled till late summer at earliest.

Yes, send the lady occcupied with housing history: I have material in Town Planning Exhibition which may interest her to overhaul.

Yes, it's a pity that we did not make more of our meeting in 1923; but I was not cleared up enough then either. Since then however I've made some progress!

I quite see that a small Review like ours is a failure – & that books are better – & I *am* struggling to write; thus though Thomson has done most of our 1560 pages or so (in *Life*, etc., coming out early June) I have done a good many too, & some of the more difficult ones.[1]

Now at *Sociology* for Home University Series – & a big *Social Life* (to match Life for Biology) besides that long delayed *Olympus*. And though of course it has been a lifelong loss of readers & influence to go on thinking instead of writing, it has been a *great* economy of time too – since letting thinking go on faster & further & over more fields (if I don't fail in getting more written now!).

Now however, I *don't accept your criticism* that I should have been writing instead of College-making! See Indian College (enclosed postcard & note on back) & think it over. For here is the fullest & clearest conjunction of Biology & Sociology, working on parallel lines, in *the University world so far* – & so already mutually educative to their varied workers, often grasping far more clearly that the needed synthesis is of – & in – Life! – & that thus natural sciences are here incorporating the Physical, while on the opposite line the bio-physicists & bio-chemists even, never get properly to biology at all! *Similarly too for Letters & Humanities.* Thus I am trying to put clearly, & practically too, what Comte & Spencer, Ward & Fiske wished to do; but without giving concrete example.

Again a *School* of great promise is opening at Château d'Assas and so a further co-operation; and an Outlook Tower is building on my old mill at Domme (Dordogne) by my old Comrade of nearly 40 years Paul Reclus. A Swiss group with Germans & Austrians (botanists, etc.) is opening also on 1 June beside us & the Palestinians are reviving also. It only needs some live Americans, to pick up the "Collège des Americains" beside us! & so on.

As to the present rabid excess of Nationalism everywhere, what more amusing solution than this house, with St. Andrews Cross & Ruddy Lion,

Thistles, etc., as architectural outside decoration – & International folk & thought inside! So this is *not one bit* like the rich new foundations – whether in east or West, but the definite, & concentrated endeavour to *work out more & more* of *the ideas & ideals of University progress*, for Juniors & for Seniors than heretofore: as from Examination replaced by Estimation, analytic & *post-mortem* studies re-related to Life, & so on: & at every point with attack on the verbalistic empaperment of mis-instruction, to replacing this – by realistic observant studies of Life organic & social, & as in Evolution: so in short from *3 R's* to *3H's, Hearty*, & *Handy* re-Education of *Head*. And so producing at least a few Socians, who would otherwise have remained too simply Individualists (much as of old my University Hall at Edinburgh helped Thomson, Branford, and many more to their vital concerns – as Branford did you!). Thus our one bright painter girl I've got starting with Hospital Decoration – & infecting our School of Art – & by & by others: at every point in short the endeavour of the leaven & mustard-seed & so far of the Kingdom coming on Earth as in the Ideal. *I can't accept, in short, the idea of sociological writing unassociated with action* – and the greatest of writers, the dramatists – have written *for the stage*! I submit it is for us to go beyond them – by writing for *Eutopia* – & as *Eutopitects*! Here are the beginnings of the Cité Universitaire on $\frac{1}{60}$ of scale of that of Paris. Yet many times more educationally ambitious – since they remain simple *residences* still – but here with *Epidauros, Theleme*, etc., on our plans as well, as this Experimental College *grows*.

So next time you are good enough to mention me, pray give some idea of this policy, of *Vivendo discimus*!

I have written about gardens, and it passed: but this fullest of Flower-Paradises of the seasons will survive me – & always be more exuberant with time – Life more abundantly – Life in Evolution – thus in fact and deed, not merely words! Our improving village points are already showing results – & by & by more – first towards the many of this department & much as regional stimulus more & more widely (with Reclus as above said, our Outlook Tower at Domme).

Thus again our Outlook Tower at Edinburgh has to stage a *Scott Masque* for his centenary, 1932, as lately for the Edinburgh's six-centenary of Bruce's charter.

This is more vivid (and even of large a circulation) than a new paper or book on Scott! (Yet I do want to write too! and am trying.)

I could *talk* one of these, rapidly, & many more books & papers could I find any one even as good as Miss Defries, though you saw her limitations! – but the prevalent hard-boiling of schools & Colleges has given me no simple capable secretary these many years – else I could talk out books – had I again a person of understanding, such as one long ago for Dumfermline book & *Evolution of Cities*. You remember the *other task* of Paul Kellogg: machine reporter of my lectures: well, they are all more or less like that, & in Branford's experience as well as mine!

I have at last one student botanist to take over my large accumulations & reinterpretations which would have taken ten years to write & to help make small new type Botanical Garden. But no one yet for any field of Sociology!

Nor for my book begun 52 years ago – on *Universities Past Present & Possible*: though here far more material than Flexner, etc., much less merely historic writers!² Yet to be at this and in the *working it out* & in this social & educational laboratory (this synthetic endeavour of University progress in practice) is worth the while of any younger man – even yourself! for your proposed month or so! And this would sell all right on both sides of the water (& with translations too!). So do think of it.

Here only for a month longer. Then a strenuous time in London till near end July – when the little *Sociology* is due. Then a fortnight or more in Edinburgh. Then back in London for September Centenary of British Association, at which something can be done. So if you can come here in *October or later*, till following June 15th 1932 or so (for now I doubt my health would allow visit to U.S.A. next spring & sumer) – we could have some peace! Why not that Universities book together – beyond goose step, a Pallas owl-flight! With this we could reach many universities! – & *rouse* the "University Militants," & some of their old corporations [...] too. *There then is a business proposition! Why not consider it?*

Yours

P.G.

Best regards to wee laddie, whom I hope to see over here. Mrs. Mumford too of course, & you too, will thrive in this place – of old the most famous health-resort of Europe – & going stronger than ever, so towards renewing that, as it really deserves to do, & is beginning now that the Riviera is so overcrowded – & with so much of commonplace luxury & pleasure hunting!

1. Geddes and J.A. Thomson, *Life: Outlines of General Biology*, 2 vols. (1931).   2.   Abraham Flexner (1866–1959), author of *The American College* (1908); affiliated with the Rockefeller Foundation where he directed support of medical education.

Is there any prospect of your again coming over to Europe? If so, I very much hope you will visit us at Montpellier. For there I am really getting towards beginnings of the *University Militant*, and it would profit by your encouragement, and *counsels*. Beside our Scots College, we have now an Indian College building: an American College is on plan, and even others; while the neighbouring French College, on a larger scale, for 120 students, is also filling up.

After planning (in part or whole) for *14* Universities, of East and West, I have settled this last of my initiatives at the Montpellier University, not only because of its great historic past – as from the very foundations of modern medicine and natural science teaching – whence, for instance, the Anatomy, the Botany, and Botanic Garden of *every other medical school anywhere*! And much else as well; for thence also, for instance, not only Rabelais, but Sydenham – the founder of medicine in England – as Sibbald for Scotland. And so on to Pinel, of whose recent monument at Edinburgh the enclosed booklet (of my Goodwill Series) has its verse of commemoration.²

All this since occupying the nearest point of contact between Mediterranean and Northern Universities: whence active sparks, and even currents;

[1931]¹

[Ts UP]

as from Petrarch onwards. It is thus also the most illuminating school I can find for *History*.

I am striving to make it a link with the East also: whence our Indian College. Towards this our University Geographer has produced the best of European books on India so far: and Sylvain Levi, at the Sorbonne, also helps us.[3]

Similarly for Education: in touch with the admirable Institutes at Geneva, Brussels, and others.

In many other ways I might explain its values, and our incipient endeavours to carry these further. Thus, returning to my own lifelong interest in Botany, we have now so eminent a leader in our Laboratory at College, that, in the month before I left, we had visits from the Professors of Botany even of Berlin and other German Universities (not usually very sympathetic); as also those from Prague and Amsterdam, Jerusalem and Chicago; and all promising to send us research students. Their work starts with the thorough Ecology of our flower-rich heath, and thence extends onwards; at length to the Geo-Botanic Survey of the Mediterranean Region, for Alps and Pyrenees to the African desert; by and by from Tangier to Palestine. Is not that a good example of reunion of specialism and generalisation?

The same, too, for other subjects; as for Prehistoric Archaeology: and then History – from Gauls and Romans onwards to to-day; and all first traced in direct open-air Surveys – for which our region is the most favourable of any, since e.g. we have better monuments of old Rome than those which remain in Rome herself! And when it comes to learning and reading History, our students are not, like most others elsewhere, confined between Lecture-Room, Library, and lodging, but feel this more real to themselves in our fine old country-house, the Château d'Assas. For this, with its village, affords a microcosm of history, since Post-Roman days, with origin in the Dark Ages. They have thus before their eyes the ramparts of Medieval times – up to the Renaissance, and its decline, and with its 18th Century revival, and ruin. And of that we have visible reminders, alike of Marie Antoinette, with her Trianon fashion, and its damages in the "Terror." And so on; to our own time, with its problems as of the re-adjustment of its past, with the present, and so outwards towards a better future – for Village and for College together – and so with influence and example beyond.

Again, everybody is familiar with the phrase "Peripatetic Philosophy." But this merely recalls how the Greeks talked philosophy, walking up and down under their plane-trees – much as Indians – like Tagore and Bose – still teach, sitting in the shade. But through our large (six acre) garden, I have a scheme of peripatetic teaching, by laying out paths, and view-points, vividly expressive of the changing philosophies through history; as from early thought to Greek, and at its highest, as even to Gods and to Muses' inspirations in so many fields these opened to them – and they have handed on to us. Thence towards Medieval and Renaissance philosophies. Beyond these to modern ones; as (for instance) Kant and Hegel, each made clearer than he was, or now usually is. And so on, as to Comte and Herbert Spencer etc.: to Le Play and to Bergson and so on. So now onwards again, towards such fuller syntheses as we can see opening, and try in our turn to advance, yet thus utilising all of preceding thought we can. All this, of course, is as yet far from complete: but it is already helpful, and I am always trying to develop it further.

Again, we have no Examinations, in the usual sense, but Estimation; as from term's work, character, etc.: and with increasing guidance towards finding of career, and preparation for it. With all this the best of psychology and pedagogy we can muster, again with help from University and abroad.

In such ways we begin to do better for junior students than usually elsewhere and heretofore: and we thus prepare them to profit more fully by their later studies in their Home Universities. Furthermore, as to senior students, since our beginnings, over six years ago, we have had a series of Theses for the Doctorate, in Letters or Science. And for both juniors and seniors, we try to help their specialised preparation, to keep in touch with general ideas, synthetic, evolutionary, and thus social also. In this way, too, we increasingly attract University Lecturers and Professors, as often an American on his sabbatical year.

Furthermore I am now fortunate in no longer being alone – as "a one man show" (and an old one) but am geting a staff of first-rate Fellows and Tutors at their age of retirement – 55 and onwards – and these without the annual salary for capital foundation, which is elsewhere necessary; because our optimum climate, which has made Montpellier for so many centuries a metropolis for health-seekers, is also the best of places for us senescents; since at once offering good conditions for the completion of one's life-work, and continued interest in life and leisure, as in helping the young folks of College; all with the resources and cultivated society of University and City.

1. This undated, typed copy follows Geddes' letter of 17 May 1931 in the UP file. 2. Distinguished physicians who had studied at Montpellier: François Rabelais (1494–1553), classic satirist; Thomas Sydenham (1624–1689), known as the "English Hippocrates"; Philippe Pinel (1745–1826), pioneered humane treatment for the insane; Robert Sibbald (1641–1722), who became President of the Edinburgh Royal College of Surgeons, studied medicine at Paris and Angers. 3. Sylvain Levi (1863–1935), author of *L'Indie et le monde* and other works on India.

Dear Master,

Amenia, New York

27 June 1931

[Ts NLS]

We came up here six weeks ago, and will remain here till the end of September; although, if you mislay this letter, the city address is always safe. In the act of finishing up my book on *The Brown Decades*, a study of the landscape, architecture, and painting in America from 1865 onward, I exhausted by doing no other work the money I had hoped to use for a European trip: so I am now definitely applying for a fellowship for next spring, and in the meanwhile am busily gathering lecture engagements for the winter. Last year, we were able to do little to either the land or the farmhouse: but this spring we threw ourselves into it, and have painted and carpentered and gardened to our heart's content, and the place's great improvement. We have put in a large vegetable garden and have planted an assortment of annuals: waiting until we get the feel of the place a little more deeply before we plant any new perennials: fortunately, there are rose bushes, lilacs, and lilies of the valley left over from the efforts of previous occupants. Your little namesake has a garden of his own, of course: and not merely weeds it and waters it thoroughly, but works on it before breakfast without a hint from us. He is a good fisherman, is quick to distinguish birds, and not merely knows most of the common plants but can distinguish them when they are only an inch or so above the ground. He may be a better credit to you than his father!

Your opus came last week, and I have been reading it steadily since, with the same thrill and delight that I remember originally when I came upon your part of the little *Evolution*. It is a masterly work, and I can't begin to congratulate you sufficiently upon it; for I see your hand all through it, and am particularly glad that you have definitely put all the philosophic part of it in order. I had scarcely dared to hope for so much. As merely a physical feat it would be a credit to both Thomson and you: but as an intellectual feat, it is little less than a miracle: everywhere, the knowledge is seasoned with your common experience of life, and the experience is widened into a philosophic vision. If the *Outlines of Sociology* is even half as good it will be a landmark. I am trying to arrange a double review of it for the *New Republic*, the biological portion to be handled by a biologist like H.S. Jennings or Wheeler, and the philosophical part by myself.[1] I have still to hear from the editors about this. I shall hope to have absorbed it well, by the time we meet.

I was glad to read your obituary on Branford: there was much in it that I had never known, for he was a very reticent man. Your verses to him are, I think, the best you have done: how warmly he would have felt them![2] As for your ingenious insertions of the sections of my letter into the article, I have nothing but praise: it was very skillfully done – praise from Sir Hubert! Farquharson writes me that nothing can be done about Victor's works until the court of Chancery hands down a decision; and perhaps it is just as well, then, that no New York publisher prematurely accepted the works. The book market is so depressed here that I have told Farquharson that it would probably be cheaper to print a uniform edition in England, in the hope that one or two of the books might, with a good introduction, be taken in sheets in the United States, and find a public for themselves here. Harry Elmer Barnes asked me last winter to do an article on Branford and yourself for a book he was editing on contemporary sociology: I have not been able to do it yet and if you have any data, therefore, which I may not have in my possession, I wish you would send them along. When did you take over Sociology? What was your and Branfod's exact share in founding the Sociological Society? What were your contacts, if any, with Durkheim, De Tourville, Demolins, Elisée Reclus, Fouillée?[3]

Next week I go to the University of Virginia to attend a conference on Regionalism. The Southerners, particularly the younger intellectuals, have lately become conscious of themselves as the repositories of the agricultural and regional traditions of the country: a group of them recently published a book, *I'll Take My Stand*, to uphold these traditions against the financial and mechanical standardization of the rest of the country; and though they tend to be slightly reactionary, still dreaming of the past instead of shaping a more integrated future, they may prove valuable allies. John Gould Fletcher, whom you may have met in London, will be there: our New York group, Stein, Wright, MacKaye, are engineering it.[4] We shall miss you. In Oklahoma a group at the university have published these last two years a Regional Miscellany, chiefly literary, called *Folk-say*: an interesting straw in the wind. The writer of the article on Meiklejohn's college, which I enclose, is another one of the young people who have studied you well: he did a study of the French bastides last summer, but unfortunately missed you in London, though he saw Farquharson. Your disciples are coming along now rapidly; the younger generation that is, those now under twenty-five, are much more concrete-minded than their elders were: architecture begins to share interest with literature in the critical journals! Miss Bauer, the girl who is going to write a history of the House, is a very adept pupil: she has gotten much out of your *Biology*. These young people are more ready for Graphics than you perhaps realize.

I am asking the publisher to send you a book called *Living Philosophies*: I happen to be one of the miscellaneous rabble of contributors.[5] . . . I trust that your health improves and that your summer will not tax you too much.

Affectionately,

Lewis

1. Herbert S. Jennings (1868–1947), author of *The Biological Basis of Human Nature* (1930). William M. Wheeler (1865–1937), author of *Foibles of Insects and Men* (1928). 2. "Obituary Notice," *Sociological Review*, 23 (January–April 1931): 2–5. 3. Alfred Fouilée (1838–1912), author of *La science sociale contemporaine* (1880). 4. John Gould Fletcher (1886–1950), poet, member of the Agrarians, and contributor to *I'll Take My Stand* (1930). 5. Mumford contributed the essay "What I Believe," *Living Philosophies* (New York: Simon and Schuster, 1931), 205–219.

Dear Mumford,

I hear with great regret of the difficulties at Wisconsin of President Glen Frank & Professor Meiklejohn, and their Experimental College – but I am not surprised. Neither the governing Bodies of our English-speaking Universities – nor the professors who have been trained in their respective specialisms, apart from such synthetic endeavours, & in subjection also, for very bread, to silence if not convinced conformity – are ready for experimental freedom – and still less for (Pro-)Synthetic endeavours of educational adequacy, as befits the times, and as these Wisconsin endeavours are pioneering.

Here at Montpellier however, we have, as far as I know, the only such free grouping as yet – save *at* the University of Montpellier but not *of* it (a position not permitted by French law since the Revolution) & thus freely & fully utilising all it can do for us (and that is much); yet also completely independent of its formal *authority* – which means not simply that of Rector and Deans of Faculty, but that of their subordination to the Ministry of Public Instruction at Paris, which keeps strict authority over all, Paris included. But our foundation stone is marked *"Rector me posuit"*[1] – & we have all the help & counsel we could wish, since recognized as loyal and helpful to the University, & even as supplementing its resources & lecturing, on our small scale, on the same principle as Oxford & Cambridge Colleges do for their Universities, with their larger means. Thus we are already housing (as per enclosed postcard) the *"Institut Géo-Botanique Méditerranéenne et Alpine"* (which had not room at the University Institute of Botany), and also the *"Survey" de Montpellier et sa Région*, with the *"Institute de Sociologie de Montpellier"* (of which society I am president, as at London & at Edinburgh).

Collège des Ecossais
Plan des Quatre Seigneurs
Montpellier (Hérault)
(but at Outlook Tower Edinburgh in August – & 3 Netherton Grove Chelsea SW10 London in September)

22 July 1931

[Ms NLS; incomplete and unposted]

So too our archeology & History with Geography in Dordogne & here. Similarly too for the further Synthetic studies – Outlines of Sciences & Philosophies – in literally peripatetic & graphic presentments. And so on, with concerns also for social students & workers – and with Occupational Courses, specially for juniors, seeking to find their profession – yet also of use to all, since uniting practice with theory on our principle of learning by living (Vivendo discimus).

Despite its nationalistic name, this College is peculiarly international – in fact is nucleus of Cité Universitaire Internationale; so it has already an Indian College partly built, a Swiss (& Austro-German) group beginning with October, & others on plan & in project; while at a little distance a big French house is just finished for 120 students, & with plans for as many more next year if possible. Paris, with its 15000 foreign students (!) now realizes it has gone too far, & is thinking of insisting on previous stay in one of the provincial universities (of which no less than 17) – & of which Montpellier by common consent as well as long traditions on the whole stands first. (For though Strasbourg is as yet in some respects more fully equipped, foreign students do not understand Alsatian German, nor like Alsatian French.)

Now as to the American College here – this was planned five years ago with one of your "Sabbatical Year" visitors, whom we have occasionally. (This year the Professor of French from Colgate University has been with us, & done a good thesis for Dr. of Letters.) This American College will be an expensive affair – for 50 or more residents: but in these hard times it need not be proposed, but can wait till they mend. My point however is that just as each & all of the above-mentioned Colleges, going or beginning, have sprung from this one, so we can now also invite American students, graduates & professors, & put them up here, and near by: so with no outlay for buildings, etc. The only expense (beyond that of journeys and pocket-money for excursions, etc.) is our very modest charge for academic year (of £110, *inclusive* of university fees & tuition!). A resident of two years back – since at Cambridge (England) & graduating also there, has returned here, saying he spent just four times our amount there, but got less for it; & though his £440 was high expenditure, we are certainly far less expensive than London or even Scotland, let alone Cambridge & Oxford, yet with longer session – 8 months instead of 6 – & in various respects even more facilities.

The real question however is of course the educational one – at once *free* & *synthetic*. You will see on reflection how the Geologic–Botanic Survey organizes the study of the synthetic & the geologic & the physico-chemical conditions of life more intensively than ever – and similarly how our *Surveys* (in Dordogne in vacations, & at Montpellier throughout the academic year) similarly unify all the "Humanities." Our Latin (& even Greek) are more vital with the finest Greco-Roman temple in existence beside us at Nimes; & the amphi-theatres of Nimes & Arles are far better preserved than is the largest Colosseum: while the first main Roman road – "*Via Domitiana*," for the north & England – soon forking off the *Via Herculeana* to Spain (Hannibal's road too before that) runs just below us, while its predecessors – the Bronze Age Road – from Philista (Palestine) to Cornwall, runs along the Ridgeway of our grounds here. (In such ways how Prof. Meiklejohn would be in his element – for here Roman movements & traditions around are essentially *Greco*-Roman.)

There then in brief is my proposal. Let us start the nucleus-group of the American College here – since not only at this foremost meeting-point of Mediterranean & Northern Culture, but that now beginning to serve also (as indeed likewise from antiquity) for meeting of West with east, both near and far – Greek & Palestinian, Indian too, (& by & by Chinese as well).

This self-governing American group, in collaboration with us also as free from control of any external governing body, such as we have in British Universities – to our general inhibition, as with you for yours – can thus rapidly work up, & with us as far as they please, these needed freely & experimentally vital & synthetic endeavours, re-educational accordingly.

So I ask speedy co-operation – if possible by 1 October (so as to give time for tutorial polishing up of French before University opens 1 November). Near though this time is, a few to begin with may surely be found whose numbers might be recruited for next semester, beginning 1 March 1932 – and still more at Easter; and the following October & November.

I should perhaps repeat that the teaching of "Nature & Civilisation" as fundamental preliminaries and which so many Universities & colleges with you are adopting, are in our two above-named Institutes of (1) Botanical & (2) Social Surveys carried out in regional detail as well as in general outlines. The former (1) is actively co-ordinating our physical & natural sciences studies in its laboratory: & the latter (2) our prehistoric & historic ones are similarly going through pre-Roman, Graeco-Roman, Barbarian to Medieval, & Revolution to Modern times – & thus even something of the incipient ones towards which we are all feeling & searching our way.

Furthermore, the fine endeavour of my Edinburgh friend, Professor Stoughton Holbourn, along with his President at Carleton College, Minnesota – of co-ordinating University studies with those of Fine Arts of Music & drama are also beginning with us: and on the educational basis of Occupational opportunities from the outset. In Montpellier, of course as elsewhere, the University, the Ecole des Beaux Arts, & the Conservatoire de Musique are all independent as elsewhere, & with no students in common, as similarly the schools of Agriculture & Horticulture, and of Commerce: whereas we have, in germ at least, the coordination of them all – so something of Synergy as well as of Synthesis.

Pray ask any questions you think fit: but meantime I should be glad of such publicity as you can help us to, so as to make a beginning of this Collège des Americains without longer delay.

1. The rector laid me.

Dear Master,

Amenia, New York

6 September 1931

[Ts NLS]

I trust the summer has gone well with you: the last letter received was dated May: but I remember that you had planned a busy summer, and I hope that your silence has only been an indication of your absorption and busy-ness. I have, as I wrote you last July, been reading the opus on *Life* with great joy and exhilaration and pride: some of which I hope has been comunicated to the review which I am herewith sending you.[1] It was an almost impossible book to review, despite the fact that I gave myself more than a month to think it over after I had finished with it; and I shall be mightily surprised if the ordinary reviewers do it anything like justice. Harper's is selling it here for the fabulously low price of eight dollars for the two volumes; so it looks as if they were counting on a wide sale. In July I attended the conference on Regionalism we had arranged at the University of Virginia: when I get back to the city, I hope to be able to send you copies of the papers that were read, although, of course, the best of the conference was not contained in the speeches that were written in advance. The southerners are all regionalists and ruralists at heart: so we had a sympathetic audience, and although the intolerable heat enervated us all, and although we needed another week, really, to bring matters even intellectually to a head, the conference at least marked a beginning. An American poet, John Gould Fletcher, who lives in London, attended the conference, and I trust you have seen him since he returned to England. He has a very fine mind, and is a natural ally; and he wished to consult with you further about the educational theories you advance in *Life*. Since coming back from Virginia I have been working mightily on a new book which at last begins to take form: it is a sort of conspectus of modern civilization and its possibilities – social, esthetic, moral – beginning with the part played by the machine and carrying the integration upward through buildings, cities, regions, and civilization as a whole.[2] I shall have a semi-final draft to show you next spring, I think; and after observation and reflection, verification and correction through European travel next year, I hope to return to America and ship the mss. finally into shape. The life up here is very healthy: we have lived off our own garden since the end of June, and we plan to have an even richer one, both floral and vegetable, next year. Little Geddes is a good gardener: his first move after breakfast is inspection of the vegetables in his own plot, and he hoed it and weeded it assiduously. He has a sharp eye, and can identify the plants he knows in even their young or withered forms. The horsetails and the clubfoot moss grow in different places up here: the first is familiar to him, but the second we had never discovered till the other day. Geddes looked at the segmented stem and said: This is just like the horsetail! In the lake, he is a little porpoise: he taught himself this summer to swim, dive, and float, and he has no more fear of that element than he has of the land. In addition, he is a passionate fisherman and has a pretty good anatomical knowledge of the fish he catches. In short he is in many ways a credit to you as well as to his parents! We are returning to the city at the end of September, provided that the infantile paralysis epidemic has abated then. Our address remains: 4002 Locust Street, Long Island City. I shall send you a copy of *The Brown Decades* at the end of the month when it will appear. With warm greetings from us all –

Yours gratefully,

Lewis

1. Mumford's review of *Life: Outlines of General Biology* appeared in *New Republic*, 68 (28 October 1931): 303–304.   2. Eventually published as *Technics and Civilization* (1934), the first volume in the Renewal of Life series.

Dear Master,

I am overwhelmed with work at the moment; but I hope to have a breathing space next week & I will write you there. The enclosed criticism – by a young philosopher & historian of science – may interest you.[1] It is even better than my review, and none the worse for being completely disinterested!

Warmly,

Lewis

4002 Locust Street
Long Island City, New York

11 November 1931

[Ms Strathclyde]

1. Benjamin Ginzberg, "Life and Continuity" (review of *Life: Outlines of General Biology*), *Nation*, 133 (11 November 1931): 523–524. Ginzberg writes: "Indeed, as far as the reviewer is concerned, he frankly confesses that he finds it difficult to overpraise the book, so ideal are both its conception and its execution."

Lewis Mumford AND *Patrick Geddes*

THE CORRESPONDENCE

1932

Dear Mumford – (no – Lewis my Son!)

5 January 1932

Still 3 N. Grove
S.W.10
(& here if you reply
more or less soon –
otherwise Collège
d'Ecossais,
Montpellier, France
within this month)

[Ms UP; copy typed by
Sophia Mumford at Oslo]

Herewith a bit of personal news – which has not touched my pulse one bit[1] – since life-habit of quiet meditation on one side & of quiet experimental practical learning on other has had *no touch of interest in popular appreciation*! (Alas, no expectancy left, but government Honours List – but herewith something.) & good or e.g. hardly or not at all bad reviews, not even the *one* venomously malignant-impish one, of *Life* did not turn a hair – so that I could ask him to lunch when I met him at History of Science congress, though there he again made more (*insulting*) refusal, taken without being touched – but made me really angry by next insulting old Haldane (& in *unique canaille fashion*!) – I could not but see to good reviews of his recent book by Thomson since it has points of (unexpected) merit & value![2] This story would amuse you, but already has more space than it matters – though helpful too towards speculation in psychologies.

Returning to point of this "honour" – I declined from old friend Lord Pentland some 20 years ago when he was member ("Secretary") for Scotland, on democratic grounds to permit thinking peace, & avoidance of "Society," etc. (and again would have been offered it in India but for personal animosity of last Governor over his detection & exposure, which though not public could not but reach him). But now this time & from Ramsay Macdonald – after advice from Thomson & Hartog – who replied that it did help to increase their usefulness and influence with wider public, & for help of good movements & organizations too.[3]

So already I find it! What shoals of congratulatory letters – what photographers! & next interviewers! There is no doubt that the public are more impressed by honours than I knew – & that the proportion of honours obviously not politically earned or purchased are surprisingly approved by intelligent public – & of *great value even in bluffing the Philistines*! who have despised us all as cranks, but have now to accept us! Thus a real help for instance towards getting them to see after public approval that they were mistaken in thinking us merely "dreamers," etc. One can now much more readily get it understood that Sociology is not crass socialism nor Education, the mere misinstruction in parrot-memory, through the three R's, of print-habit, scribble-habit and memory-habit! – but that there may be something in the three H's after all – as of *true memory* of *vital impressions* & developments as "Mother of the Muses" – so of Heart & Hand to Head (Vivendo discimus!).

The above too as suggestion to yourself when you are ready – as already so largely – to use your publicity towards *Purpose*.

You know something of how it injured Branford in the City to be thought "man with red tie" (though he never had one) though Balfour our Sociological Society President was also Conservative Member for the City! Surprising idiots of prejudice – as in Oxford, etc., too commonly still!

I hope I may last 2–3 years more, though health[4] found by last able physician as more, & deeper seated, than previous also able ones had found! And of course I've only gone to them from increasing weakness, only too fully felt & as hindrance. *Per contra*, with best past year of brain freshness & fertility (*approfondissements*).[5]

Yet clarity too. Pro-syntheses & Sub-synthesis clearer ("Day by day to her darlings –"), but never quite clear – if that can ever be! Yet *practical* progress in applications is now needed. Thus for simple single instance, you know of my old interest in Student *Hostels* – in Edinburgh since 1886, & in Chelsea with Crosby Hall, etc. Well, I've a scheme for London, with which this will help. In *Synthetic* side, I told you of Exhibition of Graphs for British Association, & in support of Smuts' appeal as President, for more of synthesis co-ordinating to specialisms. There was no room to continue this at School of Economics (itself *too arid*). (Sir W. Bragg – a great fellow![6] – would have given me room at *Royal Institute* (the foremost scientific lighthouse of Britain within its physical range) but had no room, every one already with one or more of his researchers. University London had *no* room, nor could find it in any College! All full up!) But fortunately I have made great friends with *Adler Society* – the most active I have ever found, in London or elsewhere. Look into them & their developments! One of these, the nascent New Europe group – insisted on my Presidency – but I reduce that (and strengthen it!) to Editorship! For their part, they've not only taken in my Exhibition for storage – but out again, into a good room up the street close by – so nearer University College, etc.

Thus an informal yet real *germ of Synthetic Institute* & etc., Outlook Tower, Sociological Society, & *LePlay Society* (now separated from Socio-logical Committee) Miss Tatton from Farquharson – each free now for British & foreign Survey (& much better so). And all close to University College, Indian Students, Public Health Institute, Dramatic & Visual Arts Groups, British Museum, etc., etc. And in fact the *incipient needed University Department* too.

(This may encourage you to the like in the range of your University *Influence*.)

Did I tell you a Students' organization in Philadelphia cabled me last spring to come over to plan Outlook Tower for their building? Paul Reclus & I are making a tiny one at Domme – as Valley Type!

Cordial wishes to Mrs. Mumford and to Geddes and to yourself in '32.

Yours,

P.G.

P.S. A *business point – asking your help*! This fall of exchange wipes out a third of our capital – mine all in France, & much of my wife's too, and so of our Colleges' *incomes* too – & *personal* incomes as well! It was hard before, but touch & go with ruin now for College schemes; fewer students, etc., etc.

But American Exchange yields more! So do you think you could place my articles for me in *U.S.* papers & periodicals? if I manage to write a few on return to Montpellier? (If so, suggestions welcomed! And can you suggest any outlet for more of these College circulars – etc. – towards the American College. If so, I can send them.)

This occasion of recent *Life* Book well reviewed, & now of title – for my many congratulations are of satisfaction with the unusual recognition of *ideas* – might let you say a word in one of the good papers – so helping to introduce me to your public? P.G.

Did I send you my M.K. Fraser booklet in *completed* Edition *64 pages?*
(Stir up Scots to it – when her Iona re-burial is mentioned, if you see it.)

1. Geddes was awarded a knighthood in the 1932 New Year's Honours List. 2. The reference is probably to Julian Huxley (b. 1887), biologist and prolific author, who reviewed *Life: Outlines of General Biology* in *Discovery*, 12 (July 1931): 207–208. John Scott Haldane (1860–1936), physiologist and philosopher whom Geddes had undoubtedly known at Edinburgh University, author of *The Sciences and Philosophy* (1929). "Canaille": vulgar, scoundrel-like. 3. John Sinclair, Baron Pentland (1860–1925) served as Secretary for Scotland (1905–1912) and governor of Madras (1912–1919); in 1914 he invited Geddes to bring the Cities Exhibition to India. James Ramsay Macdonald (1866–1937), Labour party leader who had a particular interest in India; author of *The Government of India* (1919). 4. Mumford's annotation on Sophia Mumford's typed copy of the handwritten letter: "sic!" 5. "Appro-fondissements": deepening through research. 6. Sir William Bragg (1862–1942), physicist, Nobel Prize laureate (1915).

Dear Sir Patrick,

4002 Locust Street
Long Island City

16 February 1932

[Ms NLS]

Our warmest congratulations: it sounds almost as inevitable to say Sir Patrick Geddes as to say Sir Patrick Spens,[1] and our republican hearts warm to John Bull for honoring himself by bestowing the title on you. Even the communists have found it necessary to institute the Order of Lenin. The recognition of merit has a different effect, as you say, from the recognition of money or political power; and I am sure that it will widen the sphere of your work in England and make it easier for your conquests. Your letter crossed one I wrote to you from Dartmouth, and I am sending this one promptly in order to get from you your probable itinerary in April, May, and June. I hope to get abroad some time during that period to complete my preparations for my book; and if you would indicate the time and place in which it would be most convenient for you to see me, I shall try to do so, for at least a week. Economically, we are now as badly prostrated as Europe: perhaps worse than anywhere else save Germany. This makes it hard to place articles here: most of the magazines have cut down the number of pages, and are using up old stores of manuscript: I myself have been unable to write for either *Scribner's* or *Harper's* for a whole year. I will do my best to place anything you may write; but I cannot be sanguine about its prospects. Had it not been for my university work this year and my outside lectures, I should have had a pretty hard time of it. In New York, fortunately, the winter has been exceptionally mild: our Japanese honeysuckle leafed freshly in November and has stayed green and kept growing ever since: the irises and tulips are already inches above the ground – alas! for the latter, they will be frostbitten before spring sets in – and the hedges have not lost all their leaves, while the lilac bush is swelling and ready to leaf! Later in the week I shall send you a catalog of an Exhibition of Modern architecture we have been showing here. In the housing section, we showed photographs of an East side lower class slum and a Park Avenue upper class slum: and drew the parallels between the two. I wish I could send you a photograph of that particular chart: it was very good. At the top of it was a quotation from you: "Slum, semi-slum, and super-slum, to this has come the evolution of cities."

Sophie joins me in warm greetings to you.

Affectionately,

Lewis

1. Eponymous hero of an early Scottish ballad.

Montpellier

2 April 1932[1]

[Ms UP]

Dear Lewis,

I have last got home here from London to recruit – & so am at length beginning to do so. What a difference from the long depression (yet overwork) of the London winter – the last, I hope, I shall ever have to spend there – I feel that another would kill me; but I may last a few (a very few) more years here – with only visits to London & Edinburgh in pleasant summers to keep my work alive there.

Here however *only till end May* – then Geneva for ten days to renew & extend contacts with its many activities. Then further touch with Synthetists at Paris & old friend Otlet at Brussels. Could we travel from here together? Also to similar touches & developments in London? (A. Farquharson, etc., have *paralyzed* Sociological Institute; but your visit (*after being here* if possible) may help them.)

The new *Le Play Society* around Miss Tatton (& for which I am President) is doing better. Also for *Adler Society* & its New Europe Group. So try reserve your London visit till after here in May or June – when I'll be there too – in later portion – whence on to Edinburgh & back to Nice – Edinburgh Congress July 25–Aug. 12, then Montpellier again.

Now your *Architectural Booklet* – I don't quite accept that booming of particular architects – & especially without reference to real origins! If you could come or go by *Glasgow* – & see Charles Macintosh's *Glasgow School of Art*, you'd have this "New" movement thereafter in better perspective than as of all from Le Corbusier, etc.[2] (I'll value your criticism of this house – as free and circular – & as going a step further, as I believe you'll *come* to admit.) Still more my proposed *schemes for revalorizing* the existing London quarters, from within (& with aid of small *University Hostels* & local cultural life e.g. of Chelsea, Bloomsbury, etc., to begin with – but increasingly adaptable to other quarters – & in fact to paleotechnic areas of towns generally – why not American too?).

Here in this scheme I claim to have a real advance – & of vast potentialities – & with real contribution to re-employment – from bricklayers to architects & artists (co-operation of writers too). The like also in outlying suburbs – & out into villages – & Regions.

In short a far further development beyond my old *Cities in Evolution*, but too long for letter – & needing you to look into our plans – now elaborating with Mears & others. (You will then at once see developments possible in U.S. cities.)

Correspondingly for *Synthetics* – developed from outlines & graphs in *Sociological Review* – 1929–30–31 – & with problem of *Crisis – Factors & Treatments*: (cf. *arbor saeculorum*[3] – as on covers of *Making of the Future*

Series – & window in Outlook Tower since my Summer Course of 1895 or 6 – put in by its students, & anticipating both development of Crisis then manifest – & something of its Solution too). Hence towards *Synergetics*.

You have never answered my demands for criticism of these papers but I take it you see something in them. Also in *Science & Sanctity*. I look for your help in continuing Victor Branford's vitally synthetic thought (in its Re-Religious way further advanced than mine).

Here too you will find endeavours towards incorporating this – and all else of best I can find – and no doubt get helpful stimulus & contributions from you. (I'm more active – despite physical ebb to weakness – than when I was with you – eight years ago & when I was merely observing & speculating & striving towards Synthesis of IX–9 – as per *Sociological Review* – which indeed first had expression in a Cornell Lecture after our New York discussions.)

I am very anxious for your visit here in May – for either I am quite mad, or I have got at last this College of the Future into beginnings of working Order – & for sciences & humanities appreciably advanced, with sociology & psychology unifying these. So remember I'm rising 78 (in Oct.) & have none like you, or but you, to be my heir – as a bright young Parsi botanist hopes to do for my 40 years accumulation of botany carried further, etc., than in *Life Book*.

So you must take over much of my further Sociology & Education, etc.

Yours, with regards to each of triad & their friends also,

P.G.

1. Mumford's annotatation dated April 1969: "His last letter to me." Geddes died on 16 April 1932.   2. Charles Rennie Mackintosh (1888–1928), architect who designed the Glasgow Art School in the art nouveau style. Le Corbusier, pseudonym of Charles Jeanneret (1887–1965), prominent French architect influential on the development of the International Style in modern architecture.   3. The "tree of life" was a symbolic diagram designed by Geddes (see reproduction in Meller 313).

APPENDIX 1:
NOTE WRITTEN AFTER VISIT
WITH GEDDES IN EDINBURGH

Geddes drives me to tears; almost he does. I have been with him since Monday; and though I anticipated the worst, the first night and day were happy; he took me about the city, showed me the hundred improvements that he had made or initiated; waste spaces become gardens, courts tidied, tenements renovated, students' hostels built, splashes of color introduced by red blinds on windows; fountains and urns designed: a great achievement in itself, all these things. On the second day, Mabel Barker came, and some other visitors arrived on the same day, and then we were back again in the old cruel mess and chaos: engagements broken, time wasted on trivial idiots, and, in the interim, an increasing volume of anecdote, suggestion, and diagrammatic soliloquy. The weakness and the strength, the steadfastness and the impatience, the effacing humility and the ruthless arrogance of this great man emerged from all this with an effect upon me that is still mingled. He is perfectly loveable in his human moments; in fact he is enchanting; a portrait of him at thirty, a bad portrait but a sufficing one, showing a black-bearded, rather chubby man, with red cheeks almost choked me with emotion: here was a man I might have worked with and merged myself with. But what can I do with the man whose muffled soliloquy spreads over hours, the man who is caught in his "thinking machines," as one who had invented decimal notation might perhaps spend his life by counting all possible objects in tens; what am I to do with a pathetic man who asks for a collaborator and wants a supersecretary, who mourns the apathy and neglect of a world that he flouts by his failure to emerge from his own preoccupations and to take account of other people's; this man who preaches activity and demands acquiescence, who requires that one see the world completely through his spectacles, and share, or make a murmur as if sharing, every particular and personal reaction. I have still to have an hour's conversation with him. What an affectionate, loyal relation he could have, if he would permit it to exist: how much one would get out of him if he did not give one so much! But he wants all or nothing; and without trying to get more deeply into one's own life, he sets before one the ambitions and ideas of his own. Once and again he returned to the notion of my getting a doctorate at Montpellier with a year's residence; for he would like me to be a professor, a

19 September 1925

Saturday

[UP f. 8181]

president; he even, amazingly, hinted something about being an ambassador! In short, everything but what I myself, consciously and unconsciously, have been driving at. I squirmed out of his presence to get into the train; I wouldn't let him and Mabel Barker stay to see me off. The incessant soliloquy, like the insistent noise of a radio I simply had to run away from. And yet I love him; I respect him; I admire him; he still for me is the most prodigious thinker in the modern world. His arrogance and his weakness have frustrated him; he lacks some internal stamina, in spite of all his strength and energy; and is not merely a discouraged old man, but, as I found out from things he has dropped, he fell often into black discouragement as a young man. What is responsible for all these incomplete endeavors, all these unverified hypotheses, whose rejection irritates him, and whose proof, when others have done it, does not interest him? Why is there that streak of feebleness in that greatness? For all that, the greatness is indisputable; and if I have perhaps seen him for the last time – and how sadly incomplete these days were, and how fine they might have been – I shall retain in him, not the memory of the stern, sorrowed old man, interminably talking, demanding what one cannot give and forgetful of all one could; no, I shall retain the memory of the comrade I found too late, the great companion who, he and I, might have made over the world, gaily and exuberantly, in our double image!

APPENDIX 2:
"THE DISCIPLE'S REBELLION"

———————

1

As the reader knows, I did not actually meet Patrick Geddes in the flesh until the spring of 1923, but there were many premonitory tremors and quakes before we met, for our correspondence became more frequent. As early as 1919 he had suggested that I collaborate with him in writing a book about contemporary politics: one of a dozen volatile suggestions for projects that never came to fruition either in my mind or in his.

In 1920, early in my stay at Le Play House, Geddes had cabled Victor Branford of his imminent arrival; his coming, in a matter of days, filled everyone there, from Branford down, with eager trepidation and anticipatory anguish. But his work in Palestine, where he was already employed on various planning schemes by the Zionist organization – it was he who selected Mount Scopus as the site for the University and, with Mears, drew up the original plan for it – kept him in Jerusalem.

Then, at long last Geddes decided in 1923 to come to the United States, a place he had not visited for a quarter of a century. He had then left behind, at least in the mind of my old 'Dial' friend, Robert Morss Lovett, the memory of his marvelously vivid talks at the University of Chicago. In fact, everyone who knew Geddes in his prime – including Auguste Hamon, the French biographer of Bernard Shaw – was wont to couple him with Shaw for his brilliant conversation, satiric wit, and quick, sometimes savage repartee. Verbally he was master of the disconcertingly unexpected: so one can easily guess why the gentlemanly Sir Edward Lutyens, finding his imperially monumental plan for New Delhi severely criticized by this unconventional professorial nobody, turned on him with a furious contempt, as Lutyens's biographer would later disclose.

Now Geddes, full of years and experience, felt drawn back to America, partly in the spirit that had governed so much of his life, that of the wandering scholar seeking fresh intellectual contacts, partly to chat with old academic friends, partly stirred perhaps by the hope that his young American disciple would turn out to be the person who, as "collaborator," would manage to transform the accumulated thoughts of a lifetime into an orderly, readable form. Then, too, he doubtless looked forward to the refreshment of a new scene and a new "école libre," though the purlieus of the New School for Social Research, then on London Terrace in West Twenty-third Street, with an old red-brick house serving as faculty hostel backing it on West Twenty-fourth Street, could hardly be called a sufficient stimulus, that hot and humid summer, even though an ailanthus-shaded garden stretched between the two street rows.

My friend Delilah Loch, still Branford's part-time secretary, anxiously warned Sophia and me about how we should take him.

> Geddes must be accepted [she wrote], as a good Catholic accepts grief, with an open heart and no reserves, *if* he is to benefit those whom his presence scourges. He will brook no reserves. . . . I have lost much of Geddes from my flights, fears, and reserves. Stop dealing with him when you (or if you) must, but when you may, go with him fully. Don't forget he is an old man and lonely, and the very-most-vicious-cave-barbarian when sad, angered, or thwarted.

Delilah, in the last characterization, was doubtless remembering the terrible moment when she had interrupted a conversation between Geddes and Branford in order to find a much-needed ashtray. Geddes had seared her with contempt for this housewifely respect for ashes and this philistine indifference to the meed of thought, which must not be broken by concern for material arrangements. Delilah's advice was excellent; but it did not make for easier intercourse, nor did it prevent reserves, frustrations, and repeated flights. But in the matter of the ashtray incident I can't resist adding that Geddes had the backing of Emerson, who noted in his 'Journals' (1861): "The vice in manners is disproportion. 'Tis right that the hearth be swept and the lamps lighted, but never interrupt conversation, or so much as pass between the faces of the inmates, to adjust these things."

In the months before Geddes was to arrive, I had tried at his suggestion to arrange a full-fledged lecture tour for him, not realizing how impossible this was in late spring season or summer. But offers from the distant Midwest in May, at fifty dollars a lecture, with expenses, did not attract Geddes, so it all simmered down to a course of specially announced lectures at the New School, and a few dashes to other places: to Washington for a social workers' conference, to Worcester to see his old friend, Stanley Hall, to Cambridge to lecture for the landscape architecture students and compare notes with George Sarton on the history of science, and – for he never lost touch with biological research – to Woods Hole for a week with his scientific colleagues.

My first glimpse of Geddes was at the White Star Pier near Twenty-third Street, on the other side of the customs barrier: a little, narrow-shouldered man, frail but wiry, with a flowing, gesticulating beard and a head of flaring reddish-gray hair, parted in the middle: hot and impatient that warm morning, vexed that I had not gotten a ticket to take me past the barrier, talking in a rapid stream whose key sounds were muffled by the gray thicket of mustache and beard. To my still adolescent horror I noticed, as he turned his head, that he wore no necktie. I feared it might be his regular costume; but discovered later that in the haste of a belated packing that morning he had dumped all his neckties into his bag before he realized that he had not put one on. Still, Geddes's head, with its great crown of hair and the equally bulging crown of the brain case, was reassuring. Even at first glance he seemed thought incarnate.

He spoke with a fine Edinburgh diction, perhaps the clearest and purest English that is spoken anywhere in educated circles: and except when he deliberately imitated braid Scots, there was not, disappointingly to me at first, even a burr in his speech. Yet for a man who relied so heavily on the spoken word, Geddes was guilty all his life of a singular oversight: he had never mastered the art of making himself audible, either on the platform or in closer quarters. He never realized that his beard muffled every sound, and that he couldn't be followed, even if one were sitting close by, without the greatest difficulty. On walks, when he talked in profile, whole paragraphs and chapters were lost on the air. Everyone complained of this obstacle, and there were certainly difficulties enough in following his agile thoughts without this extra handicap: yet no one apparently had dared to bring it to Geddes's attention – or at least could persuade him to change his ways.

I would not say that there was anything willful in this: the sound of one's voice is so much a part of one that few people have the faintest notion of their own speech till a tape recorder, to their consternation, plays it back; and Geddes came before the day of tape recorders. Perhaps the explanation is that he shared his beloved Carlyle's contempt for mere verbalism; but if so, no one except Carlyle ever put that contempt in so many words – or made so little effort to see that those words, when uttered, should be effective. Not heeding my warning, the editors of the 'Survey Graphic' hired a steno-typist to take down Geddes's New School lectures: but the weird product of that effort, though it finally got into print, exacerbated both Geddes and Paul Kellogg. It certainly didn't justify the expense.

Somehow, our companionship got off on the wrong foot; and we never managed to fall in step afterward, though we tried more than once to begin all over again. Geddes commanded all my time that summer, for he had sent me a *pour boire* in advance so that I might devote myself to him during the next few months, instead of deviling at reviews and articles for "The Freeman" or "The New Republic." But neither of us knew how to make use of our opportunity. Partly, no doubt, this was due to my own awkwardness and ineptitude; but also it came from his never taking the time to get acquainted, in the way that Victor Branford had tactfully done in 1920 on my first visit to him in Hampshire. Instead Geddes started out by making a direct demand on me I was so unprepared emotionally to fill that I put up my guard and never thereafter fully lowered it.

On the day after his arrival, in the basement lounge of the New School, which gave out on the garden, he took me squarely by the shoulders and gazed at me intently. "You are the image of my poor dead lad," he said to me with tears welling in his eyes, "and almost the same age he was when he was killed in France. You must be another son to me, Lewis, and we will get on with our work together." There was both grief and desperation in this appeal: both too violent, too urgent, for me to handle. The abruptness of it, the sudden overflow, almost unmanned me, and my response to it was altogether inadequate, not so much from shallowness of feeling as from honesty. For I knew enough of his son, Alasdair, through Brandford's memoir and other people's memories of him, to know that there was no essential likeness between us, in temperament or in background, in our modes of life or in our potentialities. If Alasdair's schooling, thanks to Geddes's theory that he could provide a better education at home than anyone could get at school, had been more irregular than mine, he was also inured to a far more varied and adventurous kind of life, for he had shipped for a cruise on a fishing boat, had tramped with a fiddle through Europe, making friends with simple people through his music, and had served as a balloon observer in the War, before a chance shell, dropping behind the lines when he was off duty, killed him.

Though Alasdair had indeed served as P.G.'s assistant in arranging his Town Planning Exhibitions, he definitely had a mind and a will of his own: so that the father's hopes of making him a docile junior partner who would carry through his hundred unfulfilled projects and tasks were unfounded. Those who knew both father and son realized that in this respect the imperious old man was willfully dreaming. He had not actually taken Alasdair's measure any more than, at this moment of paternal adoption, he was taking mine. The pathos of that fearful assumption and adoption was not lost on me at the time; nor have I ever recalled that moment without a shrinking within over my own stilted response.

The effect of this encounter was unfortunately to cover over and freeze up some of the natural warmth I felt toward him. In his need to falsify our relations and warp them in accordance with his own subjective demands, he also gave me a clue to a certain blind willfulness that in some degree, and often at the most inopportune moments, had undermined not a little of his own work. But the lesson I learned then carried into my own life: for when I was faced with a similar grief I was careful not to seek from my students, even by an indirect appeal, the response I would no longer get from my own son.

2

From the beginning our collaboration proved abortive: in fact, it never reached the initial stages of its conception. Yet my sense of Patrick Geddes's "greatness" survived the whole summer I spent in his company, and if my criticisms of this or that aspect of his thought seem captious, or my admiration stinted, I beg the reader to modify that impression by this broader admission. Much of what I must say about Geddes applies equally, I find, to not a few other men of genius I have known: to Frank Lloyd Wright, for one, and to Adelbert Ames; for in one way or another one must pay for their extraordinary gifts: the very self-absorption that sustains their work, along with a godlike self-confidence, breaks down normal social attachments.

Genius, just because of its originality, tends to be self-isolating; and the less its departures are understood and accepted, the more self-protectively inviolable becomes the resulting solitude, and the more difficult it is to overcome the solecisms that result from this isolation. My mature experience with Adelbert Ames, in the forties, parallels my youthful relations with Geddes.

Back in the thirties, at Dartmouth, when Ames's pioneering researches in optics were taking shape, he had demonstrated to me some of his ingenious experiments that showed – conclusively, I believe – that pure sensations do not register automatically, by a reaction similar to the chemical changes in a photosensitive film; that every sensation is a perception that draws on the past experience and the present purposes of the organism. Whatever the physical conditions, they register only through interpretation, and through the total response of the whole organism. Human value and purpose have played as great a part in forming the "objective world" as number, sensation, and abstract pattern. All these "objective" phenomena are refined products of human history, not aboriginal components. On that basis we were in full agreement.

After the Second World War, when I lived in Hanover for a few years, I saw much of Ames, and he then wanted desperately the help of someone who would put his observations and theories into viable literary form: so more than once he tentatively suggested that I might join him in this task. But by this time, Ames had worked his own interpretation into a kind of solipsism: since he could produce illusions at will under experimental conditions, the "external" world had come to seem an illusion, wholly dependent upon the observer. Reality, no less than beauty, seemed to lie in the eyes of the beholder. The very exquisite shapes of his experiments kept him from realizing that the apparent uniqueness of each person's perceptions resulted from the abstraction of the experiment itself, for by isolating the eye – and usually a single eye at that – he prevented the whole organism from coming into operation as it normally would.

When I tentatively broached some of my doubts to Ames, he shrank back into his private shell: the collaboration he begged for did not in fact allow for any criticism and rectification. Though I asked for the privilege more than once, he never invited me to go through his series of experiments a second time. Despite our underlying agreement, he was not prepared, I realized, to examine my criticism. The criticism that might have helped him should have come at a far earlier stage. Now it was necessary for him to go it alone, or at least accept aid only from those who would take his findings unreservedly. So it had been earlier with Geddes.

I used to come to Geddes at the New School before ten o'clock in the morning, and would leave at four or five to call for Sophia at "The Dial" for a later dinner, myself usually in a stage of fatigue approaching exhaustion; for this Ancient, who lived, like the ultimate creatures in Shaw's "Back to Methuselah" – a play published that year – in a vortex of thought, had enough energy to tire out "fifty fast-writing youths" like myself, even if the strain of listening so long to smothered, only half-caught sentences was not enough by itself to drain one's vitality. But unfortunately this daily intercourse did not bring us closer; and I was never alert enough at the end of the day to put down even a fragment of all that he had told me and taught me, though I had had the ambition, before we met, of using this period to gather materials for his biography. He did, of course, tell me much about his past, as he might tell some chance acquaintance in a railroad train or a restaurant; but he was averse to making any systematic effort to review his life: "Time for that later," he would say, and by later he meant after he was dead.

To show how closely Dewey's educational influence had paralleled his own doctrines, I took him one day to my friend Caroline Pratt's City and Country School in West Twelfth Street. I had expected him to be enthusiastic about a plan of education that shied away from abstractions and verbalisms and even delayed reading till need and interest turned the pupil toward it: a plan, moreover, that sent groups of children out to various parts of the city to explore its daily life and recapture it in paintings, blocks, and stories. Was this not Geddes's Regional Survey realized in education?

To my surprise, Geddes fastened, instead, upon what he considered a radical weakness – the absence of any firsthand contact with nature, in plants, animals, museums, and gardens, and made such swift scarifying criticisms that after this I hardly dared to introduce him to any of my other friends. Fortunately, his contact with my fellow members of the Regional Planning Association of America, on his long weekend at the Hudson Guild Farm, had proved memorable. We all retained a picture of him, sitting cross-legged, like an Indian guru, under a great oak, which almost turned, under the spell of his stories, into a banyan tree, telling us about his town planning in India – and how as "Maharajah for a day" he had banished the plague from Indore.

Somehow, it is symptomatic of the strain and tension that existed between us that, after promising to spend a whole Sunday afternoon with my wife and me in our little apartment on Brooklyn Heights, visiting our home for the first time, he entirely forgot the engagement and never even phoned. Characteristically, he was always forgetting engagements, or making two or three for the same hour: but he and I both realized that this particular oversight, on a day when he had nothing else on his calendar, was a Freudian slip. I was annoyed, too, that the one day we visited the American Museum of Natural History together, it was for the sake of two young sons of a chance acquaintance he had met at a dinner party, not for my sake. Had I given up a summer of my own time, just to take the leavings of his days? As you see, there were two demanding, self-absorbed egos to reckon with.

3

Frustrated, indeed exasperated, I felt the time had come for a show-down. So finally, early in July, I stayed away and wrote him, not without trepidation, the following letter: [Here Mumford quotes a long passage, "In one sense ... Mohegan Colony," from his letter of 6 July 1923, pp. 177–179 above.]

In his reply Geddes rose to the occasion, as he so often did at critical moments, with overwhelming insight and magnanimity: not angry, but sorry over his own apparent failure to "take me in"; realizing "only too sadly" the gap between our generations: his, heady even now with the hopes of endless progress through science, technics, and education: mine, war-shocked, disillusioned, discouraged, unconsciously anticipating the even more formidable barbarisms that were to follow. At the end he emphasized his interest in my development "as no mere exponent of mine." How that latter admission reawakened my love and admiration!

When we again came together, we had one illuminating, almost ecstatic day, which I recorded in my notes. But then, alas! he fell back once more into the old soliloquy, and repeated once more the old irrelevant demands, wanting me to take over this or that half-finished theme as a "legacy" from his "mingled heaps," ideas he himself, he knew in his heart, would never put together in any viable form no matter how long he might live. As the moment for his departure came nearer, our relations got even worse: I kept asking for the living kernel of his wisdom, and he insisted that I should swallow a heap of husks. Though I tried hard for a while to treat the latter as food, I could not deceive myself. He himself had taught me the difference.

Only now, looking back, do I realize all we might have done together. Had we spent some part of our days rambling in Central Park, dropping into the menagerie, revisiting the Museum of Natural History, or exploring the historic images and symbols in the Metropolitan Museum of Art, these shared experiences might easily have evoked by association the richness of Geddes's own mind, once he allowed it to roam freely outside his graphic enclosures. But from having hoped for too much from me as his "collaborator," he still had, even after meeting me, the same curiously constricted view of my interests and capabilities, which he had exposed in a letter I discovered when sorting Branford's files in 1957.

"I hope," he wrote Branford then, "Mumford does not feel I want to make a planner of him or biologist: no such idea: what I suggested was that with his vivid writing he might easily get the United States public to understand the use of Civic Exhibitions." When I came upon this letter, I was dismayed, indignant – and saddened. Civic exhibitions indeed! It was emphatically in biology and city development that he had something unique and personally valuable to pass on to me. What I have learned about cities did not come from his Cities Exhibition, for I never saw any part of it. Nor did it spring from conning his graphs or memorizing his categories. How blind Geddes was not to realize that from the beginning it was in biology that he had made his most lasting contribution to my education – and even more profoundly to my life.

For all this, that summer had some good moments, brief but memorable. Though my direct contact with Geddes was confined to such a short period, nothing that he had written or correlated in graphs had an influence on my thinking nearly as profound as he in his own person had on my life. Among books, Plato's "Republic," Emerson's "Journals," Whitman's "Democratic Vistas," Melville's "Moby Dick," Ruskin's "Munera Pulveris," Zimmern's "The Greek Commonwealth," Whitehead's "Science and the Modern World," to mention only a handful, left a deeper impression on me than any single work of Geddes's. But the impact of his person shook my life to the core.

The most impressive thing about Geddes, even at sixty-nine, was the sense he still conveyed, as Branford did, too, of what it is to be fully alive; so it was not a surprise to me, though it was a scandal to some of his friends, to find him at seventy-three, taking as his second wife a woman considerably younger. He was seeking something more, I have reason to believe, than mere companionship and care. And it was by this magnificent aliveness that Geddes towered above those around him.

Gladys Mayer was right. Blake would have understood Geddes, but Bentham would not have; for even Geddes's old-fashioned rationalism – and he had not a little of that – was too impassioned to satisfy its classic later exponents. "I and mine do not convince by arguments: we convince by our presence." If Geddes had only accepted that dictum of Walt Whitman's more fully, he might have conserved for better uses some of the time he actually frittered away in his endless graphic demonstrations and arguments.

His presence was what counted; and if there had been long periods of silence, interspersed with longer periods of listening, it might have counted more!

It was true that the very intensity of his vitality, its exorbitance, made impossible demands upon those about him. Even the New School itself, under the shrewdest of directors, Alvin Johnson, found it impossible to contain this explosive personality. Johnson, I should explain, had rather grudgingly offered Geddes the temporary use of a guest room, till he should find more permanent quarters; but since his bedroom would remain vacant, Geddes saw no reason why he should leave it: he was quite content with the house and the rear garden and Professor Harry Dana's companionship, and had no mind to change them for a less congenial settlement house or a hotel.

So Geddes had not merely stayed on all summer, but had taken possession of the academic quarters of the school, room by room, opening up the bales of charts and diagrams he had sent over in advance, and spreading out his papers. For Geddes loved to work on large tables – he said this was the aristocracy's secret of mastery! – and here he had as great a space to work on as in the Outlook Tower. In short, Johnson found that he had a scientific Bartleby on his hands, as loquacious as Bartleby was silent, but just as difficult to dislodge. Yet with all this paraphernalia of preparation, nothing came of Geddes's graphic demonstration: it was mere gymnastic. The unformed theses, the unwritten books, mostly ramained unformed and unwritten.

It would have been flattering to think that Geddes had brought his boxes of graphs and charts over for my special benefit: but the truth is that they were a standard part of his travel equipment, like the cabinet of medicines an invalid dares not leave behind, even if he never touches one. But since I was present, Geddes took advantage of the opportunity; and it was in one of the New School classrooms, hung with charts, piled with memoranda, that my sense of frustration came to a climax, just a few days before Geddes sailed. I still smart over the feeling of rebellious humiliation in my breast – I was then almost twenty-eight – when he demanded, while he was busy elsewhere, that I spend the morning setting down on the blackboard, one after another, all the graphs and charts of his that I had mastered, including not a few that I rejected entirely.

That exercise made me feel as if I were back in elementary school again! But I had lost my old docility, and though I went through this exercise grimly, and did it thoroughly enough to earn his approval on his return, I never felt further from him than at that hour. This was the complete antithesis of the bold, challenging, insurgent, vitalizing mind that I had been drawn to: as autocratic, as tyrannously Teutonic, as that of Carlyle. Secretly I still looked forward, as reward for this final submission, to at least one day of his undivided attention. But it never came.

The final evening P.G. was in New York – he was to sail at midnight – he had made an engagement to have dinner with Lillian Wald and the sisters Lewisohn, as I remember; and he left to me the dreary task of packing his bags and papers – those heaps of clothes, those middens of notes and charts, those shelves of new books! An English artist, Stephen Haweis, who had long been influenced by Geddes, dropped in for a last chat, and he, too, was disappointed at finding the master absent. Years later Haweis recalled to me the remark I had made then, in the midst of my packing: that it was like putting the contents of Vesuvius back into the crater after an eruption.

By the time eleven o'clock came, I was tired, and too sick at heart to retain the reproachful fury I had felt earlier in the evening. When Geddes came back, I found a taxi and put his bags into it, but I refused to accompany him the few blocks to the pier. For a moment, he began to protest and urge me, but he must have sensed my bitter disappointment, for he let me go. We shook hands hastily on the sidewalk and the taxi rolled off. Though we met briefly in 1925 and corresponded at irregular intervals up to the month of his death in 1932, this parting was really our final one.

APPENDIX 3:

"THE GEDDESIAN GAMBIT"

[Mumford's headnote]

3 March 1976

I am still not satisfied with this for many reasons. The tone is wrong, to begin with, and it is over-written – clumsily over-written at that.

Shall I use a few good pages for the second miscellany? Or try all over again from a fresh point of view – if I can find it?

12 March 1976

Another reading has not yet satisfied me. This has everything but inspiration.

4 August 1976

Rewrite or Destroy?

1

Up to this point I have treated Geddes' life and thought very largely in terms of the influence they had on me. And yet in my account of what happened when we finally spent a summer in almost daily companionship, it might seem it was this personal encounter that first opened a breach between us, a breach which came psychologically to a climax at that moment I recorded on the terrace of the Outlook Tower. The actual record shows something different. From the beginning there was a fissure in our intellectual masonry which, so far from being sealed by time, only opened wider and went deeper; and sooner or later, if only to explain my own career and my own philosophy, I must dwell upon our differences. This is something that cannot be properly handled without doing what no one has so far adequately done, or even attempted: namely to make a close critical examination of Geddes' achievement as a systematic thinker. If this does nothing else, it will at least throw a light on why I never heeded Geddes' testamentary injunction to be his biographer.

Over this problem I have spent many uneasy hours in a attempt to deal fairly with my obligations, my doubts, and our unresolvable differences. Not the least part of my problem has been to decide at what point this particular chapter should be introduced. Since this should be a summation of my mature judgment on Geddes' life and thought at the end of half a century, it might seem fitting that it should come at a later state of the narrative, or even be tacked on as an appendix. This decision is all the harder to make because I realize that my earlier presentments of Geddes' life and work, as late as that in *The Condition of Man*, tend to be idealizations, not in the sense that they falsify his achievements, but that, as in an old-fashioned Victorian biography, they deliberately overlook personal faults and intellectual defects. So let me explain at once that I have no wish to besmirch any of these favorable judgements; nor shall I say anything by way of criticism that I did not say to Geddes openly, almost from the beginning of our correspondence; although, naturally, I can now say with greater confidence what I originally put forward in a more respectful and tentative fashion.

Fortunately I kept carbons of some of my earliest letters to him: a practice I long ago abandoned, except in the case of either business letters, or letters written, often in futile indignation, to the *New York Times*. What is even luckier for me, almost from the beginning Geddes preserved all my letters, and, even more miraculously, they were finally deposited in the National Library of Scotland through whose offices I have obtained Xeroxes that now recall many ideas and incidents I had quite forgotten.

Even my earliest letters to Geddes reveal that I had reserves about some of his central themes, along with the graphic categories he had employed to express them. Branford's preparatory exposition in 1920 of Geddes' central graph, known to the initiates as the "36," had already set my intellectual teeth on edge. From the beginning that kept me from swallowing Geddes whole. The failure of our intercourse during his term at the New School in 1923 could hardly be understood without allowing for the fact that in 1920, when Geddes invited me to collaborate on a book, I already intuitively realized that for him collaboration meant unquestioning acceptance of his clichés as well as his many vivid original insights. As I explained to him then, I could not absorb his ideas without transforming them; and even at that early stage I realized how impermissible that kind of co-operation would be.

As late as April 1932 – indeed in his last letter to me – Geddes, speaking of a series of papers his had published during the previous decade in *The Sociological Review*, wrote: "You have never answered my demands for criticism of these papers, but I take it you see something in them." What I saw, alas!, was something quite different from what Geddes wanted me to see. In point of fact, all through our correspondence, sometimes in unmistakable words, more often by my silences and my abstentions from his urgent proposals, I had gently, perhaps too gently, indicated my criticisms – yet always with the deference due to an older mind. But since no one else has yet made the long overdue examination of Geddes' abstract framework of ideas, the time has come to break down this wall of silence. Whatever was valuable in Geddes' work will survive this examination. But the worst possible service to his genius would be to pass over in polite silence a method that seemed to him so revolutionary in its potentialities as if it were unworthy of critical attention. Much of Geddes' published thought will stand out more strongly if one exposes those parts of his intellectual structure where inadequate, untested, or obsolete materials had been used, or where the several parts of the structure do not, even when forced together, form a functioning whole.

Paradoxically, no one will properly understand Geddes' influence on my thinking who does not realize how stimulating even these negative contributions were; for some of my most original work had come through supplying certain components left out of Geddes' systematic thought. And if the present critique prove faulty, I shall be content if this chapter nevertheless turns the reader back to Geddes himself, for independent verification – or rebuttal.

2

Though Geddes' career was to encompass a variety of occupations and activities, so that he was at home in many other fields than biology, he came to associate his ability to pass freely from one area to another with his development of his new system of graphic logic, suitable for dealing simultaneously with all manifestations of organic complexity. In the course of his thinking, he had invented a series of graphs for different purposes: some extremely illuminating, some useful in making shortcuts, some too private to be put in words – at least convincing words. With his elevation of visual graphs over verbal language, his exercises with graphs became a standard feature of his working day. This served as a kind of private intellectual gymnastics, often to the exclusion of any further active application to the fields or events they were supposed to bring under intelligent direction.

Geddes' "thinking machines," as he called his graphs, were the compensatory byproduct of a period of blindness he had undergone on a solitary scientific expedition in Mexico, when he was twenty-five. The traumatic threat gripped him all the more because a tendency to blindness ran in his mother's family. Confined to a darkened room on a local physician's dubious advice, believing that the visible world might be permanently closed to him, Geddes in his desperation sought to open a way out by creating a notation for marshalling and integrating ideas within a network of squares suggested by the barely visible window frame he could feel with his fingers. He made these squares visible to himself by folding a sheet of paper into compartments, to each of which he sought to assign a definite function and meaning. All through life he continued to use folded paper, rather than ruled cartesian graph paper, for his exercises, for he held that the slight manual activity involved in carefully creasing the paper was helpful to thought by keeping all his organs in play.

When Geddes' sight came back, the method itself still seemed valuable to him: doubly so, naturally, because of its healing value during his ordeal; and the fact that he whiled away the tedious voyage home by playing chess incessantly perhaps reinforced the image of the chessboard in which his graphic system of thirty-six squares eventually culminated. The original appeal of Geddes' squares increased rather than diminished with repetition, especially in the emotionally disrupted years marked by his elder son's death and that of his wife. Repetition itself, as normally practiced in religious ritual, or even in today's Greek "worry beads," is a long-used method for reducing anxiety – having its pathological counterpart in the tics and the automatisms of a "compulsion neurosis."

When still a student of Ernst Haeckel's at Jena, as Geddes recalled to me, he had become acquainted with a fellow boarder, witty, intelligent, scholarly, the very ideal of a thinking man, who greatly attracted Geddes. One day this man took Geddes aside and told him that he must tell Geddes his sad secret: the truth is that he was crazy! For he had spent years attempting the art of simultaneous thinking, and just as he was on the point of achieving it, his mind had broken down. When Geddes finally began putting his elementary categories on squared paper, he remembered the talented lunatic and realized that he, too, was now engaged in "simultaneous thinking." My friend Lucie Zimmern's musical name for this method, polyphonic thinking, was even better; but perhaps when it draws in other sciences one ought to conceive it as contrapuntal. In time Geddes came to perceive that by laying out his categories on squared paper, he could disclose at a glance inter-playing functions and relations that non-graphic methods could present seemingly only in successive words. Though his wife's talent for music might have freed him from his one-sided commitment to graphs, he was never able to appreciate the complementary method of language and music.

As Geddes developed this skill, he became a kind of ideological map-maker, attempting to do for every area of human culture what the geographic map did for the physical environment. If Geddes called these diagrams "machines," that epithet contained the latent suggestion that they were supposed to do work, and what is more save labor and time which the slower tempo of verbal discourse did not permit. It was the same Victorian labor-saving penchant that caused Charles Babbage to design the first computer. Yet perhaps I should add, as the term polyphonic happily implies, that music has other ways of achieving simultaneity in dealing with complexity than those spatial symbols which Geddes centered all his attention on.

Before going further, let me say a word about Geddes' more obvious graphs, usually of earlier date. His early diagrams took a variety of forms: not the least interesting was his use of the classic "Tree of Life," which was worked into a window of the Outlook Tower. Perhaps his most easily grasped diagram was his profile of an "ideal" longitudinal river valley – the "Valley Section" in his terminology. In this graph he correlated the basic primitive occupations with the physiographic profile from mountain top to sea, descending from the hunter, the miner, and the woodman to the herdsman, the farmer, the fisherman. These, he pointed out, were the environmental types from which the more complete technologies and occupations had developed. I found that graph useful in my first attempt at systematic survey of *Technics and Civilization*.

Not less viable was Geddes' correlation of the sciences and humanities with their corresponding practical applications in the arts and technologies, from agriculture and medicine to education and architecture. I used this graph in the introductory lecture I gave at Stanford in a new basic course in the Humanities back in 1942. And in some ways I took Geddes' chessboard model more seriously than he did: for whereas he confined himself to the vertical and horizontal relations between the sciences, the arts, and the humanities, I pointed out how the varied characters of the chessmen exposed other interrelations and cooperations by making knight's moves and bishop's moves, to say nothing of offering the freedom that the queen enjoyed in covering the whole board. A religious myth accordingly might open up new observations in physics, or an improvement in contraceptives might effect a moral change. In action, the formal spatial barriers between all these spatial arrangements dissolved. Certainly, this desegregation of specialized and isolated functions was the core of Geddes' own behavior – when his eyes were not fixed on the graphs themselves.

Even before that exposition, in preparing the two final volumes of the "Renewal of Life" series, I spent a whole month in laying out my further studies in terms of my own quasi-Geddesian logic, before going further in active research. Though the reader would not detect this framework in my book, it helped give form to the complex structure I build around it. Later thinkers, like the psychologist J.L. Moreno the inventor of the "psychodrama" – independently developed such graphs, which added the necessary spatial dimension to all descriptions of human behavior.[1] (Geddes had invented the term "psychodrama" earlier and used it in one of his University of London lectures.) Such graphs, obviously, may perform a service much as the architect's working drawings do: facilitating the construction of a building better than constant verbal instruction could do.

From all this it should be plain that I had no innate bias against Geddes' graphs, though some of those he treasured most and kept on "perfecting" (repeating!) never took hold of me. For the first few years I even tried hard to overcome my initial set against the "36." Recently I discovered in the Geddes file in Edinburgh an essay of mine whose very existence had dropped out of my mind. That was an exposition of Geddes' graphic method, contrasting his kind of visual two-dimensional thinking with auditory thinking, which depends upon a seemingly linear succession of sounds and words. In this analysis, I favored Geddes' concentration of spatial graphs over my own native aptitude for words, which exist in time and become simultaneous only through memory. For some reason that paper never was criticized or printed, but the fact that it was discovered among the Geddes and Branford papers would indicate that I must have sent it to one or the other, perhaps hoping for publication in "The Sociological Review." Strangely I have as yet found no reference to it in any of our correspondence.

This pious explication, undated but probably written between 1923 and 1925, marks the high point in my role as a faithful Geddesian disciple. Though in letters to Branford or Geddes I continued to use the familiar Geddesian terminology, I realized that I needed a more ample and flexible vocabulary than his, no longer limited to Geddes' private definitions, if I was to draw on other ideas and systems. But well before this I had put to Geddes my reservations about some of the generalizations he regarded as fundamental: notably Comte's historic "Law of Three Stages," or the formal division of society into Chiefs, People, Intellectuals and Emotionals – as if these were biological traits and denoted permanent roles and castes. In March 1923, before Geddes and I had met, I wrote him: "I seriously doubt the existence of any sort of linear progress in human society such as the law of three stages presupposes; and it seems to me rather that all three are implicit or potential in each society and that in the rhythm of history one or the other may be expressed or subordinated as society develops, only to return again."

If I now proceed to examine the whole Geddesian system, I have at least alerted the reader to my original youthful doubts and criticisms. It was the living man and his formative example, not the regimented assemblage of his thinking machines, that had drawn me to him and that still attract me. Had I stumbled upon the finished system first, I possibly would never have been tempted to come nearer. In short, P.G.'s rigid cartesian graphs of nine or thirty-six squares, each compartment occupied by categories with fixed, unreplaceable meaning, some functional, some arbitrary or fanciful, every term calcified by incessant repetition, put me off even before I had listened to his lively but too often unconvincing expositions and impromptu elaborations. From the outset, Geddes' overvaluation of his graphic methods was what made anything that could be called collaboration impossible. And the closer Geddes advanced toward me, like a Roman gladiator with his trident and net, ready to snare me in his cunning ideological net, the more warily I was inclined to retreat and preserve my freedom of movement: indeed my very life!

Even Geddes' oldest and closest colleagues had faced this difficulty; and though some of them could repeat his graphs and explanations from memory – more or less! – they did not in fact make use of them except in paying homage to Geddes. Like Branford they were often at their best when they had shaken themselves free. In my one meeting with Geddes' old colleague Professor J. Arthur Thomson, a gentle, ruddy-faced man with an erratic heart, he told me with a deprecating smile that, in writing their final two-volume biological opus together (*Life: Outlines of Biology*) he had wished to turn Geddes' noun "function," in "organism: function: environment," into the participle "functioning," in order to emphasize their activity and interaction. But Geddes for long stubbornly opposed Thomson's overdue innovation. So it was, but even more, with Geddes' personal interpretations of the subjective life. For one half of the thirty-six squares no other terms, no other interpretations but his own were admissible.

3

In the course of time two graphs came to occupy a central place in Geddes' thinking. One was his "bookcase" diagram for the logical ordering of the biological sciences. This graph presented organisms in both their static and kinetic, their passive and their active aspects, arranged in a time order that embraced the past, the present, and the possible; that is, the latent or potential future. It had a place equally for the individual and for the species: for both externally and internally determined events, with all their possible modifications. Geddes summarized this interpretation in a neat equation: Life equals OFE over EFO, where O stands for organism, F for function, and E for environment. As one increases the denominator, the forces of life shrink and retreat: as one increases the numerator, the organism actively dominates or overcomes the circumstances that condition it and increasingly determine its behavior.

Not being a professional biologist, I am not competent to estimate the value of this graph for biologists. Speaking only in terms of philosophy, I believe that Geddes made an important contribution in restoring the Aristotelian concept of potentiality and purpose, as necessary categories in the interpretation of life-processes. Here I reserve comment for only one fact: namely, two essential aspects of life that properly find a place in the biological bookcase diagram, Time and History, are strangely absent from Geddes' final "Chart of Life."

Now it was this "Chart of life," this "Chessboard of Life" as Geddes often called it, that he regarded as his central contribution to the reorganization of both thought and practice: a method for restoring the unity that had been disrupted by the multiplication of specialisms; and the increasing volume of detailed knowledge. The "Chart of Life" was the elaboration of Geddes' basic graph of nine squares, itself based on the triad of organism–function–environment, in which the terms on the diagonal interacted directly upon each other and generated the surrounding field of interaction and interpenetration. As both a biologist and a field naturalist, Geddes understood that none of these aspects of organic existence could be isolated from the other two except in thought, for some temporary immediate purpose. Once one allowed the organism to be separated even in the mind from its other two aspects, and likewise from the fields of other organisms (ecology) even the most rigorous scientific observations would prove under-dimensioned, and their conclusions would be inadequate, if not false. Witness Lord Solly Zuckerman's once famous experiments on the behavior of apes, which established only how such animals behaved in scientific laboratories, and as ethnologists have since revealed, told almost nothing about their behavior in the wild. The appreciation of Geddes' basic nine squares which applied to all organic activity would have guarded against such a *faux pas*.

Geddes, coming in contact in Paris with the sociological investigations of Frédéric Le Play, a mining engineer whose detailed studies of working class family budgets were concerned with the interaction of work, place, and family, became impressed at an early date with the universal nature of Le Play's triad: so he extended his own biological outline by adding Folk, Work, and Place to his more familiar Organism, Function and Environment. This was an important unification of the natural world, as given, with the emergent cultural world of man, which has, like the earth itself, been continuously modified for at least half a million years by the agency of man, with his objective explorations and selections, and his even more decisive subjective contributions. Geddes' bold recognition of this dynamic interaction – so constant and "obvious" that it had passed unobserved – was a central contribution. In this respect, Geddes' fascination with the graph of nine squares was justified, and by his constant playing with it, he naturally encouraged his own hope of further discoveries.

Indeed one may go further. If I am right, Geddes' cartesian framework was the first systematic application of "field theory," a whole generation before Kurt Lewin and his successors. By presenting the area of interaction in this systematic fashion, he provided a framework for handling many other relationships, however complex – provided that the central term in the triad could be described as an activity, a response, or a function – a condition that Geddes in some of his own later graphs unfortunately did not observe. There was nothing in this logical structure that was peculiar to any one kind of organic relationship; nor was there anything in these fundamental terms that was peculiar to Geddes or that depended upon his private definitions for their successful application.

Once one had the clue, one came upon the triad constantly, and in many guises. In the structure of language it was noun–verb–predicate or sentence; in a corporate organization chart, it was agent, activity, situation; and as a radio operator in the United States Navy I discovered that once a day the commander of a naval vessel was obliged to report on the condition of his ship with respect to "Personnel–Operation–Material" – OFE all over again. Through his use of the nine squares Geddes sought to overcome the illusion that we live in a world of discrete objects in empty space, manifested in separate unrelated events, neatly describable by equally formalized and isolated abstractions.

In the course of elaborating his fundamental graph, Geddes had in fact exposed the archetypal drama of life: and what was even more important had restored the missing factors of time and change. In the scenario for this drama the actors, the plot and the scenery, the dialogue, the performance, the setting, actively bring into existence an interwoven sequence of events whose meaning and purpose no single part, however clearly presented, can possibly convey. Geddes' conception of life as essentially an enacted drama – a cosmodrama, a biodrama, a technodrama, a politodrama and autodrama – all terms that Geddes himself had coined – lay at the bottom of his own immense potential, yet insufficiently realized creativity. It was in this conception of the dynamic role of the drama in human evolution, not in his static "Chart of Life," that Geddes not merely expressed his own unique insights, but likewise did justice to life's realities: Time, Organic Change, Memory, Insurgence, Purpose, Significant Action, and Social Intercourse – life with its unpredictable exuberance, its spontaneous inventiveness, its increasingly meaningful transformations as "the plot thickens."

If Geddes had not been diverted at first by a dozen other claims, this new trail that he himself had blazed would alone have penetrated much unexplored territory which many others by now, perhaps, would have opened up and begun to cultivate. And the very characteristics that are absent from his Chart of Life – Time, History, Language, Dialogue – are, it turns out, the central components of the human drama. Had he recognized the nature of this cosmic drama he would have banished from his own mind the illusion that he had achieved an ultimate synthesis: a boast comparable to Kant's dubious claim to having written a prolegomenon to all future systems of metaphysics.

Both the reader and I are handicapped at this point, since without the "Chart of Life," with Geddes' own interpretations before him, the reader will be unable to follow my verbal exposition. For the reader's later guidance, let me say that, after publishing tentative outlines of the "36" elsewhere, Geddes first presented this encompassing graph in a communication to the *Sociological Review* in 1923; and he turned it over in even more complete form to Amelia Defries for publication in her semi-biographic book, *Geddes the Interpreter*, parts of which were actually written by him. Geddes then presented his "Charting of Life" in a long article in the *Sociological Review* in 1927; and finally re-stated and amplified it, but without any essential changes, in the two volume opus he and J. Arthur Thomson published in 1931, just the year before his death. The "Chart of Life" is perhaps more accessible to library readers in the end-papers used in Philip Mairet's biography of Geddes – though Mairet's own emphasis was unfortunately upon Geddes' contribution as "pioneer sociologist." But the fullest exposition of Geddes' ideas remains that in the second volume of *Life*.

Even in its bare skeleton form, this "Chart of Life" was not easy to navigate by the derivative categories he regarded as equally fundamental: Feeling–Experience–Sense: Emotion–Ideation–Image: Etho–Polity–Synergy–Achievement – were all open to challenge once one got past the first quadrant. As if this were not sufficient, Geddes superimposed upon this framework perceptions and intuitions and highly personal interpretations that were equally arbitrary and sometimes irrelevant. Thus in his seeking to summarize the ultimate expressions of creative life in terms of the Grecian Nine Muses, he even limited life's ideal possibilities to those of a particular moment in a particular culture; and I uncharitably suspect that Geddes was lured into this extravagance by the fact that the Nine Muses fitted so neatly into his top right hand quadrant. If the Greeks had recognized only five muses, he might not have succumbed to that mischievous temptation!

Up to a point, Geddes might well have supported his aim in elaborating the nine squares by citing Descartes' observation: "To him who perceives the links of the sciences, it will prove no more difficult to retain them in the mind than to retain a number series." That aim, of bringing all the actors of the human drama together simultaneously on the stage, was worthy of his great mind. It was to Geddes' credit that instead of confining his graphs to science, he sought audaciously to include every aspect of human experience, in so far as it could be reduced to ideographs. This, one must admit, was a pretentious ambition; yet in actual use it might be justified as a kind of mnemonic system, like the Roman memory training, which used an imaginary theater to aid in the recall of long, complex speeches.

In his terminal Montpellier days, in fact, Geddes created a series of gardens in which his principal graphs actually took concrete form in the layout of his plants; as if by repeating his graphs often enough he could imprint his system in the mind of even a passerby. Ironically, Geddes, who castigated the examination system in education, as based only on cramming and memorizing, relied upon his own repetitive drill to gain adherence to his ideology. Characteristically – but how shockingly! – in all the thirty-six squares there was no place for the specific human trait Alexander von Humboldt[2] correctly regarded as the prime index of man's emergence from his animal state: Language.

Unfortunately, once one got past the nine squares which dealt with the common facts of daily experience, the game of intellectual chess Geddes had invented could not be played except according to the fixed rules he had laid down, with pieces he himself had carved and given his own special values to. In fact, it had become a form of intellectual solitaire. From his personal intercourse with Geddes, for example, Bernard Berenson, the critic of art, had picked up the term "ideated emotion." Berenson had appropriated this as the happiest possible description of the Renascence paintings he admired. But in Geddes' dictionary, "ideated emotion" was identified as "Doctrine"! Not merely, then, did one have to accept Geddes' private definition of his terms and not question their fitting into this or that compartment: one soon discovered that on this tightly occupied and over-crowded chessboard whole tracts of life had been excluded, if for no better reason than that he had allowed no room for them. In this regard it was as bad as the Dewey decimal system of library classification, which originally had no place for sociology.

In short, unless one gave each piece the specific place and function that underpinned Geddes' interpretation, the game could not be played; for Geddes had left no open spaces, as in the orthodox chess boards, for manoeuvre or riposte. One might call this chess opening the Geddesian Gambit. Worse even than in a Fool's Checkmate, the opponent was checked before he could make the first move. In short, this was not a public "Chart of Life," but a private chart of Geddes' mind.

Thus what seemed to Geddes an ultimate "Opus Syntheticum," universally valid on all occasions, was in fact only an elaboration of his own too-neatly fabricated thesis, without the benefit of any encounter with its Hegelian anti-thesis or counter-statement – or even with any specific criticisms or modifications by other competent minds of his own special terms. Though he had been an avid reader of Emerson, he overlooked Emerson's observation that a true doctrine must embrace the heresies as well as the beliefs of mankind. The possibility of constructing such a "final" synthesis was, in terms of his own most vital insight, a delusion. Synthesis is not a goal: it is a process of organization, constantly in operation, never finished. Any attempt to produce a single synthesis good for all times, all places, all cultures, all persons is to reject the very nature of organic existence. Geddes' elaborate formula for synthesis was no sounder than Herbert Spencer's once-plausible formula for evolution itself, as the change from indefinite homogeneity to definite heterogeneity. While life goes on no "final synthesis" is thinkable.

Geddes never realized how static his system had become or how limited and stultifying was his graphic vocabulary. He relied on the interaction of his selected categories, one with another, to automatically produce new categories and open the way to new areas of knowledge, new fields for action. But such dynamism as his system exhibited – symbolized on his "36" graph by the legs of a swastika! – could occur only within the limits that the system imposed. All further experience had somehow to be squeezed in or lopped off. To overcome the specialization and compartmentalization of post seventeenth century thought Geddes had in fact only substituted his own private set of compartments, concealed behind the public facade of Folk–Work–Place. Nothing, certainly could have been further from his original insights, or further from the viable lessons of his life and thought. But as Geddes presented his graph of life, he left nothing to chance and nothing to the correction or elaboration of other minds. Though he persistently asked for "criticism" and "collaboration," what the system itself demanded was converts.

Though by training Geddes accepted the neutral disciplines of science, he had the imperious ego that usually goes with great positive talents and keeps them from wilting with discouragement when their most original contributions are misunderstood or unheeded. Just to the extent that Geddes met with rebuffs and dismissals from his more bovine contemporaries, his ego sought compensation in self-reinforcement. This is what happened to Freud, too, and it kept both thinkers from absorbing ideas from other quarters which did not fit into their system. Such egoism must not be confused with a low kind of vanity; for one finds it in personalities seemingly full of humility and free from petty self-assertions, witness Einstein.

As a young man Einstein aspired to be no more than a science teacher in a high school, almost at the moment he was advancing his original views of relativity. But actually, if one examines Einstein's own statement toward the end of his life, one discovers that his ego was more exorbitant than either Freud's or Geddes'; for he said that if he were compelled to believe, as quantum theory suggested, that the ultimate nuclear particle could behave as if it had a choice to move or not to move in a predictable course, he would abandon physics for the trade of a shoemaker. In other words, he was not prepared to accept a universe that was not ordered on his own determinist terms – which perhaps explains why to the very end he kept trying to find a mathematical formula that would hold true for all possible changes in physical states – and died still seeking it.

When one remembers these examples, one must view more charitably Geddes' insistence on his having blocked out within his thirty-six squares all the essential components of human existence. Had he actually performed this feat, it would be only reasonable that he should seek at all costs to implant such valuable knowledge in the minds of his disciples and followers. So much must be said for him in absolution.

5

What I had already discovered in my first encounter with the "Chessboard of Life" under Branford became even plainer when I want further into it with Geddes, though it is only now, on returning to it after many years, that I have come to recognize fully its most unacceptable assumptions. Geddes, the sworn enemy of all "abstract" or "metaphysical" thinking, apparently never realized that his own ideographs were abstractions, too. And though he was a merciless foe of all closed systems he had, in fact, put together a tightly closed system of his own, which left no openings for time, chance, experience, feedback, or future emergents.

Now Geddes, as an evolutionary biologist, was fully aware of the factor of time: astronomical time, biological time, historical time: time registered in the genetic inheritance, in the social heritage, in individual memories, time embedded in the modifications of the landscape made by geological and meteorological process, and, not least, the changes made by man. Yet somehow time plays no part in his "Chart of Life": time is represented only in the person of the muse of History, history sporadically recorded, not history as lived, experienced, remembered. I cannot explain this oversight. For in the Outlook Tower Geddes showed me a whole series of "Time Charts" he and his colleagues had once worked on, attempting to represent the flow of historic events in colored streams, widening, thinning, sometimes blending. And he himself when going over his graphic chessboard with me once even speculated on the possibility of creating a kind of four dimensional chess in a box, composed of graphs whose movements through space would enable one to express changing relationships in time.

But how impossible it is to express time in such spatial terms! It is only in the symbols of language and music, or – as Henri Bergson saw – in motion pictures that the flow of living experiences through time can be symbolized. Unfortunately all these aspects of historic culture remained more or less closed to him; and though he vividly responded to the written word, especially in poetry, he too easily defended his own failure to develop his thoughts in this accessible public form by castigating the free use of writing as a mode of "verbalistic empaperment." Despite his own mountainous middens of graphs, he did not recognize how his own thought had been smothered in *graphic* empaperment.

Yet in the end Geddes could hardly help realizing that some block had kept people from taking over the system *in toto*, or even freely using the parts. Though he never lost an occasion to present his graphs, at the end he had to face the fact that he had made no headway in getting them accepted and used. For a time he might cover up this failure to gain acceptance by treating it as the result of their inhibiting formalistic education, or of insufficient preparation on the part of his listeners, his readers, his potential followers. But this is not a plausible defense against his palpable inability to win over to his graphic methods competent, sympathetic minds like Bergson's or D'Arcy Thomson's. The thinkers who were most loyal to him, Arthur Thomson and Victor Branford, were those he had captured at an early stage, before his graphs had finally crystallized.

On this matter, Delilah Loch's observation not merely accounts for my own difficulties with Geddes-and-Branford expositions, but also for this more general resistance. "Unless," she observed, "thou believe *just* this thou canst not be saved." The "Chart of Life" had become something to *believe in*: it called for converts. Thus the further use of the chart was not as a fresh incentive to observation and reflection: its use was a a devotional exercise, a daily ritual, which admitted no deviation or alteration. As with any theological dogma, whether Buddhist, Christian, Marxian, or Freudian, every alteration, even in the interest of a more adequate presentation of experience and truth, became a heresy. Yet nothing of course could have been a greater betrayal of Geddes' own insurgent mind than the lifeless finality of this final chart. He, who had so often sardonically characterized other thinkers as being "imprisoned in their own thought cage," had fabricated a super thought cage of his own. Geddes had in fact thrown away the key that would have opened his thirty-six cells: the key of language. Yes: language with its multiplicity of terms, its flexibility, its delicacy of shading, and its precision, its use of a vast store of collective experiences common to all men, its historic accumulation of meanings and – through its very nature – its fluent expression of time and change.

Not that Geddes' graphs were intelligible without verbal explanation. Para-doxically, it was only after his ideograms were authoritatively explained by him that even their most obvious statements could be verified, as in some geometric proposition, "by inspection." But once past Folk–Work–Place the only language that could be used with these graphs was a kind of Geddesian basic English, in which a limited number of fixed terms always meant the same thing, whenever and however used: precisely what Geddes himself meant, and only that. And whereas I.A. Richards' basic English used some eight hundred words,[3] the vocabulary of the "Life" graphs, taken alone, numbered only a hundred and fifty. As with that more sophisticated thinking machine, the computer, Geddes' students could get nothing out of the Chart of Life that he had not already programmed into it. What is more, without the sound track he himself supplied afresh each time he utilized it, the machine did not work.

Geddes seems never to have realized how open to challenge many parts of his "36" were when cut off from his dominating, attention-absorbing presence. Nor did he realize how his own resistance to even the most polite and gingerly dissent dampened the interest of his more mature listeners, when they realized that Geddes' insistent monologue never opened into a dialogue. Again one thinks of the computer. One cannot argue with a computer! By theoretically spurning language, by its nature the most social and truly human of the arts, Geddes had seriously limited the sphere of his own influence, not only upon his contemporaries, but among the generations to come. Unfortunately, the more adept that Geddes became in using his graphs, the more recondite were his categories, the more magical they seemed to him. The language of ideographs, going back to the ancient Egyptians and Chinese and a wide variety of religious cults, early and late, seemed to him so superior to the language of the spoken and especially the written word that he increasingly disparaged the need for more lengthy verbal explications.

So committed to his thinking machines did Geddes become that he felt that the graphic shortcuts he made actually dispensed with the need for more detailed observations over a more winding road: the shortcuts, he believed, would enable even beginners to arrive at the ultimate destination more quickly. Philip Boardman even quotes Geddes as insisting in his last years that with the help of his graphs an intelligent student might in ten hours of study grasp the essentials of a complete intellectual synthesis – Geddes' synthesis. Unfortunately Geddes left no instructions as to what the student should do with this synthesis once the ten hours had passed. And I suspect the reason for this omission: the student could do only what Geddes himself kept on doing: repeat the demonstrations to other docile eyes and ears – if he could find them!

Here I must note parenthetically that though Geddes' basic vocabulary was restricted when used in his central graphs, he himself found a need for many other terms: indeed he was as fertile in inventing them as was Jeremy Bentham,[4] and thus left the imprint of his thought on many areas. Some of Geddes' neologisms, like paleotechnic, geotechnic, conurbation, psychodrama have already found a place in the dictionary, though like all words that enter into common speech they cannot be confined to their original definition; while others like "eupsychic," "politology," or "thematametrics" – in place of logic – have not yet turned out to be viable. Of course in this effort to match new words derived from Greek or Latin roots to new needs, Geddes was only following common practice in the sciences: a practice that has in time provided each special discipline with a secret language whose private passwords serve to keep out intruders who have not undergone the requisite ideological initiation. As experience and knowledge pile up, many new terms are indeed indispensable: witness early coinages like "biology" and "sociology" themselves, not to mention Bentham's "international." But Geddes' prolific production of neologisms called for further indoctrination of his hearers before he could gain acceptance for his ideas – even for proposals that might have been readily entertained if expressed in plain language without the intervention of graphs.

In the loneliness of his old age, after his wife died in India in 1918, Geddes turned back more and more to going over his graphs to relieve his anxiety and solace his griefs.

In sense, they were his agnostic substitute for prayer. As the habit grew upon him, no new disciple was permitted to cast doubt on his categories or suggest amendments and additions: what was worse, he himself had no second thoughts of his own. In fact, he turned his graphs into a kind of solitaire: a game he could play by himself, if no one else would join him. Even without an audience he would go through one or another set of graphs, morning after morning, often hours before breakfast, as a pianist might go through his finger exercises, in preparation for a concert. Only the concert was never given; for he was unable to compose the new music. Not being able to see the extent to which this preoccupation had stultified his thought in later years, Geddes would justify this monotonous routine by insisting that he was "clearing up" this idea or carrying that one further, though in fact his mind was now standing still. Only when his attention shifted from his inner world to problems and demands that came from the outside, as in the magnificently comprehensive report on the planning of Indore, or his radical proposals for the new University of Jerusalem, did Geddes succeed in bringing together the concrete proofs of his whole life experience, as student, researcher, teacher, planner, activator, thinker. The later buildings – dubbed the Collège des Ecossais – he put up at Montpellier were the empty substitutes for the books he had never disciplined himself to write.

Though I have not minimized the genuine, if only potential, contributions that Geddes's own insistent use of graphs so early made to "field theory," most of the decisive innovations in thought and practice that he had made had been made by him between 1880 and 1910, before his graphs had been elaborated and completely assimilated in the "Chart of Life." Yet the more resistance Geddes encountered in getting his "final" synthesis accepted, the more addicted he became to his daily ritual, even when at Montpellier the students he hoped to win over to his ideological system fled from his presence when he threatened them with still another exposition. He had exchanged the living word, which carried the electric charge of his whole personality, for a neatly articulated skeleton, to be pulled out of the closet on any and all occasions for anatomical demonstration.

When I said earlier that Geddes had thrown away the key of language which could have unlocked the prison cells that his graphs had become, I was using no far-fetched metaphor of my own. The crux of Geddes' error was his belief that his graphs, without the aid of words, actually provided the indispensable key. Geddes used indeed to compare his own success in accumulating and utilizing a vast store of knowledge, drawn from a score of specialized fields, with the depredations of the infamous Deacon Brodie, seemingly a respectable citizen, who burglarized the houses of his Edinburgh friends and neighbors by using a key that unlocked every front door.[5] To understand Geddes' fixation – nay, his obsession! – one must realize that with his graphs, Geddes actually believed he had shaped such magic keys, and he mistakenly attributed to his thinking machines the broad mastery that followed from his own insatiable appetite for detailed scientific knowledge and personal observation.

What Geddes could legitimately claim for his graphic method was that it provided abstract guidelines for the orderly study of all organic events and activities: it brought together in the mind what had hitherto too often been left dispersed, isolated, unrelated. Though many other philosophers had made a similar effort, Geddes, with his biological background, had gone further in laying the ground for a systematic ecology of human culture, which would counterbalance the purely analytic and reductive approach that had so largely dominated the physical sciences and had placed biology and sociology under embarrassing restraints – as if they were ultimately to be studied only as minor branches of physics. Going further one might suggest that Geddes' fundamental graphs established a kind of order similar to that Mendelejev had effected through the Periodic Table in chemistry. Like Newland's original "Law of Octaves" Mendelejev's table had been at first flatly rejected by his contemporaries on the ground that Nature was inherently more disorderly than that. Such a rhythmic pattern in the arrangement of a long series of elements presumed a kind of esthetic harmony that the physical universe, still often described dogmatically in ancient Greek terms as a "fortuitous concourse of atoms" did not supposedly reveal.

When I pointed out this encouraging parallel to Geddes in 1923 – for he had never referred to it – I was both surprised and disappointed over his reaction. That happy feat of graphic thinking had made no impression on him, despite its fertile applications. He explained that he had learned his chemistry before the Periodic Table had been established or was generally accepted: so that he knew the 92 elements only in alphabetical order by their atomic weights – and he began reciting them rapidly, to show how effectively he had been drilled. This seemed to me only an added argument for Geddes' innovations, since the mere re-arrangement from a serial linear order to a multi-dimensional framework, with all the components simultaneously visible, revealed significant vertical and horizontal associations that had hitherto been hidden. What better proof of the capacity of graphs to generate new ideas? For a man as insatiable in the pursuit of fresh knowledge as Geddes, it seemed to strange that he should disdain such a potent confirmation of his own methodology.

By now I think I understand the unconscious reason for his indifference. For one thing, it was impossible to superimpose the Periodic Table on either the "Bookcase Diagram" of the biological sciences or the "Chart of Life"; for it followed a valid but quite different logic of its own, proper to the physical (pre-organic or proto-organic) sciences' observations of the constitution of "matter." But even more: if Geddes had accepted the precedent of the Periodic Table, he would have also been forced to face the radical flaw in his own graphs; and still more in the over-ambitious claims he made for them. His comprehensive "Chart of Life" promised more than any graph could perform. Preoccupied with the fabrication of his graphic skeleton key, he forgot that unlocking the door is only the first step. Such easy entry does not dispose of the time-consuming task of penetrating the interior and carrying off the contents. As with Deacon Brodie's successful raids, most of the valuable interior furnishings would still be left behind.[5]

The genuine contribution of the Periodic Table was that in effect it outlined a strategic plan of campaign. But the strategist who had prepared that plan had not saved the army from the effort to take possession of the new territory, step by step; he had only given them a map that made it easier to advance. Merely memorizing the Periodic Table did not make one a chemist. Thus it was with Geddes' graphs. So far from claiming that his graphs would reduce the need for mastering specialized knowledge, Geddes might properly have pointed out that his synthesis helped to bring out the significant patterns underlying a multitude of otherwise confusing details. Was this not in fact the lesson of Geddes' own life before his graphs were perfected? – his insatiable search for first-hand acquaintance with existential realities through personal experiment and communal activity. Of what use were all these swift mental break-ins and shortcuts unless one "took one's time" and recognized at every moment that *all organic activities take time*. Without enlisting the arts of time, symbolized in language, one could not do justice to life itself or express more than a small part of its significance.

So attached was Geddes to pictures and pictographs that one of the many unfulfilled projects still nagging him in his old age was for an "Encyclopedia Graphica," in which the explanatory letterpress could be reduced to a minimum or even omitted. This world have been the equivalent, in terms of today's technical resources, of the *Biblia Pauperum* of the late Middle Ages, which presented the Bible to the illiterate by means of pictures. Here again Geddes was something of an educational pioneer. And if audio-visual methods of teaching have never been entirely lacking in schools since John Amos Comenius presented to teachers his *Orbis Sensualis Pictus*, the current enlargement of these methods owes not a little, at least in Britain, to Geddes' original initiatives.

But if I am right, Geddes' crucial experience of blindness caused him, naturally enough, to stress the function he had nearly lost. He distrusted words as against images, and if he overvalued the images, it was because the visual world had almost been closed to him. (Once, at the New School, in 1923, when a bowl of liquid soap exploded in his face and temporarily stung him with blindness, I found him close to panic.) Just because of Geddes' favoring the eye, he no doubt perceived better than most of his academic contemporaries how much can be known, directly and swiftly, through the eye; for an animal, a forest, a city, a human being, all tell many things about themselves directly that the most meticulous description by words cannot convey. What Geddes overlooked in theory, though not in practice, was that images are no less abstract than sounds unless they are further vivified by smelling, touching, tasting, fondling, chewing, grasping, lifting, handling.

6

No graphic system or encyclopedia can do justice to all the dimensions of life. And who had demonstrated this better than Geddes or had said it more succinctly? "We learn by living."

Geddes, by confining so much of his thought to his limited graph-bound vocabulary, was incapable except in spontaneous speech of drawing on the fullness of his own life experience. No wonder his thought became costive when the time came to express himself in such writing as a larger audience might share: no wonder he postponed that effort and knew no purgative for overcoming it! But let me be brutally honest: behind the inability to put his Opus Syntheticum into adequate words was his reluctance to expose to others – or even critically examine for himself – the limitations of his favorite terms or his mode of assembling them in what was too often an arbitrary pattern. Brazenly he scorned written expression as – this again was his own epithet – verbalistic empaperment. When ideas are embalmed in textbooks, he would say, they are dead! Admitted. But ideas that are embalmed in graphs are also dead – and besides, the language that counts is not the language of textbooks. It is through sharing the spoken and written and printed word that ideas in time come to life in other minds. In language there are no cartesian squares to impede the flow of ideas or limit the area of their circulation and interaction.

As soon as Geddes translated his graphs into words their weakness became apparent: his sentences, as long as they were tethered to his graphic symbols, reflected far too small a part of his immense life-experience. Paradoxically, his "synthesis," so far from expanding the reach of his thought, actually contracted it; for the very squares that had once rescued him from blindness had compartmentalized his own thinking and arrested the free play of his mind. Yet, unlike the Cambridge philosopher, George Moore, who confessed that he kept silent, and so imposed silence upon others, because he had nothing to say, Geddes had much to say. Even so little of Geddes's direct reflections as had been published in his letters, in reports of his conversations by others, and in those vivid extracts that are to be found in "Patrick Geddes in India" – not least perhaps in his all-too-little-known *Masque of Learning* – give one a measure of what the world has lost by his fatal addiction to his thinking machines.

In practice, Geddes recognized this fact; for fortunately he did not try to carry his thinking machines around with him. When asked to replan an historic city, he would first ramble about it for days at a time, soaking in the entire milieu, without reading about its history, or getting current information about its economic or political institutions, or even consciously directing his thoughts anywhere. In this mood he did not make the error of associating place primarily with sense-relations, nor did he confine experience solely to work, or feelings to folk. On the contrary, he allowed feelings, emotional urges, ideas, institutional ideas, remembered images of other villages and towns to float into his consciousness by the most random route possible. And if actions speak louder than words, Geddes' actions likewise spoke louder than his graphs – and often more to the point.

Once Geddes shook loose from his graphs and his increasingly standardized proposals to transform Comte's three historic stages into the new Geddesian order – geotechnic, biotechnic, eupsychic – he usually had fresh personal observations to make and could make them without embarrassment in quite straightforward prose. Witness some of the letters that Mairet reprinted: such as Geddes' reflection in October 1914 on passing through the Suez Canal. Let me here quote Geddes' reaction on his first visit to Benares:

"Here we are," he wrote, "in this wonderful old city of religion, and after an afternoon of rambling through its lanes crowded with temples and stairs running down to the bathing ghats – which are of all kinds and degrees, from the simplest and rudest of old stairs to the palaces of state, and these again standing stately as of old, or overwhelmed and collapsed, shattered and half-buried, by the tremendous violence of the great stream in its floods. Contrast of cosmic indifference and human religions can go no further. Yet the religions too are right; and in the main they hold back the cosmic stream and make it the human one it is. Yet how simply! – that nature is sacred, that

sex is sacred, that creatures are sacred, and that even out of destruction comes new life
– these seem to be the main teachings; while the burning of the dead and the sweeping
of the ashes of the pyre into the river we have just been watching are the fitting and frank
expression of the great fact of death ... !"

All through Geddes' published work, meagre though it is, paragraphs as meaty as
this can be singled out, even many pages, such as the chapter in *Cities in Evolution*,
unfortunately left out of the Tyrwhitt edition, wherein Geddes condenses the historic
change from medieval to recent industrial culture by retelling the story of Cinderella.
Tyrwhitt more than atoned for this omission, I hasten to add, by ransacking Geddes'
Indian reports and extracting passages equally memorable. And in my Introduction to
Patrick Geddes in India I underlined the fact that his selection emphasizes the non-
systematic side of Geddes' thinking: "His insights, his gift for swift and penetrating
observation, and the life-wisdom he brought to each fresh situation. The Aristotle and
the Socrates are equally important in the living personality; but, paradoxically, in things
of the spirit the living flesh of wisdom outlasts the bony skeleton of abstractions; and
Geddes the teacher properly takes precedence over Geddes the systematic thinker. In
these living fragments ... one finds Geddes following Bergson's dictum that the thinker
should think as a man of action, and that the man of action should act as if he were a
thinker, though that unity of thought was engrained in Geddes' whole philosophy."

Let me close this analysis of the constraints that Geddes' over-addiction to his
thinking machines placed upon the full expression of his experience, by describing the
contraceptive device that kept him from gestating a commissioned book. One of the
books he had long promised himself to write was his biophysic interpretation of the
Olympian Gods of the Greeks. As early as 1923, as I recall, arrangements for publishing
this work had been made with the Yale University Press. As a biologist, Geddes had
made a simple but interesting discovery: the Gods and Goddesses of the Greek
pantheon, beginning with Eros and Hebe and ending with Zeus and Hera, are ideal
representatives of the successive stages of human development: the traditional seven
stages of man. The one exception perhaps is Hephaistos, the Blacksmith God, whose
lameness, on Geddes' apt interpretation, stood for the crippling effect of vocational
specialization, which so often marks middle age, both physiologically and psychologically.

That was, no doubt, a suggestive *aperçu* for the mind to play with, though once
stated the idea seems almost too neatly self-explanatory to be carried further, except by
a comparative study of these stages in other religions. Geddes nevertheless returned to
this Olympic model again and again, each time arranging the deities with the opposite
sexes facing each other in the same double ellipse, always recapitulating their characters
in the same set terms, as described briefly in his 1912 *Masque of Learning*. Even at the
end of Geddes' life, Boardman remembers him at Montpellier laying out such a
Pantheon in the open, and getting the students of his College to stand in positions on
pedestals at appropriate intervals in order supposedly to vivify the idea. Needless to say,
nothing else happened; nothing *could* happen on Geddes' terms; and needless to say the
promised book never got written. Once Geddes has laid out the graphs and had
substituted these concrete symbols, the idea was finished. Without a long *excursus* into
Greek mythology and anthropology, what was there further to say?

Had it not been for Geddes' graphic fixations, his germinal impulse need not have remained so sterile. Since I had long before escaped from a life sentence in Geddes' graphic prison, nothing prevented me in 1938, in presenting the City as a favorable habitat for human growth, from writing a paper on "Planning for the Phases of Life." Without mentioning the Olympian Gods, I showed that each phase of growth, from babyhood onward, demanded an appropriate habitat, corresponding to the needs and functions of that stage, from the child's sandpile, immediately under the mother's eye, to the elderly person's need for visual stimulus, human companionship, and medical attention. This was such a paper as Geddes might easily have written at a much earlier date, and with far greater richness of illustration, had he not been fettered to his graphs. Though I was not conscious of this Geddesian background when I wrote my paper, I have no doubt that his early presentation of these Olympian phases had been quietly germinating in my mind.

A few decades later Erik Erikson, without having any acquaintance with Geddes' ideas, gave a course on the Stages of Life at Harvard University: a course to which the students flocked eagerly with good reason, for he brought to it both his insights as a psychologist and his wisdom as a mature man who himself had passed through most of the stages of life. By all reports, this was one of the crowning moments of Erikson's intellectual career. And again, this was the sort of course that Geddes himself might well have given and developed into a book, for in outline it already existed in one of his early University of London lecture syllabuses. The idea need only a fuller rounding out through application of the additions to psychology that had already been made by Geddes' close contemporaries, Stanley Hall, Freud, and Jung, along with Greek scholars like Gilbert Murray and Jane Harrison. But for all Geddes' wide-ranging scholarship, for all his varied personal experience, the old lion went back to his thought cage for his stale cut-up food, instead of ranging abroad for fresh game over the unfenced fields he had once so freely explored.

By now, I trust, I have said enough to illuminate both the glory and the tragedy of Geddes' career, and to define quite incidentally the extent of his influence over my own thinking. My response to Geddes the man of vision, the liberated and insurgent spirit, was what had first brought me close to him, and evoked a personal loyalty in me more intimate than with any other teacher. Once we had met, I had no difficulty in addressing him by letter, European fashion, as "Dear Master," and thirty years later, in telling my students some anecdote drawn from him, I still easily would refer to him as "my master" – which of course seemed as quaintly Victorian, as if I had called trousers "unmentionables." Even after we had drawn apart both in space and in thought, that original underlayer of empathy and sympathy – call it rather love! – remained. But at the time we met, Geddes' graphic system, with its desiccated ideology, had taken possession of him and displaced the personality I had been drawn to. Though I noted these reservations in my introduction to *Patrick Geddes in India* it had taken me many years to piece the facts together and examine the illuminations they throw on both our lives.

Both Geddes and Branford regarded Geddes' collection of Graphs, most particularly the "Chessboard of Life," as not merely the culmination and quintessence of Geddes' own thinking: it was for them a new organon, a veritable philosopher's stone, that would bring order into a disorderly intellectual world and cast a light into its most obscure and distant corners. After rejecting all earlier attempts by others at such a complete ideological synthesis, Geddes nevertheless claimed success for his own. He gave it, indeed, a completeness, a finality, he would not have accorded to Plato or Aristotle, to Aquinas or Ibn Khaldun, to Leibniz or Kant, to Vico, Hegel, Marx, Spencer, or Croce. But what an upshot! The most damning thing one could say about Geddes' system was that its spirit was profoundly un-Geddesian; for it denied the most vital lesson he exemplified in person – the uncageable and inexhaustible creativity of the life force itself.

Fortunately for me, what had drawn me to Geddes at the beginning was what had attracted him in his youth to Bastian: something that rested on a deeper foundation than any conscious feats of thought. A delight in every manifestation of life, a sense of its wonder and mystery, was what issued forth from Geddes' personality, a kind of radiant emanation, with a halo of visible and usable wisdom, not essentially different from what emanated from William Blake. No graph, nor yet I confess any written words, could have captured fully Geddes' response to the *élan vital*. Except in the gnomic forms that great poets use, words are ultimately as mute as Geddes' graphs and charts proved to be. As for his "Chart of Life," his "Opus Syntheticum," what was this but the leaden residue left in his mind when the original radium had disintegrated?

For me, Geddes' greatest service was to open up the House of Life, from rooftop under the open sky to labyrinthine cellar. But I cannot forget that there are many chambers in that house we shall never, to the end of our days, penetrate; and that no single life, no single culture, no single philosophic or religious outlook, no single period or epoch, nor yet all the assembled products of science and technics, however condensed and computerized, will ever exhaust life's boundless and unpredictable manifestations of creativity. Nothing less than the total effort of all creatures and all minds, aided by the stars in their courses, is necessary to convey even faintly the meanings and values of life. And from whom did I first learn this lesson? From Patrick Geddes. But I did not find it in any of his graphs.

1.   Jacob L. Moreno (1892–1974), Austrian psychiatrist who introduced the technique of psychodrama; author of *Psychodrama* (1946).   2.   Alexander von Humboldt (1769–1859), German scientist; author of *Cosmos* (1845–1862).   3.   I.A. Richards and C.K. Ogden created "Basic English", consisting of 830 words, in 1930.   4.   Jeremy Bentham (1784–1832), British philosopher associated with Utilitarianism.   5.   William Brodie (d. 1788), leading citizen and "deacon of the incorporation of Edinburgh Wrights and Masons" who headed a gang of thieves.

INDEX

Note: Page references in *italics* indicate diagrams.

lot by Dri & bat. Then quiet slum
+ fully refreshed. Try it! — a
(full day, hard.
the ^ working man is a wonder
d at — but was just as inclined to
lean to it so far — If But two
± sedentary
d for us ^ fellows in most cases. (

best Child's education up to going a
of 11-12) was from trotting aft
Mother among her flowers — at the
tt I really was, a little — & has
do that with Geddes? (the no
4
But new your " Modern Synthes
Review — never saw it before
a dollar gradually, me posting me was